Essentials of Optoelectronics

Optical and Quantum Electronics Series

Series editors

Professor G. Parry, University of Oxford, UK
Professor R. Baets, University of Ghent, Belgium

This series focuses on the technology, physics and applications of optoelectronic systems and devices. Volumes are aimed at research and development staff and engineers involved in the application of optical technologies. Advanced undergraduate and graduate textbooks are included, giving tutorial introductions to the many exciting areas of optoelectronics. Both conventional books and electronic products will be published, to provide information in the most appropriate and useful form for users.

Essentials of Optoelectronics

With applications

Alan Rogers

Head of Department
Department of Electronic Engineering
School of Physical Sciences and Engineering
King's College London
UK

CHAPMAN & HALL

London · Weinheim · New York · Tokyo · Melbourne · Madras

Published by Chapman & Hall, 2–6 Boundary Row, London SE1 8HN, UK

Chapman & Hall, 2–6 Boundary Row, London SE1 8HN, UK

Chapman & Hall GmbH, Pappelallee 3, 69469 Weinheim, Germany

Chapman & Hall USA, 115 Fifth Avenue, New York, NY 10003, USA

Chapman & Hall Japan, ITP-Japan, Kyowa Building, 3F, 2-2-1 Hirakawacho, Chiyoda-ku, Tokyo 102, Japan

Chapman & Hall Australia, 102 Dodds Street, South Melbourne, Victoria 3205, Australia

Chapman & Hall India, R. Seshadri, 32 Second Main Road, CIT East, Madras 600 035, India

First edition 1997

© 1997 Alan Rogers

Printed in Great Britain at T.J. International Ltd, Padstow, Cornwall

ISBN 0 412 40890 2

A catalogue record for this book is available from the British Library

Library of Congress Catalog Card Number: 96-70805

♾ Printed on permanent acid-free text paper, manufactured in accordance with ANSI/NISO Z39.48-1992 and ANSI/NISO Z39.48-1984 (Permanence of Paper).

To my wife and family, past and present

Contents

Preface

Optoelectronics is a science and a technology based on the conjunction of optics with electronics. The special advantages of such a conjunction should become clear as the reader progresses through the book but, broadly, they result from the large speed and information-carrying capacity of light coming into alliance with the ready controllability of electrons.

The great speed of light is well known ($\sim 3 \times 10^8\,\mathrm{m\,s^{-1}}$ in a vacuum); it is, in fact, the greatest speed which is possible for any form of physical energy, as was elucidated by Einstein in his special theory of relativity (1905); and it is not so very long ago (c. 1640) that many scientists held the view that the speed of light was infinite.

The large information-carrying capacity of light derives from the high frequency of its oscillations. The implication is that optical processes occur on a time-scale of order one optical period ($\sim 10^{-14}\,\mathrm{sec}$), and thus that they can absorb and convey information on the same kind of time-scale, leading to information rates of the order of the optical frequency.

Electrons, unlike photons, carry electrical charge. As a consequence they respond to electric and magnetic fields, which thus can be used to control and manipulate them. This is a significant advantage, since modern electronics provides finely adjustable fields at a sophisticated level of flexibility.

Since there are well-understood processes by which electrons can give rise to photons and photons to electrons, the association of the two sciences was a natural and convenient one, once certain conditions were satisfied. These conditions related to the nature of optical sources which, until 1960, were noisy and imprecise. The light which these sources emitted had ill-defined properties, implying that it was difficult to impress information clearly upon it, and that it was difficult to control to an extent which allowed even a reasonable match to the control that was possible with electrons.

In 1960 the laser was invented. The laser provides light with the required sharp definition of properties, and the subject of optoelectronics effectively began in that year.

Optoelectronics was referred to earlier as both a science and a technology. It is a science because it has contributed a great deal to our understanding of photons and electrons and of the inter-relationships between them. It is technology because, based on the science, a wide range of applications has begun to affect our daily lives: optical communications, compact audio-discs, laser printers, liquid-crystal displays on watches and calculators, laser surgery, supermarket laser check-outs, to name just a few.

The prospects for the future are remarkable. It is almost inevitable that optoelectronics will come to dominate our material lives in the early 21st century.

The book seeks to provide the essential understanding of the principles and applications of the subject necessary for comfortable familiarity with it, and for an appreciation of its many possibilities. It does not employ detailed mathematical analyses, but aims to provide the firm basis of understanding necessary to underpin more advanced studies of the specialist topics, if desired.

Furthermore, all important results are proved (albeit some in appendices!) so that no external references are necessary to follow the full logical development. Those external references which are included are for background and interest, and are not crucial to the flow.

Hence, an important aspect of the book is that it comprises an entirely self-contained development of the essentials of the subject, and will thus be useful for early students of optoelectronics, and for those who wish to gain deeper physical insights.

Above all the book seeks to interest, enlighten and stimulate the reader to the point where he or she may enjoy the subject, seek to apply it, and wish to pursue it further.

1
Photons and electrons

1.1 INTRODUCTION

In this first chapter we shall take a quite general look at the nature of photons and electrons, in order to gain a familiarity with their overall properties, insofar as they bear upon our subject. Clearly it is useful to acquire this 'feel' in general terms before getting immersed in some of the finer detail which, whilst very necessary, does not allow the interrelationships between the various aspects to remain sharply visible. The intention is that the gentle familiarity acquired by reading this chapter will facilitate an understanding of the other chapters in the book.

Our privileged vantage point for the modern views of light has resulted from a laborious effort by many scientists over many centuries, and a valuable appreciation of some of the subtleties of the subject can be obtained from a study of that effort. A brief summary of the historical development is our starting point.

1.2 HISTORICAL SKETCH

The ancient Greeks speculated on the nature of light from about 500 BC. The practical interest at that time centred, inevitably, on using the sun's light for military purposes; and the speculations, which were of an abstruse philosophical nature, were too far removed from the practicalities for either to have much effect on the other.

The modern scientific method effectively began with Galileo (1564-1642), who raised experimentation to a properly valued position. Prior to his time experimentation was regarded as a distinctly inferior, rather messy, activity, definitely not for true gentlemen. (Some reverberations from this period persist, even today!) Newton was born in the year in which Galileo died, and these two men laid the basis for the scientific method which was to serve us well for the following three centuries.

Newton believed that light was corpuscular in nature. He reasoned that only a stream of projectiles, of some kind, could explain satisfactorily the fact that light appeared to travel in straight lines. However, Newton recognized the difficulties in reconciling some experimental data with this view, and attempted to resolve them by ascribing some rather unlikely properties to his corpuscles; he retained this basic corpuscular tenet, however.

Such was Newton's authority, resting as it did on an impressive range of discoveries in other branches of physics and mathematics, that it was not until his death (in 1727) that the views of other men such as Euler, Young and Fresnel began to gain their due prominence. These men believed that light was a wave motion in a 'luminiferous aether', and between them they developed an impressive theory which well explained all the known phenomena of optical interference and diffraction. The wave theory rapidly gained ground during the late eighteenth and early nineteenth centuries.

The final blow in favour of the wave theory is usually considered to have been struck by Foucault (1819-1868) who, in 1850, performed an experiment which proved that light travels more slowly in water than in air. This result agreed with the wave theory and contradicted the corpuscular theory.

For the next fifty years the wave theory held sway until, in 1900, Planck (1858-1947) found it mathematically convenient to invoke the idea that light was emitted from a radiating body in discrete packets, or 'quanta', rather than continuously as a wave. Although Planck was at first of the opinion that this was no more than a mathematical trick to explain the experimental relation between emitted intensity and wavelength, Einstein (1879-1955) immediately grasped the fundamental importance of the discovery and used it to explain the photoelectric effect, in which light acts to emit electrons from matter: the explanation was beautifully simple and convincing. It appeared, then, that light really did have some corpuscular properties.

In parallel with these developments there were other worrying concerns for the wave theory. From early in the nineteenth century its protagonists had recognized that 'polarization' phenomena, such as those observed in crystals of Iceland spar, could be explained if the light vibrations were transverse to the direction of propagation. Maxwell (1831-1879) had demonstrated brilliantly (in 1864), by means of his famous field equations, that the oscillating quantities were electric and magnetic fields.

However, there arose persistently the problem of the nature of the 'aether' in which these oscillations occurred and, in particular, how astronomical bodies could move through it, apparently without resistance. A famous experiment in 1887, by Michelson and Morley, attempted to measure the velocity of the earth with respect to this aether, and consistently obtained the result that the velocity was zero. This was very puzzling in view of the earth's known revolution around the sun. It thus appeared that the medium in which light waves propagate did not actually exist!

The null result of the aether experiment was incorporated by Einstein into an entirely new view of space and time, in his two theories of relativity: the special theory (1905) and the general theory (1915). Light, which propagates in space and oscillates in time, plays a crucial role in these theories.

Thus physics arrived (ca. 1920) at the position where light appeared to exhibit both particle (quantum) and wave aspects, depending on the physical situation. To compound this duality, it was found (by Davisson and Germer in 1927, after a suggestion by de Broglie in 1924) that electrons, previously thought quite unambiguously to be particles, sometimes exhibited a wave character, producing interference and diffraction patterns in a wave-like way.

The apparent contradiction between the pervasive wave-particle dualities in nature is now recognized to be the result of trying to picture all physical phenomena as occurring within the context of the human scale of things. Photons and electrons appear to behave either as particles or as waves to us only because of the limitations of our modes of thought. We have been conditioned to think in terms of the behaviour of objects such as sticks, stones and waves on water, the understanding of which has been necessary for us to survive, as a species, at our particular level of things.

In fact, the fundamental atomic processes of nature are not describable in these same terms and it is only when we try to force them into our more familiar framework that apparent contradictions such as the wave-particle duality of electrons and photons arise. Electrons and photons are neither waves nor particles but are entities whose true nature is somewhat beyond our conceptual powers. We are very limited by our preference (necessity, almost) for having a mental picture of what is going on.

Present-day physics with its gauge symmetries and field quantizations rarely draws any pictures at all, but that is another story...

1.3 THE WAVE NATURE OF LIGHT

In 1864 Clerk Maxwell was able to express the laws of electromagnetism known at that time in a way which demonstrated the symmetrical interdependence of electric and magnetic fields. In order to complete the symmetry he had to add a new idea: that a changing electric field (even in free space) gives rise to a magnetic field. The fact that a changing magnetic field gives rise to an electric field was already well known, as Faraday's law of induction.

Since each of the fields could now give rise to the other, it was clearly conceptually possible for the two fields mutually to sustain each other, and thus to propagate as a wave. Maxwell's equations formalized these ideas and allowed the derivation of a wave equation (see Appendix I).

This wave equation permitted free-space solutions which corresponded to electromagnetic waves with a defined velocity; the velocity depended on the known electric and magnetic properties of free space, and thus could be calculated. The result of the calculation was a value so close to the known velocity of light as to make it clear that light could be identified with these waves, and was thus established as an electromagnetic phenomenon.

All the important features of light's behaviour can be deduced from a detailed study of Maxwell's equations. We shall limit ourselves here to a few of the basic properties.

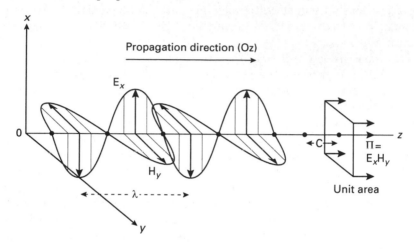

Fig. 1.1 Sinusoidal electromagnetic wave.

If we take Cartesian axes Ox, Oy, Oz (Fig. 1.1) we can write a simple sinusoidal solution of the free-space equations in the form:

$$E_x = E_0 \exp[i(\omega t - kz)]$$
$$H_y = H_0 \exp[i(\omega t - kz)]$$
(1.1)

These two equations describe a wave propagating in the Oz direction with electric field (E_x) oscillating sinusoidally (with time t and distance z) in the xz plane and the magnetic field (H_y) oscillating in the yz plane. The two fields are orthogonal in direction and have the same phase, as required by the form of Maxwell's equations: only if these conditions obtain can the two fields mutually sustain each other. Note also that the two fields must oscillate at right angles to the direction of propagation, Oz. Electromagnetic waves are transverse waves.

The frequency of the wave described by equations (1.1) is given by:

$$f = \frac{\omega}{2\pi}$$

and its wavelength by:

$$\lambda = \frac{2\pi}{k}$$

where ω and k are known as the angular frequency and wave number, respectively. Since f intervals of the wave distance λ pass each point on the Oz axis per second, it is clear that the velocity of the wave is given by:

$$c = f\lambda = \frac{\omega}{k}$$

The free-space wave equation shows that this velocity should be identified as follows:

$$c_0 = \frac{1}{(\varepsilon_0\mu_0)^{1/2}} \tag{1.2}$$

where ε_0 is a parameter known as the electric permittivity, and μ_0 the magnetic permeability, of free space. These two quantities are coupled, independently of (1.2), by the fact that both electric and magnetic fields exert mechanical forces, a fact which allows them to be related to a common force parameter, and thus to each other. This 'force-coupling' permits a calculation of the product $\varepsilon_0\mu_0$ which, in turn, provides a value for c_0, using equation (1.2). (Thus was Maxwell able to establish that light in free space consisted of electromagnetic waves.)

We can go further, however. The free-space symmetry of Maxwell's equations is retained for media which are electrically neutral ($\rho = 0$) and which do not conduct electric current ($\mathbf{j} = 0$). These conditions obtain for a general class of materials known as dielectrics; this class contains the vast majority of optical media. In these media the velocity of the waves is given by:

$$c = (\varepsilon\varepsilon_0\mu\mu_0)^{-1/2} \tag{1.3}$$

where ε is known as the relative permittivity (or dielectric constant) and μ the relative permeability of the medium. ε and μ are measures of the enhancement of electric and magnetic effects, respectively, which is generated by the presence of the medium. It is, indeed, convenient to deal with new parameters for the force fields, defined by:

$$\mathbf{D} = \varepsilon\varepsilon_0\mathbf{E}$$
$$\mathbf{B} = \mu\mu_0\mathbf{H}$$

where $\mathbf{D}$ is known as the electric displacement and $\mathbf{B}$ the magnetic induction of the medium. More recently they have come to be called the electric and magnetic flux densities, respectively.

The velocity of light in the material medium can (from 1.3) also be written as

$$c = \frac{c_0}{(\varepsilon\mu)^{1/2}} \tag{1.4}$$

where c_0 is the velocity of light in free space, with an experimentally determined value of $2.997925 \times 10^8\,\mathrm{ms}^{-1}$. For most optical media of any importance we find that $\mu \approx 1, \varepsilon > 1$ (hence the name 'dielectrics'). We have already noted that they are also electrical insulators. For these, then, we may write (1.4) in the form:

$$c \approx \frac{c_0}{\varepsilon^{1/2}} \qquad (1.5)$$

and note that, with $\varepsilon > 1$, c is smaller than c_0. Now the refractive index, n, of an optical medium is a measure of how much more slowly light travels in the medium compared with free space, and is defined by:

$$n = \frac{c_0}{c}$$

and thus

$$n \approx \varepsilon^{1/2}$$

from (1.5).

This is an important relationship because it connects the optical behaviour of the optical medium with its atomic structure. The medium provides an enhancement of the effect of an electric field because that field displaces the atomic electrons from their equilibrium position with respect to the nuclei; this produces an additional field and thus an effective magnification of the original field. The detailed effect on the propagation of the optical wave (which, of course, possesses an electric component) will be considered in Chapter 4 but we can draw two important conclusions immediately. First, the value of the refractive index possessed by the material is clearly dependent upon the way in which the electromagnetic field of the propagating wave interacts with the atoms and molecules of the medium. Second, since there are known to be resonant frequencies associated with the binding of electrons in atoms, it follows that we expect ε to be frequency dependent. Hence, via (1.5), we expect n also to be frequency dependent. The variation of n (and thus of optical wave velocity) with frequency is a phenomenon known as optical dispersion and is very important in optoelectronic systems, not least because all practical optical sources emit a range of different optical frequencies, each with its own value of refractive index.

We turn now to the matters of energy and power in the light wave. The fact that a light wave carries energy is evident from a number of its mechanical effects, such as the forced rotation of a conducting vane in a vacuum when one side is exposed to light (Fig. 1.2). A simple wave picture of this effect can be obtained from a consideration of the actions of the electric and magnetic fields of the wave when it strikes a conductor. The electric field will cause a real current to flow in the conductor (it acts on the 'free' electric charges in the conductor) in the direction of the field. This current then comes under the influence of the orthogonal magnetic field of

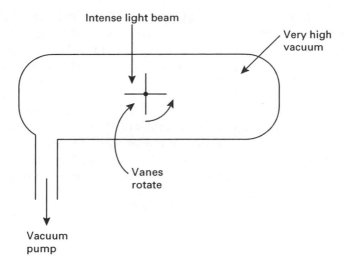

Intense light beam

Very high vacuum

Vanes rotate

Vacuum pump

Fig. 1.2 Force exerted by light falling on a conducting vane.

the wave. A current-carrying conductor in a magnetic field which lies at right angles to the current flow experiences a force at right angles to both the field and the current (motor principle) in a direction which is given by Fleming's left-hand rule (this direction turns out to be, fortunately, the direction in which the light is travelling!). Hence the effect on the conductor is equivalent to that of energetic particles striking it in the direction of travel of the wave; in other words, it is equivalent to the transport of energy in that direction.

We can take this description one stage further. The current is proportional to the electric field and the force is proportional to the product of the current and the magnetic field, hence the force is proportional to the product of electric and magnetic field strengths. The flow of energy, that is the rate at which energy is transported across unit area normal to the direction of propagation, is just equal to the vector product of the two quantities:

$$\mathbf{\Pi} = \mathbf{E} \times \mathbf{H}$$

(the vector product of two vectors gives another vector whose amplitude is the product of the amplitudes of the two vectors multiplied by the sine of the angle between their directions (in this case $\sin 90° = 1$) and is in a direction orthogonal to both vectors, and along a line followed by a right-handed screw rotating from the first to the second vector. Vectors often combine in this way so it is convenient to define such a product).

Clearly, if $\mathbf{E}$ and $\mathbf{H}$ are in phase, as for an electromagnetic wave travelling in free space, then the vector product will always be positive. $\mathbf{\Pi}$ is known as the Poynting vector. In Appendix I we also find that, in the case of a

propagating wave, **E** is proportional to **H**, so that the power across unit
area normal to the direction of propagation is proportional to the square
of the magnitude of either **E** or **H**. The full quantitative relationships
will be developed in Chapter 2, but we may note here that this means
that a measurement of the power across unit area, a quantity known as the
intensity of the wave (sometimes the 'irradiance') provides a direct measure
of either **E** or **H** (Fig. 1.1). This is a valuable inferential exercise since it
enables us, via a simple piece of experimentation (i.e., measurement of
optical power) to get a handle on the way in which the light will interact
with atomic electrons, for example. This is because, within the atom,
we are dealing with electric and magnetic fields acting on moving electric
charges.

The units of optical intensity, clearly, will be watts.metre^{-2}(Wm^{-2}).

1.4 POLARIZATION

The simple sinusoidal solution of Maxwell's wave equation for E and H
given by equations (1.1) is, of course, only one of an infinite number of
such solutions, with **E** and **H** lying in any direction in the xy plane, and
with ω taking any value (except zero!).

It is customary to fix attention on the electric field for purposes of general
electromagnetic wave behaviour, primarily because the effect of the electric
field on the electrical charges within atoms tends to be more direct than
that of the magnetic field. But the symmetry which exists between the **E**
and **H** fields of the electromagnetic wave means that conclusions arrived at
for the electric field have close equivalence for the magnetic field. It simply
is convenient only to deal with one of them rather than two.

Suppose that we consider two orthogonal electric field components of
a propagating wave, with the same frequency but differing phases (Fig.
1.3(a)).

$$E_x = e_x \cos(\omega t - kz + \delta_x)$$
$$E_y = e_y \cos(\omega t - kz + \delta_y)$$

From Fig. 1.3 we can see that the resulting electric field will rotate as
the wave progresses, with the tip of the resulting vector circumscribing (in
general) an ellipse. The same behaviour will be apparent if attention is
fixed on one particular value of z and the tip of the vector is now observed
as it progresses in time. Such a wave is said to be elliptically polarized.
(The word 'polarized', being associated, as it is, with the separation of
two dissimilar poles, is not especially appropriate. It derives from the attempt to explain crystal-optical effects within the early corpuscular theory
by regarding the light corpuscles as rods with dissimilar ends, and it has
persisted). Of notable interest are the special cases where the ellipse degenerates into a straight line or a circle [Fig 1.3(b),(c)]. These are known as

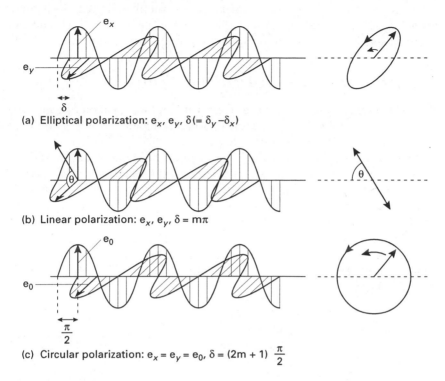

(a) Elliptical polarization: e_x, e_y, $\delta\,(=\delta_y-\delta_x)$

(b) Linear polarization: e_x, e_y, $\delta = m\pi$

(c) Circular polarization: $e_x = e_y = e_0$, $\delta = (2m+1)\,\dfrac{\pi}{2}$

Fig. 1.3 Linear and circular polarization as special cases of elliptical polarization.

linear and circular polarization states, respectively, and their importance lies not least in the fact that any given elliptical state can be resolved into circular and linear components, which can then be dealt with separately. The light will be linearly polarized, for example, when either $e_x = 0$ or $e_y = 0$, or when $\delta_y - \delta_x = m\pi$; it will be circularly polarized only when $e_x = e_y$ *and* $\delta_y - \delta_x = (2m+1)\pi/2$, where m is a positive or negative integer: circular polarization requires the component waves to have equal amplitude and to be in phase quadrature. A sensible, identifiable polarization state depends crucially on the two components maintaining a constant phase and amplitude relationship. All of these ideas are further developed in Chapter 3.

The polarization properties of light waves are important for a number of reasons. For example, in crystalline media, which possess directional properties, the propagation of the light will depend upon its polarization state in relation to the crystal axes. This fact can be used either to probe crystal structure or to control the state of the light via the crystal. Furthermore, the polarization state of the light can provide valuable insights into the restrictions imposed on the electrons which gave rise to it.

Wherever there is directionality in the medium in which the light is travelling the polarization state of the light will interact with it, and this is an extremely useful attribute, with a number of important applications.

1.5 THE ELECTROMAGNETIC SPECTRUM

Hitherto in this chapter we have dealt with optical phenomena in fairly general terms and with symbols rather than numbers. It may help to fix ideas somewhat if some numbers are quoted.

The wave equation allows single-frequency sinusoidal solutions and imposes no limit on the frequency. Furthermore, the equation is still satisfied when many frequency components are present simultaneously. If they are phase-related then the superposition of the many waveforms provides a determinable time function via the well known process of Fourier synthesis. If the relative phases of the components are varying with time, then we have 'incoherent' light; if the spread of frequencies in this latter case exceeds the bandwidth of the optical detector (e.g. the human eye) we sometimes call it 'white' light.

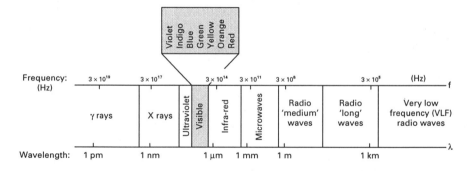

Fig. 1.4 The electromagnetic spectrum.

The electromagnetic spectrum is shown in Fig. 1.4. In principle it ranges from (almost) zero frequency to infinite frequency. In practice, since electro-magnetic wave sources cannot be markedly smaller than the wavelength of the radiation which they emit, the range is from the very low frequency ($\sim 10^3$ Hz) radio waves ($\lambda \sim 300$ km) to the very high frequency ($\sim 10^{20}$ Hz) gamma radiation, where the limit is that of the very high energy needed for their production.

The most energetic processes in the universe are those associated with the collapse of stars and galaxies (supernovae, black holes), and it is these which provide the radiation of the highest observable frequencies.

Visible radiation lies in the range 400 to 700 nm (1 nm = 1 nanometre = 10^{-9} m), corresponding to a frequency range of 7.5×10^{14} to 4.3×10^{14} Hz.

The eye has evolved a sensitivity to this region as a result of the fact that it corresponds to a broad maximum in the spectral intensity distribution of sunlight at the earth's surface: survival of the species is more likely if the sensitivity of the eye lies where there is most light!

The infrared region of the spectrum lies just beyond 700 nm and is usually taken to extend to about 300 000 nm ($\equiv$ 300 μm; we usually switch to μm for the infrared wavelengths, in order to keep the number of noughts down).

The ultraviolet region lies below 400 nm and begins at about 3 nm. Clearly, all of these divisions are arbitrary, since the spectrum is continuous.

It is worth noting that the refractive index of silica (an important optical material) in the visible range is $\sim$ 1.47, so the velocity of light at these wavelengths in this medium is close to 2×10^8 ms^{-1}. Correspondingly, at the given optical frequencies, the wavelengths in the medium will be $\sim$ 30% less than those in air, in accordance with the relation: $\lambda = c/f$. (The frequency will, of course, remain constant.)

It is important to be aware of this wavelength change in a material medium, since it has a number of noteworthy consequences which we shall later explore.

1.6 EMISSION AND ABSORPTION PROCESSES

So far, in our discussions, the wave nature of light has dominated. However, when we come to consider the relationships between light and matter, the corpuscular, or (to use the modern word 'particulate'), nature of light begins to dominate. In classical (i.e. pre-quantum theory) physics, atoms were understood to possess natural resonant frequencies resulting from a conjectured internal elastic structure. These natural resonances were believed to be responsible for the characteristic frequencies emitted by atoms when they were excited to oscillate, by external agencies. Conversely, when the atoms were irradiated with electromagnetic waves at these same frequencies, they were able to absorb energy from the waves, as with all naturally resonant systems interacting with sympathetic driving forces. This approach seemed to provide a natural and reasonable explanation of both the emission and absorption spectral characteristics of particular atomic systems.

However, it was soon recognized that there were some difficulties with these ideas. They could not explain why, for example, in a gas discharge, some frequencies were emitted by the gas and yet were not also absorbed by it in its quiescent state; neither could they explain why the energy with which electrons were emitted from a solid by ultraviolet light (in the photoelectric effect) depends not on the quantity of absorbed light energy but only on the light's frequency.

We now understand the reasons for these observations. We know that atoms and molecules can exist only in discrete energy levels. These energy levels can be arranged in order of ascending value: $E_0, E_1, E_2...E_m$ (where m is an integer) and each such sequence is characteristic of a particular atom or molecule. The highest energy level corresponds to the last below the level at which the atom becomes ionized (i.e. loses an electron).

Fundamental thermodynamics (classical!) requires that under conditions of thermal equilibrium the number, N_i, of atoms having energy E_i is related to the number N_j having energy E_j by the Boltzmann relation:

$$\frac{N_i}{N_j} = \exp\left[-\frac{(E_i - E_j)}{kT}\right] \tag{1.6}$$

Here k is Boltzmann's constant $(1.38 \times 10^{-23}\,\mathrm{JK}^{-1})$ and T is the absolute temperature.

The known physics now states that light frequencies ν_{ij} can be either emitted or absorbed by the system only if they correspond to a difference between two of the discrete energy levels, in accordance with the relation

$$h\nu_{ij} = E_i - E_j$$

where h is Planck's quantum constant $(6.626 \times 10^{-34}\,\mathrm{joule.seconds})$. The more detailed interpretation is that when, for example, an atom falls from an energy state E_j to E_i, a 'particle' of light with energy $h\nu_{ij}$ is emitted. This 'quantum' of light is called the photon; we use the symbol ν to denote frequency rather than f (or $\omega/2\pi$) to emphasize that light is now exhibiting its particulate, rather than its wave, character.

Thus the relationship between light and matter consists in the interaction between atoms (or molecules) and photons. An atom either absorbs/emits a single photon, or it does not. There is no intermediate state.

The classical difficulties to which reference was made earlier are now resolved. First, some lines are emitted from a gas discharge which are not present in the absorption spectrum of the quiescent gas because the energetic conditions in the discharge are able to excite atoms to high energy states from which they can descend to some lower states; if these states are not populated (to any measurable extent) in the cold gas, however, there is no possibility of a corresponding incoming frequency effecting these same transitions and hence being absorbed. Second, for an incoming stream of photons, each one either interacts or does not interact with a single atom. If the photon energy is higher than the ionization energy of the atom then the electron will be ejected. The energy at which it is ejected will be the difference between the photon energy and the ionization energy. Thus, for a given atom, the ejection energy will depend only on the frequency of the photon.

Clearly, in light/matter interactions, it is convenient to think of light as a stream of photons. If a flux of p photons of frequency ν crosses unit area in unit time then the intensity of the light (defined by the Poynting vector) can be written

$$I = ph\nu \qquad (1.7)$$

It is not difficult to construct any given quantity in the photon approach which corresponds to one within the wave approach. However, there does still remain the more philosophical question of reconciling the two approaches from the point of view of intellectual comfort. The best that can be done at present is to regard the wave as a 'probability' function, where the wave intensity determines the probability of 'finding' a photon in a given volume of space. This is a rather artificial stratagem which does, however, work very well in practice. It does not really provide the intellectual comfort which we seek, but that, as has been mentioned earlier, is a fault of our intellect, not of the light!

Finally, it may be observed that, since both the characteristic set of energy levels and the return pathways from an excited state are peculiar to a particular atom or molecule, it follows that the emission and/or absorption spectra can be used to identify and quantify the presence of species within samples, even at very partial concentrations. The pathway probabilities can be calculated from quantum principles, and this whole subject is a sophisticated, powerful and sensitive tool for quantitative materials analysis. It is not, however, within the scope of this book.

1.7 PHOTON STATISTICS

The particulate view of light necessitates the representation of a light flux as a stream of photons 'guided' by an electromagnetic wave. This immediately raises the question of the arrival statistics of the stream.

To fix ideas let us consider the rate at which photons are arriving at the sensitive surface of a photodetector.

We begin by noting that the emission processes which gave rise to the light in the first place are governed by probabilities, and thus the photons are emitted, and therefore also arrive, randomly. The light intensity is a measurable, constant (for constant conditions) quantity which, as we have noted, is to be associated with the arrival rate p according to equation (1.7), i.e. $I = ph\nu$. It is clear that p refers to the mean arrival rate averaged for the time over which the measurement of I is made. The random arrival times of the individual particles in the stream imply that there will be statistical deviations from this mean, and we must attempt to quantify these if we are to judge the accuracy with which I may be measured.

To do this we begin with the assumption that atoms in excited states emit photons at random when falling spontaneously to lower states. It is not possible to predict with certainty whether any given excited atom will or will not emit a photon in a given, finite time interval. Added to this there is the knowledge that for light of normal, handleable intensities, only a very small fraction of the atoms in the source material will emit photons in sensible detection times. For example, for a He-Ne laser with an output power of 5 mW, only 0.05% of the atoms will emit photons in one second.

Thus we have the situation where an atom may randomly either emit or not emit a photon in a given time, and the probability that it will emit is very small: this is the prescription for Poisson statistics, i.e. the binomial distribution for very small event probability. (See, for example, ref. [1].)

Poisson statistics is a well-developed topic and we can use its results to solve our photon arrival problem.

Suppose that we have an assemblage of N atoms and that the probability of any one of them emitting a photon of frequency ν in time τ is q, with $q \ll 1$.

Clearly, the most probable number of photons arriving at the detector in time τ will be Nq and this will thus also be the average (or mean) number detected, the average being taken over various intervals of duration τ. But the actual number detected in any given time τ will vary according to Poisson statistics, which state that the probability of detecting r photons in the time τ is given by (ref. 1):

$$P_r = \frac{(Nq)^r}{r!} \exp(-Nq)$$

Hence the probability of receiving no photons in τ is $\exp(-Nq)$, and of receiving two photons is $[(Nq)^2/2!]exp(-Nq)$ and so on.

Now the mean optical power received by the detector clearly is given by:

$$P_m = \frac{Nqh\nu}{\tau} \tag{1.8}$$

and P_m is the normally-measured quantity. Hence equation (1.8) allows us to relate the mean of the distribution to a measurable quantity, i.e.,

$$Nq = \frac{P_m\tau}{h\nu} = \frac{P_m}{h\nu B}$$

where B is the detector bandwidth ($B = 1/\tau$). We need now to quantify the spread of the distribution in order to measure the deviation from the mean, and this is given by the standard deviation which, for the Poisson distribution, is the square root of the mean. Thus the deviation of the arrival rate is

$$D = (Nq)^{1/2} = \left(\frac{P_m}{h\nu B}\right)^{1/2}$$

This deviation will comprise a 'noise' on the measured power level and will thus give rise to a noise power

$$P_{\text{noise}} = \left(\frac{P_m}{h\nu B}\right)^{1/2}\frac{h\nu}{\tau} = (P_m h\nu B)^{1/2}$$

Thus the signal-to-noise ratio will be given by

$$\text{SNR} = \frac{P_m}{P_{\text{noise}}} = \left(\frac{P_m}{h\nu B}\right)^{1/2}$$

This is an important result. It tells us what is the fundamental limit on the accuracy with which a given light power can be measured. We note that the accuracy increases as $(P_m/h\nu)^{1/2}$, and it is thus going to be poor for low rates of photon arrival. This we would expect intuitively, since the 'granular' nature of the process will inevitably be more noticeable when there are fewer photons arriving in any given time. It will also be poor for large optical frequencies, since this means more energy per photon, and thus fewer photons for a given total light energy. Again the 'granular' nature will be more evident. For good SNR, therefore, we need large powers and low frequencies. Radio wave fluxes from nearby transmitters are easy to measure accurately, gamma rays from a distant galaxy are not.

Finally, it should be remembered that the above conclusions only apply strictly when the probability q is very small. For the very intense emissions from powerful lasers ($\sim 10^6\,\text{Wm}^{-2}$, say) a substantial proportion of the atoms will emit photons in a typical detection time. Such light is sometimes classed as non-Poissonian (or sub-Poissonian) for reasons which will now be clear.

1.8 THE BEHAVIOUR OF ELECTRONS

Our subject is optoelectronics, and so far we have been concerned almost exclusively with just one half of it: with optics. The importance of our subject derives from the powerful interaction between optics and electronics, so we should now evidently gain the necessary equivalent familiarity with electronics, to balance our view. We shall, therefore, now look at the general behaviour of electrons.

A free electron is a fundamental particle with negative electrical charge (e) equal to 1.602×10^{-19} coulombs and mass (m) equal to 9.11×10^{-31} kg.

All electrical charges exert forces on all other charges and, for any given charge, q, it is convenient to summarize the effect of all other charges by defining the electric field, $\mathbf{E}$, via the value of the force $\mathbf{F_E}$ which the field exerts on q:

$$\mathbf{F_E} = q\mathbf{E}$$

A magnetic field exerts no force on a stationary charge. When the charge moves with velocity **v** with respect to a magnetic field of induction **B**, however, the force on the charge is given by

$$\mathbf{F_B} = q(\mathbf{v} \times \mathbf{B})$$

where **v** × **B** denotes the vector product of **v** and **B**, so that the force is orthogonal to both the **v** and **B** vectors. Of course, a uniformly moving charge comprises an electrical current, so that **v** × **B** also describes the force exerted by a magnetic field on a current-carrying conductor. The two forces are combined in the Lorentz equation:

$$\mathbf{F} = q(\mathbf{E} + \mathbf{v} \times \mathbf{B}) \tag{1.9}$$

which also is a full classical description of the behaviour of the electron in free space, and is adequate for the design of many electron beam devices (such as the cathode-ray tube of television sets) where the electron can be regarded as a particle of point mass subject to known electromagnetic forces.

If an electron (or other electrical charge) is accelerating, then it comprises an electric current which is varying with time. Since a constant current is known to give rise to a constant magnetic field, a varying current will give rise to a varying magnetic field, and this, as we have seen, will give rise in turn to an electric field. Thus an accelerating electron can be expected to radiate electromagnetic waves. For example, in a dipole antenna (Fig. 1.5) the electrons are caused to oscillate sinusoidally along a conducting rod. The sinusoidal oscillation comprises accelerated motion, and the antenna radiates radio waves.

However, the electron also itself exhibits wave properties. For an electron with momentum p there is an associated wavelength λ given by

$$\lambda = \frac{h}{p}$$

which is known as the de Broglie wavelength, after the Frenchman who, in 1924, suggested that material particles might exhibit wave properties. (The suggestion was confirmed by experiment in 1927.) Here h is, again, the quantum constant.

The significance assigned to the wave associated with the electron is just the same as that associated with the photon: the intensity of the wave (proportional to the square of the amplitude) is a measure of the probability of finding an electron in unit volume of space. The wave is a 'probability' wave. The particle/wave duality thus has perfect symmetry for electrons and photons. One of the direct consequences of this duality, for both entities, is the uncertainty principle, which states that it is fundamentally

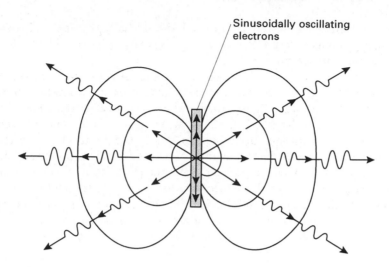

Fig. 1.5 The radiating dipole.

impossible to have exact knowledge of both momentum and position simultaneously, for either the photon or the electron. The uncertainty in knowledge of momentum, Δp, is related to the uncertainty in position, Δx, by the expression:

$$\Delta p \Delta x \approx \frac{h}{2\pi}$$

There is a corresponding relation between the uncertainty in the energy (ΔE) of a system and the length of time (Δt) over which the energy is measured.

$$\Delta E \Delta t \approx \frac{h}{2\pi}$$

The interpretation in the wave picture is that the uncertainty in momentum can be related to the uncertainty in wavelength, i.e.,

$$p = \frac{h}{\lambda}$$

so that

$$\Delta p = \frac{-h \Delta \lambda}{\lambda^2}$$

and hence

$$\Delta x = \frac{h}{2\pi \Delta p} = \frac{\lambda^2}{2\pi \Delta \lambda}$$

Hence, the smaller the range of wavelengths associated with a particle, the greater is the uncertainty in its position (Δx). In other words the closer is the particle's associated wave function to a pure sine wave, having constant amplitude and phase over all space, the better is its momentum known: if the momentum is known exactly, the particle might equally well be anywhere in the universe!

The wave properties of the electron have many important consequences in atomic physics. The atomic electrons in their orbits around the nucleus, for example, can only occupy those orbits which allow an exact number of wave-lengths to fit into a circumference: again, the escape of electrons from the atomic nucleus in the phenomenon of β-radioactivity is readily explicable in terms of the 'tunnelling' of waves through a potential barrier. But probably the most important consequence of these wave properties, from the point of view of our present discussions, is the effect they have on electron behaviour in solids, for the vast majority of optoelectronics is concerned with the interaction between photons and electrons in solid materials. We shall, therefore, need to look at this a little more closely.

The primary feature which solids possess compared with other states of matter (gas, liquid, plasma) is that the atoms or molecules of which they are composed are sufficiently close together for their electron probability waves to overlap. Indeed, it is just this overlap which provides the interatomic bonding strength necessary to constitute a solid material, with its resistance to deformation.

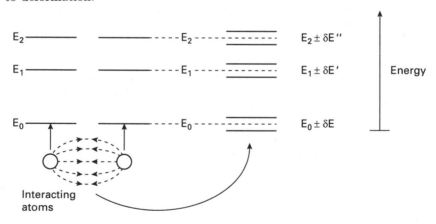

Fig. 1.6 Splitting of energy levels for two interacting atoms.

When two identical atoms, with their characteristic set of energy levels, come close enough for their electronic wave functions (i.e. their waves of probability) to overlap, the result is a new set of energy levels, some lower, some higher than the original values (Fig. 1.6). The reason for this is analogous to what happens in the case of two identical mechanical resonant systems, say two identical pendulums, which are allowed to interact by swinging them from a common support rod (Fig. 1.7). If one pendulum is set swinging, it will set the other one in motion, and eventually the second will be swinging with maximum amplitude while the second has become stationary. The process then reverses back to the original condition and this complete cycle recurs with frequency f_B. The system, in fact, possesses two time-independent normal modes: one is where both pendulums are swinging with equal amplitude and are in phase; the other with equal amplitudes in anti-phase. If these two frequencies are f_1 and f_2 we find

$$f_1 - f_2 = f_B$$

and the frequency of each pendulum when independent, f, is related to these by

$$f_1 = f + \tfrac{1}{2}f_B$$
$$f_2 = f - \tfrac{1}{2}f_B$$

i.e., the original natural frequency of the system, f, has been replaced under interactive conditions by two frequencies, one higher (f_1) and one lower (f_2) than f.

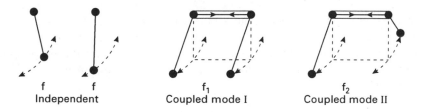

Fig. 1.7 Interacting pendulums.

It is not difficult to extend these ideas to atoms and to understand that when a large number of identical atoms is involved, a particular energy level becomes a band of closely spaced levels. Hence, in a solid, we may expect to find bands separated by energy gaps, rather than discrete levels separated by gaps; and that, indeed, is what is found.

The band structure of solids allows us to understand quite readily the qualitative differences between the different types of solid known as insulators, conductors and semiconductors, and it will be useful to summarize these ideas.

We know from basic atomic physics that electrons in atoms will fill the available energy states in ascending order, since no two electrons may occupy the same state: electrons obey the Pauli exclusion principle. This means that, at the absolute zero of temperature, for N electrons the lowest N energy states will be filled (Fig. 1.8(a)). At a temperature above absolute zero the atoms are in thermal motion and some electrons may be excited to higher states, from which they subsequently decay, setting up

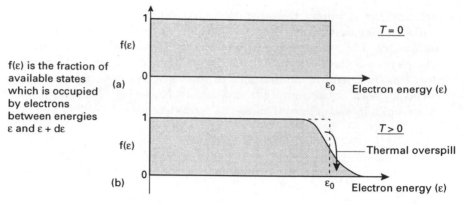

f(ε) is the fraction of available states which is occupied by electrons between energies ε and ε + dε

Fig. 1.8 The Fermi-Dirac distribution for electrons in solids.

a dynamic equilibrium in which states above the lowest N have a mean level of electron occupation. The really important point here is that it is only those electrons in the uppermost states which can be excited to higher levels, since it is only for those states that there are empty states within reach [Fig. 1.8(b)]. This fact has crucial importance in the understanding of solid state behaviour. The electrons are said to have a Fermi-Dirac distribution among the energy levels at any given temperature, rather than the Maxwell-Boltzmann distribution they would have if they were not constrained within the solid, and which is possessed by freely-moving gas molecules, for example.

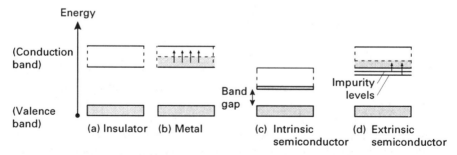

Fig. 1.9 Energy-level schematic for the three main classes of solid ($T > 0$).

Consider now the energy band structure shown in Fig. 1.9(a). Here the lower band is filled with electrons and there is a large energy gap before the next allowable band, which is empty. The available electrons thus have great difficulty in gaining any energy. If an electric field were applied to this solid it would have very little effect on the electrons, since in order to move in response to the force exerted by the field, they would need to gain energy from it, and this they cannot do, since they cannot jump the gap. Hence the electrons do not move; no current flows in response to an applied voltage; the material is an insulator.

Consider now the situation in Fig. 1.9(b). Here the upper band is only half full of electrons. (The electrons in this band will be those in the outer reaches of the atom, and hence will be those responsible for the chemical forces between atoms, i.e. they are valency electrons. Consequently the highest band to contain any electrons is usually called the valence band). The situation now is quite different from previously. The electrons near the top of the filled levels now have an abundance of unfilled states within easy reach and can readily gain energy from external agencies, such as an applied electric field. Electric currents thus flow easily in response to applied voltages; the material is a metallic conductor.

The third case (Fig. 1.9(c)) looks similar to the first, the only difference being that the gap between the filled valence band and the next higher unoccupied band is now much smaller. As a result, a relatively small number of electrons can be excited into the higher band (known as the conduction band) by thermal collisions and, once there, they can then move freely in response to an applied electric field. Hence there is a low level of conductivity and the material is a semiconductor; more specifically it is an intrinsic semiconductor. It is clear that the conductivity will rise with temperature since more energetic thermal collisions will excite more electrons into the conduction band. This is in contrast to metallic conductors in which the conductivity falls with temperature (owing to greater interference from the more strongly vibrating fixed atoms). There is a further important feature in the behaviour of intrinsic semiconductors. When an electron is excited from the valence band into the conduction band it leaves behind an unfilled state in the valence band. This creates mobility in the valence band, for electrons there which previously had no chance of gaining energy can now do so by moving into the empty state, or hole, created by the promotion of the first electron. Further, the valence electron which climbs into the hole, itself leaves behind another hole which can be filled in turn. The consequence of all this activity is that the holes appear to drift in the opposite direction to the electrons when an electric field is applied, and thus they are behaving like positive charges. (This is hardly surprising because they are created by the absence of negative charge.) Hence we can view the excitation of the electron to the conduction band as a process whereby an electron/hole pair is created, with each particle contributing to the current which flows in response to an applied voltage.

Finally, we come to another very important kind of semiconductor. It is depicted in Fig. (1.9(d)). Here we note that there are discrete energy levels within the region of energy 'forbidden' to states, the gap between bands. These are due to intruders in the solid, to 'impurities'.

To understand what is going on, consider solid silicon. Silicon atoms are tetravalent (i.e. have a valency of four), and in the solid state they sit comfortably in relation to each other in a symmetrical three-dimensional lattice (Fig. 1.10). Silicon is an intrinsic semiconductor with an energy gap between the filled valence band and the empty (at absolute zero) conduction band of 1.14 electron volts (eV). (An electron volt is the kinetic energy acquired by an electron in falling through a potential of 1 volt, and is equal to 1.6×10^{-19} joules). The Boltzmann factor [equation (1.6)] now allows us to calculate that only about one in 10^{20} electrons can reach the conduction band at room temperature; but since there are of order 10^{24} electrons per cm^3 in the material as a whole, there are enough in the conduction band to allow it to semiconduct.

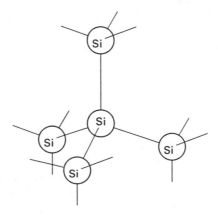

Fig. 1.10 Structure of silicon lattice.

Suppose now that some phosphorus atoms are injected into the silicon lattice. Phosphorus is a pentavalent (valency of five) atom, so it does not sit comfortably within the tetravalent (valency of four) silicon structure. Indeed, it finds itself with a spare valence electron (it has five as opposed to silicon's four) after having satisfied the lattice requirements. This electron is loosely bound to the phosphorus atom and thus is easily detached from it into one of the conduction band states, requiring little energy for the excitation. Effectively, then, the electron sits in a state close to the conduction band [as shown in Fig. 1.9(d)] and, depending on the density of phosphorus atoms (i.e., the 'doping' level), can provide significantly greater conductivity than is the case for pure silicon. Such impurity-doped materials are called extrinsic semiconductors.

As the impurity we chose donated an electron to the conduction band (as a result of having one spare) it is called an n-type semiconductor, since it donates negative charge carriers. Conversely, we could have doped the silicon with a tervalent (valency of three) element, such as boron, in which case it would sit in the lattice in need of an extra electron, since it has only three of its own. The consequence of this will be that a neighbouring silicon valence electron can easily be excited into that vacant state, leaving a positive hole in the valence band as a consequence. This hole now enhances the electrical conductivity, leading to p-type ('positive carrier') semiconductivity. It is now easy to understand why 'pentavalent' elements are said to give rise to 'donor' energy levels and 'tervalent' elements to 'acceptor' levels (in silicon).

There are several reasons why extrinsic semiconductors are so important. The first is that the level of conductivity is under control, via the control of the dopant level. The second is that p-type and n-type materials can be combined with great versatility in a variety of devices having very valuable properties, the most notable of which is the transistor: many thousands of these can now be integrated on to electronic chips.

We are now in a position to understand, in general terms, the ways in which photons can interact with electrons in solids.

Consider again the case of an intrinsic semiconductor, such as silicon, with a band-gap energy E_g. Suppose that a slab of the semiconductor is irradiated with light of frequency ν such that

$$h\nu > E_g$$

It is clear that the individual photons of the incident light possess sufficient energy to promote electrons from the valence band to the conduction band, leaving behind positive 'holes' in the valence band. If a voltage is now applied to the slab, a current, comprised of moving electrons and holes, will flow in response to the light: we have a *photo*conductor. Moreover, the current will continue to flow for as long as the electron can remain in the conduction band, and that includes each electron which will enter the slab from the cathode whenever one is taken up by the anode. Hence the number of electrons and holes collected by the electrodes per second can far exceed the number of photons entering the slab per second, provided that the lifetime of the carriers is large. In silicon the lifetime is of the order of a few milliseconds (depending on the carrier density) and the electron/photon gain can be as large as 10^4. However, this also means, of course, that the response time is poor, and thus photoconductors cannot measure rapid changes in light level (i.e., significant changes in less than a few milliseconds).

Small band-gap materials such as indium antimonide must be used to detect infrared radiation since the corresponding photon energy is relatively small. An obvious difficulty with a narrow band gap is that there will be a greater number of thermally excited carriers, and these will constitute a noise level; hence these infrared detectors usually must be cooled for satisfactory performance, at least down to liquid nitrogen temperatures (i.e., $< 77\,\text{K}$).

In order to increase the speed with which the photoconduction phenomenon can be used to make measurements of light level, we use a device consisting of a combination of n and p type semiconductor materials. The two types of material are joined in a '*pn* junction' which forms a 'photodiode' (Fig. 1.11). In this case the electron/hole pairs created by the incident photons drift in the electric field across the junction, thus giving rise to a measurable current as before; but each is quickly annihilated at the boundaries of the junction by an abundance of oppositely-charged carriers which combine with them. The reduced recombination time leads to a fast response and, with careful design, responses in times of order tens of picoseconds may be achieved. These *pn* photodiodes, in addition to being fast, are compact, rugged, cheap and operate at low voltage. They are not generally as sensitive as photoconductive devices, however, since they do not allow 'gain' in the way described for these latter devices (unless used in an 'avalanche' mode, of which more in Chapter 7).

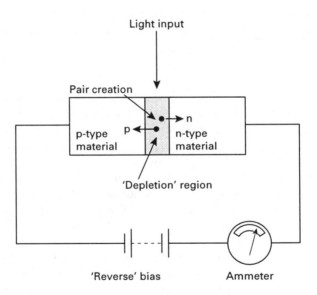

Fig. 1.11 Schematic p-n junction photodiode.

The *pn* detection process can also be used in reverse, in which case the device becomes a light emitter. For this action electrons are injected into the *pn* junction by passing a current through it using, now, 'forward' bias. The electrons combine with holes in the region of transition between p and n materials, and, in doing so, release energy. If conditions are arranged appropriately this energy is in the form of photons and the device becomes an emitter of light – a light-emitting diode (LED). Again this has the advantages of ruggedness, compactness, cheapness and low voltage operation. LEDs already are in widespread use.

1.9 LASERS

Finally, in our general view of optoelectronics, we must have a quick glance at the laser, for that is from where it really all derives.

Our subject began (effectively) with the invention of the laser (in 1960) because laser light is superior in so many ways to non-laser light. In 'ordinary', so-called 'incoherent', sources of light, each photon is emitted from its atom or molecule largely independently of any other, and thus the parameters which characterize the overall emission suffer large statistical variations and are ill-defined: in the case of the laser, this is not so. The reason is that the individual emission processes are correlated, via a phenomenon known as stimulated emission, where a photon which interacts with an excited atom can cause it to emit another similar photon which then goes on to do the same again, etc., etc. This 'coupling' of the emission processes leads to emitted light which has sharply defined properties such as frequency, phase, polarization state and direction, since these are all correlated by the coupling processes. The sharpness of definition allows us to use the light very much more effectively. We can now control it, impress information upon it and detect it with much greater facility than is the case for its more random counterpart. Add to this facility the intrinsic controllability of electrons via static electric and magnetic fields and we have optoelectronics.

1.10 SUMMARY

Our broad look in this chapter at the subject of optoelectronics has pointed to the most important physical phenomena for a study of the subject and has attempted to indicate the nature of the relationships between them in this context.

Of course, in order to practise the art, we need much more than this. We need to become familiar with quantitative relationships and with a much finer detail of behaviour. These the succeeding chapters will seek to provide; but it will all be so much easier if the broad principles introduced in this first encounter are well understood and kept firmly in mind.

Let us now proceed!

PROBLEMS

1.1 An electromagnetic wave is described by its electric and magnetic field components as follows:

$$E_x = E_0 \exp[i(\omega t - kz)]$$
$$H_y = H_0 \exp[i(\omega t - kz)]$$

Draw a sketch of this wave. What is the direction and magnitude of the wave's power flow across unit area?

1.2 Derive the relationship which connects the refractive index of an optical medium with its dielectric constant. Why are the two quantities dependent upon the optical frequency at which they are measured?

1.3 What is the essential difference between gamma radiation and infrared light? An average of 10^5 gamma-ray photons arrive at the surface of a detector every second. The frequency of the radiation is 10^{19} Hz and the detector surface area is $1\,\text{cm}^2$. What is the gamma-ray intensity? How would you relate the photon arrival rate with the gamma-ray electric field? ($h = 6.626 \times 10^{-34}$ joule.seconds.)

1.4 In question 1.3, for how long would the gamma-ray flux have to be averaged in order to measure its intensity to an accuracy of 1%?

1.5 An electron has a mass of 9.11×10^{-31} kgm and a charge of 1.6×10^{-19} coulombs. It is accelerated from rest through a voltage of $100\,\text{kV}$. What is its final de Broglie wavelength? How would you measure this wavelength?

1.6 In the hydrogen atom, an electron orbits a proton at a radial distance of 5.3×10^{-11} m. If the electron mass is 9.11×10^{-31} kg, what is the uncertainty in the momentum of the electron? What do you think are the measurable consequences of this uncertainty in the momentum? How will it relate to the uncertainty in the electron's energy?

1.7 In terms of the band theory of solids, what are the qualitative differences between metals, semiconductors and insulators?

1.8 What measurements would you make to determine whether a given light signal was from a laser source or a non-laser source?

REFERENCES

[1] Kaplan, W. *Advanced Mathematics for Engineers*, Addison Wesley, (1981) p. 857.

FURTHER READING

Bleaney, B. I. and Bleaney, B. (1985) *Electricity and Magnetism*, Oxford University Press, 3rd edn (for a readily digestible account of classical electricity and magnetism, including wave properties).

Cajori, F. (1989) *A History of Physics*, Macmillan, New York (for those interested in the historical developments).

Goldin, E. (1982) *Waves and Photons, an Introduction to Quantum Theory*, Wiley, New York (for the basics of photon theory).

Richtmeyer, F. K., Kennard, E. H. and Lauritsen, T. (1955) *Introduction to Modern Physics*, McGraw Hill (for the physical ideas concerning photons and electrons).

Solymar, L. and Walsh, D. (1993) *Lectures on the Electrical Properties of Materials*, 5th edn. Oxford University Press (for a clear treatment of general properties of electrical materials).

2
Wave properties of light

2.1 INTRODUCTION

In Chapter 1 it was noted that electromagnetic radiation, including light, exhibits both wave and particle properties, and that the type of behaviour exhibited at any one time depended upon the special circumstances.

In this chapter we shall concentrate just on the wave properties. Most of these were discovered and examined in the nineteenth century, before the advent of quantum mechanics (in 1901). The success of the wave theory was remarkable, and it led to a number of important devices, some of which are described in this chapter.

We shall begin, in earnest, in section 2.4, by looking at some aspects of the wave theory's crowning glory: Maxwell's equations for the electromagnetic field.

Before coming to that, however, there is, first, further emphasis on the range of radiation which lies within the electromagnetic spectrum and, second, a close look at the thinking behind the complex exponential representation of sinusoidal waves, since this is a most convenient and very widely used stratagem in all wave manipulations.

2.2 THE ELECTROMAGNETIC SPECTRUM

We have already noted (section 1.5) that Maxwell's equations allow the electromagnetic wave frequency ω to take any value from (almost) zero to infinity. At value zero we have electrostatics and magnetostatics. At any value other than zero we are dealing with electrodynamics and magnetodynamics.

The dynamic spectrum starts at the very low frequency radio waves and rises to the very high frequency gamma waves (Fig. 1.4). Methods of generation, interactions with material media and methods for detection all vary widely with frequency, leading to a variety of disciplines corresponding to the various frequency ranges, e.g. radio, microwave, infrared, optical, ultraviolet, X-ray, gamma rays, etc. Optoelectronics deals essentially with the optical range (wavelengths 400 nm to 700 nm) with short extensions either side, i.e., into the near infrared and near ultraviolet. It deals essentially with that part of the spectrum where the sun's light is most intense at the earth's surface, and also where efficient laser action is presently possible.

2.3 WAVE REPRESENTATION

In the study of optical waves within this chapter, we shall be examining many of the quantitative relationships between them and with other physical entities. Waves are sinusoids of the form:

$$E = e_0 \cos(\omega t + \varphi)$$

(We know from Fourier theory that any physical field disturbance can be expressed as the sum of such sinusoids).

A particular problem in the manipulation of such quantities is that the trigonometric expansions are rather cumbersome, i.e., in this case:

$$E_0 = e_0 \cos \omega t \cos \varphi - e_0 \sin \omega t \sin \varphi$$

and if, for example, we wish to add two such waves, of equal frequency, to produce a resultant, this will be another wave of the same frequency but with a different amplitude and phase. Our problem, in this case, is to find what is this resultant wave. That is to say, for

$$e_1 \cos(\omega t + \varphi_1) + e_2 \cos(\omega t + \varphi_2) = e_T \cos(\omega t + \varphi_\Gamma)$$

what are e_T and φ_T (Fig. 2.1)?

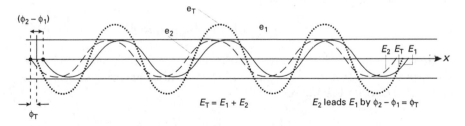

Fig. 2.1 Addition of two waves of the same frequency.

The mathematics of this, although straightforward, is tedious, and hence vulnerable to error; and the tedium increases rapidly with the number of waves!

A convenient solution to this is to express the sinusoid in its complex exponential form, for this allows factorizations which simplify the mathematics very considerably. Since this stratagem is used extensively in this book and elsewhere, it is, perhaps, worth taking some time to appreciate it more fully.

A sinusoidal wave of the form:

$$E = e_0 \cos(\omega t + \varphi)$$

is the real part of the expression

$$E' = e_0 \cos(\omega t + \varphi) + i e_0 \sin(\omega t + \varphi)$$

(E' might be expected to be a convenient mathematical entity since i (the square root of minus one) effectively is an operator which acts to rotate through $\pi/2$, and it thus brings together the cosine and sine which are, of course, $\pi/2$ out of phase.)

Now the well-known exponential expressions for cosine and sine are:

$$\cos \varphi = \tfrac{1}{2}[\exp(i\phi) + \exp(-i\varphi)]$$

$$\sin \varphi = \tfrac{1}{2i}[\exp(i\phi) - \exp(-i\varphi)]$$

hence:

$$\cos \varphi + i \sin \varphi = \exp(i\varphi)$$

and thus:

$$
\begin{aligned}
E' &= e_0 \cos(\omega t + \varphi) + i e_0 \sin(\omega t + \varphi) \\
&= e_0 \exp[i(\omega t + \varphi)] \\
&= e_0 \exp(i\varphi) \exp(i\omega t)
\end{aligned}
$$

Our original wave, $e_0 \cos(\omega t + \varphi)$, is just the real part of this, and this is sometimes expressed by writing:

$$e_0 \cos(\omega t + \varphi) = \mathrm{Re}\left[e_0 \exp(i\varphi) \exp(i\omega t)\right]$$

Now any sinusoidal wave can be written in this form, even a sine wave (as opposed to a cosine wave), since all that is necessary in that case is to subtract $\pi/2$ from the phase, i.e.,

$$e_0 \sin(\omega t + \varphi) = \mathrm{Re}\left\{e_0 \exp\left[i\left(\varphi - \tfrac{1}{2}\pi\right)\right] \exp(i\omega t)\right\}$$

The really important point about all of this, now, is that if we are dealing with a number of waves of the same frequency, then the frequency term may be factored out to leave a term which is a *complex number whose modulus represents the wave amplitude, and whose argument represents the wave's phase*. In our simple case:

$$E' = e_0 \exp(i\varphi) \exp(i\omega t) = E \exp(i\omega t)$$

where

$$E = e_0 \exp(i\varphi)$$

and E is a complex number with the properties

$$|E| = e_0; \arg(E) = \varphi$$

Similarly, for any two waves of the same frequency:

$$E_1 = e_1 \cos(\omega t + \varphi_1)$$
$$E_2 = e_2 \cos(\omega t + \varphi_2)$$

The complex exponential forms are:

$$E_1' = e_1 \exp(i\varphi_1) \exp(i\omega t)$$
$$E_2' = e_2 \exp(i\varphi_2) \exp(i\omega t)$$

Suppose, as earlier, we wish to find the amplitude and phase of the wave which results from the sum of these two waves. We may write:

$$E_T' = E_1' + E_2'$$
$$= \exp(i\omega t)\left[e_1 \exp(i\varphi_1) + e_2 \exp(i\varphi_2)\right]$$
$$= \exp(i\omega t)(E_1 + E_2)$$

The complex amplitude term can always be written in the form:

$$E_1 + E_2 = a + ib$$

where a and b are real numbers. Hence the resultant wave amplitude is given by:

$$e_T = |a + ib| = (a^2 + b^2)^{1/2}$$

and its phase by

$$\varphi_T = \arg(a + ib) = \tan^{-1}\left(\frac{b}{a}\right)$$

and hence finally:

$$e_1 \cos(\omega t + \varphi_1) + e_2 \cos(\omega t + \varphi_2) = e_T \cos(\omega t + \varphi_T)$$

So that:

$$e_1 \exp(i\varphi_1) + e_2 \exp(i\varphi_2) = e_T \exp(i\varphi_T)$$

and hence:

$$e_1 \cos\varphi_1 + e_2 \cos\varphi_2 + ie_1 \sin\varphi_1 + ie_2 \sin\varphi_2 = e_T(\cos\varphi_T + i\sin\varphi_T)$$

Here, then, we have

$$a = e_1 \cos \varphi_1 + e_2 \cos \varphi_2$$
$$b = e_1 \sin \varphi_1 + e_2 \sin \varphi_2$$

Hence

$$e_T = (a^2 + b^2)^{1/2} = e_1^2 + e_2^2 + 2e_1 e_2 \cos(\varphi_1 - \varphi_2)$$
$$\varphi_T = \tan^{-1}\left(\frac{b}{a}\right) = \tan^{-1}\left(\frac{e_1 \sin \varphi_1 + e_2 \sin \varphi_2}{e_1 \cos \varphi_1 + e_2 \cos \varphi_2}\right)$$

and the problem is solved.

This is a much more convenient mathematical technique than that which involves the trigonometric identities for the expansion of the cosines, and the convenience becomes markedly more noticeable as the number of waves increases.

For these reasons the complex exponential representation is to be used extensively in this chapter. Occasionally, however, it is convenient to revert to the real sinusoid. This usually is in those cases where the optical frequency is of direct importance, so that its removal as a separate factored complex quantity cannot conveniently represent the true physical condition.

2.4 ELECTROMAGNETIC WAVES

2.4.1 Velocity and refractive index

In 1864 Clerk Maxwell showed conclusively that light waves were electromagnetic in nature. He did this by expressing the then known laws of electromagnetism in such a way as to allow him to derive from them a wave equation (see Appendix 1). This wave equation permitted free-space solutions which corresponded to electromagnetic waves with a velocity equal to the known experimental value of the velocity of light. The consequent recognition of light as an electromagnetic phenomenon was probably the single most important advance in the progression of its understanding.

All the important features of light waves follow from a detailed examination of Maxwell's equations (see Appendix 1). Taking Cartesian axes Ox, Oy, Oz (Fig. 2.2) a typical sinusoidal solution is given by:

$$E_x = E_0 \exp[i(\omega t - kz)]$$
$$H_y = H_0 \exp[i(\omega t - kz)]$$

(2.1)

which states that the electric field oscillates sinusoidally in the xz plane, the magnetic field oscillates in the yz plane (i.e., orthogonally to the E field) and in phase with the E field, and the wave propagates in the Oz

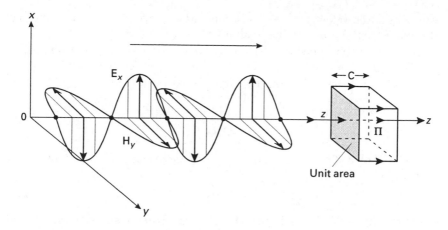

Fig. 2.2 Electromagnetic wave and energy flow (Poynting vector: Π).

direction (Fig. 2.2). The frequency and wavelength of the wave are given by:

$$f = \frac{\omega}{2\pi}$$

$$\lambda = \frac{2\pi}{k}$$

and $f\lambda = \omega/k = c$ where c is the wave velocity. This latter is related to the electromagnetic properties of the medium in which the wave propagates via the relation:

$$c = (\varepsilon\mu)^{-1/2} \tag{2.2}$$

where ε is the electric permittivity of the medium, and μ is its magnetic permeability. The relation (2.2) can also be written in the form:

$$c = (\varepsilon_R\varepsilon_0\mu_R\mu_0)^{-1/2}$$

i.e.,

$$\varepsilon = \varepsilon_R\varepsilon_0, \quad \mu = \mu_R\mu_0$$

where ε_R, μ_R are the permittivity and permeability factors for the medium relative to those for free space, ε_0, μ_0; ε_R is often called the dielectric constant. The electric displacement $\mathbf{D}$ and the magnetic flux density $\mathbf{B}$ are defined by the relations:

$$\mathbf{D} = \varepsilon\mathbf{E}$$

$$\mathbf{B} = \mu\mathbf{H}$$

(For reasons of symmetry, $\mathbf{D}$ is sometimes called the electric flux density.) We can, therefore, also write:

$$c = \frac{c_0}{(\varepsilon_R\mu_R)^{1/2}} \tag{2.3}$$

where c_0 is the velocity of the electromagnetic wave in free space, and has the experimentally determined value $2.997925 \times 10^8 \, \mathrm{m \, s^{-1}}$.

For most optical media of any importance we have $\mu_R \sim 1$ and $\varepsilon_R > 1$. These materials belong to the class known as dielectrics and they are electrical insulators. Thus we may write (2.3) in the form:

$$c \approx \frac{c_0}{(\varepsilon_R)^{1/2}}$$

and note that $c < c_0$. The ratio c_0/c is, by definition, the refractive index n of the medium, so that:

$$n \approx \varepsilon_R^{1/2} \tag{2.4}$$

where n is thus the factor by which light travels more slowly in an optical medium than it does in free space. Now ε_R is a measure of the ease with which the medium can be polarized electrically by the action of an external electric field (see section 4.2 for more details of this). This polarization depends on the mobility of the electrons, within the molecule, in the face of resistance by molecular forces. Clearly then, ε_R will depend on the frequency of the applied electric field, since it will depend on how quickly these forces can respond to the field. Thus equation (2.4) will be true only if n and ε_R refer to the same frequency of wave; hence we also note that n is frequency dependent.

2.4.2 Energy, power and intensity

Let us now consider the energy content of the wave. For an electric field, the energy per unit volume, u_E, is given by (see, for example, [1]):

$$u_E = \tfrac{1}{2} \varepsilon E^2$$

and for a magnetic field:

$$u_H = \tfrac{1}{2} \mu H^2$$

Maxwell's equations relate E and H for an electromagnetic wave according to (see Appendix 1):

$$H = \left(\frac{\epsilon}{\mu} \right)^{1/2} E$$

Hence the total energy density in the wave is given by:

$$u = u_E + u_H = \epsilon E^2 = \mu H^2 \tag{2.5}$$

Consider now the plane wave propagating in the direction Oz (Fig. 2.2). The total energy flowing across unit area in unit time in the direction Oz

will be that contained within a volume $c\,m^3$, where c is the wave velocity. Hence the power flux across unit area is given by:

$$\frac{\text{power}}{\text{area}} = c\varepsilon E^2 = \left(\frac{\varepsilon}{\mu}\right)^{1/2} E^2$$

Clearly, if the electric field E varies sinusoidally, this quantity also will vary sinusoidally; e.g., if

$$E = E_0 \cos \omega t$$

$$\frac{\text{power}}{\text{area}} = \left(\frac{\varepsilon}{\mu}\right)^{1/2} E_0^2 \cos^2 \omega t = \left(\frac{\varepsilon}{\mu}\right)^{1/2} \tfrac{1}{2}E_0^2(1 + \cos 2\omega t)$$

The average value of this quantity over one period of oscillation is called the 'intensity' of the wave (sometimes the irradiance) and clearly represents the measurable power per unit area for any device which cannot respond to optical frequencies (i.e., the vast majority!). Hence we have:

$$I = \left\langle \frac{\text{power}}{\text{area}} \right\rangle = \left(\frac{\varepsilon}{\mu}\right)^{1/2} \langle E^2 \rangle = \left(\frac{\varepsilon}{\mu}\right)^{1/2} \frac{1}{2} E_0^2 \qquad (2.6a)$$

(where $\langle \rangle$ denotes the average value) since $\cos 2\omega t$ averages to zero.

Clearly I is proportional to the square of the electric field amplitude and also, from (2.5), it will be proportional to the square of the magnetic field amplitude. The quantity I has MKS units of watts.metres^{-2}.

More generally, the intensity is expressed in terms of the Poynting vector $\mathbf{\Pi}$ (see Appendix I):

$$\mathbf{\Pi} = \mathbf{E} \times \mathbf{H}$$

where $\mathbf{E}$ and $\mathbf{H}$ are now vector quantities and $\mathbf{E} \times \mathbf{H}$ is their vector product (see Appendix I and Chapter 1). The intensity of the wave will be the value of $\mathbf{\Pi}$ averaged over one period of the wave. If $\mathbf{E}$ and $\mathbf{H}$ are spatially orthogonal and in phase, as in the case of a wave propagating in an isotropic dielectric medium, then:

$$I = \langle \mathbf{\Pi} \rangle == c\varepsilon E^2$$

as before. As is to be expected, in some more exotic cases (e.g. anisotropic media) the $\mathbf{E}$ and $\mathbf{H}$ components are neither orthogonal nor in phase, but $\langle \mathbf{\Pi} \rangle$ will still provide the average power flow across unit area. If, for example, $\mathbf{E}$ and $\mathbf{H}$ happened to be in phase quadrature, then we should have:

$$I = \langle \mathbf{\Pi} \rangle = \langle E_0 \cos \omega t \, H_0 \sin \omega t \rangle = 0$$

and thus there is no mean power flow. (This result should be noted for reference to the case of 'evanescent'waves, which will be considered later.)

In an optical medium with $\mu_R \approx 1$, equation (2.6a) can be written:

$$I = \left(\frac{\varepsilon_R \varepsilon_0}{\mu_R \mu_0}\right)^{1/2} \frac{1}{2} E_0^2 = n \left(\frac{\varepsilon_0}{\mu_0}\right)^{1/2} \frac{1}{2} E_0^2 \tag{2.6b}$$

where n is, again, the refractive index of the medium.

The quantity $(\mu_0/\varepsilon_0)^{1/2}$ is sometimes called the 'impedance of free space' and given the symbol Z_0. This is because, in free space:

$$\frac{E}{H} = \left(\frac{\mu_0}{\epsilon_0}\right)^{1/2} = Z_0$$

Since E has dimensions of volts.metres^{-1} and H of amps.metres^{-1}, Z_0 clearly has the dimensions of impedance (ohms); Z_0 is real and has the MKS value:

$$\left(\frac{\mu_0}{\varepsilon_0}\right)^{1/2} = \left(\frac{4\pi \times 10^{-7}}{8.854 \times 10^{-12}}\right)^{1/2} = 376.7 \, \text{ohms}$$

It follows that (2.6b) can be written:

$$I = \frac{n}{2Z_0} E_0^2 = \frac{n}{753.46} E_0^2 = 1.33 \times 10^{-3} n E_0^2 \tag{2.6c}$$

This is a useful relationship in two ways. Firstly, it relates a quantity which is directly measurable (I) with one which is not (E_0). Secondly, it provides the actual numerical relationship between I and E_0, and this is valuable when designing devices and systems, as we shall discover later.

2.4.3 Optical polarization

We should now give brief consideration to what is known as the 'polarization'of the optical wave. (This topic will be dealt with more comprehensively in Chapter 3.)

The 'typical' sinusoidal solution of Maxwell's wave equation given by equations (2.1) is, of course, only one of an infinite number of such sinusoidal solutions. The general solution for a sinusoid of angular frequency ω is given by:

$$\mathbf{E}(\mathbf{r}, t) = \mathbf{E}(\mathbf{r}) \exp(i\omega t)$$

where $\mathbf{E}(\mathbf{r}, t), \mathbf{E}(\mathbf{r})$ are, in general, complex vectors, and $\mathbf{r}$ is a real radius vector in the xy plane.

If, for simplicity, we consider just plane, monochromatic (single frequency) waves propagating in free space in the direction Oz, we may, for the **E** field, write the general solution to the wave equation in the form:

$$E_x = e_x \cos(\omega t - kz + \delta_x)$$
$$E_y = e_y \cos(\omega t - kz + \delta_y)$$

where δ_x, δ_y are arbitrary (but constant) phase angles. Thus we are able to describe this solution completely by means of two waves: one in which the electric field lies entirely in the xz plane, and the other in which it lies entirely in the yz plane (Fig. 2.3). If these waves are observed at a particular value of z, say z', then they take the oscillatory form:

$$E_x = e_x \cos(\omega t + \delta_x'); \ \delta_x' = \delta_x - kz'$$
$$E_y = e_y \cos(\omega t + \delta_y'); \ \delta_y' = \delta_y - kz'$$

and the tip of each vector appears to oscillate sinusoidally with time along a line. E_x is said to be linearly polarized in the direction Ox, and E_y linearly polarized in the direction Oy.

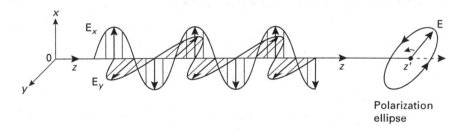

Polarization ellipse

Fig. 2.3 Electric field components for an elliptically-polarized wave.

The tip of the vector which is the sum of E_x and E_y will, in general, describe an ellipse whose Cartesian equation in the xy plane at the chosen z' will be given by eliminating ωt from the expression for E_x and E_y, i.e.,

$$\frac{E_x^2}{e_x^2} + \frac{E_y^2}{e_y^2} + \frac{2E_x E_y}{e_x e_y}\cos\delta = \sin^2\delta$$

$$\delta = \delta_y' - \delta_x'$$

This ellipse will degenerate into a straight line (and the overall polarization state of the light will thus be linear) if:

(a) $e_x \neq 0; e_y = 0$

or

(b) $e_x = 0; e_y \neq 0$

or

(c) $\delta = m\pi$

where m is a positive or negative integer. This corresponds to the condition that E_x and E_y are either in phase or in antiphase.

The ellipse becomes a circle (and the light is thus circularly polarized) if:

(a) $e_x = e_y$

and

(b) $\delta = (2m + 1)\pi/2$

i.e., the waves are equal in amplitude and are in phase quadrature.

The polarization properties of light waves are especially important for propagation within anisotropic media, in which the physical properties vary with direction. In this case the propagation characteristics for the component E_x will, in general, differ from those for E_y, so that the values of e_1, e_2 and δ will vary along the propagation path. The polarization state of the light will now become dependent upon the propagation distance, and on the state of the medium. This, also, will be covered in detail in Chapter 3.

2.5 REFLECTION AND REFRACTION

We have seen in section (2.4) that Maxwell's equations allow a set of solutions of the form:

$$E_x = E_0 \exp[i(\omega t - kz)]$$
$$H_y = H_0 \exp[i(\omega t - kz)]$$

with $\omega/k = (\varepsilon\mu)^{-1/2} = c$.

These represent plane waves travelling in the Oz direction. We shall now investigate the behaviour of such waves, with particular regard to the effects which occur at the boundaries between different optical media.

Of course, other types of solution are also possible. An important solution is that of a wave which spreads spherically from a point to a distance r:

$$E_r = \frac{E_0}{r} \exp[i(\omega t - kr)]$$

Here the factor $1/r$ in the amplitude is necessary to ensure conservation of energy (via the Poynting vector) for, clearly, the total area over which the energy flux occurs is $4\pi r^2$, so that the intensity falls as $1/r^2$. (Remember that intensity is proportional to the square of the amplitude.)

It is interesting and valuable to note that the propagation of a plane wave (such as in Fig. 2.4) is equivalent to the propagation of spherical waves radiating from each point on the propagating wavefront of the plane wave. On a given wavefront the waves at each point begin in phase (this is the definition of a wavefront), so that they remain strictly in phase only in a direction at right angles to the front (Fig. 2.4). Hence the plane wave

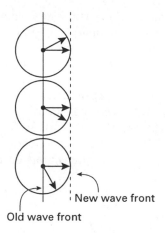

New wave front

Old wave front

Fig. 2.4 Huygens' construction.

appears to propagate in that direction. This principle of equivalence, first enunciated by Huygens and later shown by Kirchhoff to be mathematically sound [2], is very useful in the study of wave propagation phenomena generally.

The laws of reflection and refraction were first formulated in terms of 'rays' of light. It had been noticed (c. 1600) that, when dealing with 'point' sources, the light passed through apertures consistently with the view that it was composed of rays travelling in straight lines from the point. (It was primarily this observation which led to Newton's 'corpuscular' theory.) The practical concept was legitimized by allowing such light to pass through a small hole so as to isolate a 'ray'. Such rays were produced, and their behaviour in respect of reflection and refraction at material boundaries was formulated, thus:

(i) On reflection at a boundary between two media, the reflected ray lies in the same plane as that of the incident ray and the normal to the boundary at the point of incidence (the plane of incidence); the angle of reflection equals the angle of incidence.

(ii) On refraction at a boundary the refracted ray also lies in the plane of incidence, and the sine of the angle of refraction bears a constant ratio to the sine of the angle of incidence (Snell's law).

These two laws form the basis of what is known as geometrical optics, or, 'ray' optics. The majority of bulk optics (e.g. lens design, reflectometers, prismatics) can be formulated with its aid. However, it has severe limitations. For example, it cannot predict the intensities of the refracted and reflected rays.

If, in the attempt to isolate a ray of light of increasing fineness, the aperture is made too small, the ray divergence appears to increase, rather than diminish. This occurs when the aperture size becomes comparable with the wavelength of the light, and it is under this condition that the geometrical theory breaks down. 'Diffraction' has occurred and this is, quintessentially, a wave phenomenon. The wave theory provides a more complete, but necessarily more complex, view of light propagation. We shall now deal with the phenomena of reflection and refraction using the wave theory, but we should remember that, under certain conditions (apertures much larger than the wavelength), the ray theory is useful for its simplicity: a wave can be replaced by a set of rays in the direction of propagation, normal to surfaces of constant phase, and obeying simple geometrical rules.

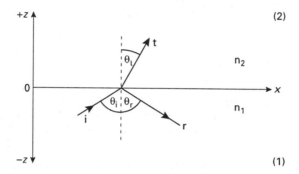

Fig. 2.5 Reflection and refraction at a boundary between two media.

Let us consider two non-conducting dielectric media with refractive indices n_1 and n_2, separated by a plane boundary which we take to be the xy plane at $z = 0$ (Fig. 2.5). Let us now consider a plane wave lying in the xz plane which is propagating in medium 1 and is incident on the boundary at angle ϑ_i, as shown in the figure. All the field components, such as (E_i, H_i), will vary as:

$$(E_i, H_i) \exp\{i\omega[t - n_1 (x \sin \vartheta_i + z \cos \vartheta_i) /c]\}$$

(see Fig. 2.6) using the exponential forms of the wave, detailed in section (2.3) and taking c to be the velocity of light in free space.

After striking the boundary there will, in general, be a reflected and a refracted (transmitted, t) wave. This fact is a direct consequence of the boundary conditions which must be satisfied at the interface between the two media. These conditions follow from Maxwell's equations, and essentially may be stated as:

(i) tangential components of **E** and **H** are continuous across the boundary
(ii) normal components of **B** and **D** are continuous across the boundary.

The above conditions must be true at all times and at all places on the boundary plane. They can only be true at all times at a given point if

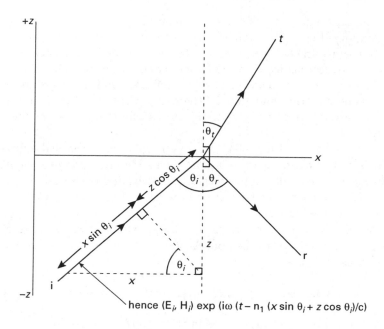

hence $(E_i, H_i)\,\exp\,(i\omega\,(t - n_1\,(x\sin\theta_i + z\cos\theta_i)/c)$

Fig. 2.6 Trigonometry of the incident ray.

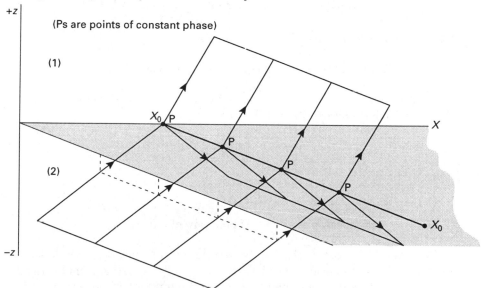

Fig. 2.7 Line of constant phase in boundary plane.

the frequencies of all the waves (i.e., incident, reflected, refracted) are the same; otherwise, clearly, amplitude discontinuities would occur across the boundary. Further, since the phase and amplitude of the incident wave must be constant on the boundary plane along any line for which x is constant (see Fig. 2.7) it follows that the phases and amplitudes of the reflected and refracted waves must also be constant along such a line, if continuity in accordance with the boundary conditions is to be maintained, and this is equivalent to saying that the reflected and refracted rays travel in the same direction and thus in the same plane (the xz plane), as the incident ray, which proves one of the previously stated laws of reflection and refraction.

To go further it is necessary to give proper mathematical expression to the waves. Any given wave is, of course, a sinusoid, whose amplitude, frequency and phase define the wave completely, and in section (2.3) it was shown that the most convenient representation of such waves was via their complex exponential form.

Suppose (Fig. 2.6) that the reflected and refracted waves make angles ϑ_r and ϑ_t, respectively, to the boundary in the xz plane. Then these waves will vary as:

$$\text{reflected}: \ \exp\{i\omega[t - n_1(x\sin\vartheta_r - z\cos\vartheta_r)/c]\}$$

(note that the reflected ray travels in the *negative* z direction)

$$\text{refracted}: \ \exp\{i\omega[t - n_2(x\sin\vartheta_t + z\cos\vartheta_t)/c]\}$$

whereas the incident wave, for reference, was:

$$\text{incident}: \ \exp\{i\omega[t - n_1(x\sin\vartheta_i + z\cos\vartheta_i)/c]\}$$

At the boundary ($z = 0$) these variations must be identical for any x, t, if continuity is to be maintained, hence:

$$n_1 x \sin\vartheta_i = n_1 x \sin\vartheta_r = n_2 x \sin\vartheta_t$$

Thus we have:

$$\vartheta_i = \vartheta_r \ \ \text{(law of reflection)}$$
$$n_1 \sin\vartheta_i = n_2 \sin\vartheta_t \ \ \text{(Snell's law of refraction)}$$

We must now consider the relative amplitudes of the waves. To do this we match the components of $\mathbf{E}, \mathbf{H}, \mathbf{D}, \mathbf{B}$, separately. A further complication is that the values of these quantities at the boundary will depend on the direction of vibration of the $\mathbf{E}, \mathbf{H}$, fields of the incident wave, relative to the plane of the wave. Therefore, we need to consider two linear, orthogonal polarization components separately, one in the xz plane, the other normal to it. (Any other polarization state can be resolved into these two linear components, so that our solution will be complete).

Let us consider the two stated linear components in turn:
(a) **E** *in the plane of incidence;* **H** *normal to the plane of incidence:*
The incident wave can now be written in the form (see Fig. 2.6)

$$E_x^i = -E_i \cos \vartheta_i \exp \{i\omega[t - n_1(x \sin \vartheta_i + z \cos \vartheta_i)/c]\}$$
$$E_z^i = E_i \sin \vartheta_i \exp \{i\omega[t - n_1(x \sin \vartheta_i + z \cos \vartheta_i)/c]\} \qquad (2.7)$$
$$H_y^i = H_i \exp \{i\omega[t - n_1(x \sin \vartheta_i + z \cos \vartheta_i)/c]\}$$

Now we can again enlist the help of Maxwell's equations to relate **H** and
E for a plane wave (see Appendix 1).
We have:

$$\frac{E}{H} = Z = \left(\frac{\mu}{\epsilon}\right)^{1/2}$$

Z is now known as the characteristic impedance of the medium. Since we
are dealing, in this case, with non-conducting dielectrics, we have $\mu = 1$
and $n = \varepsilon^{1/2}$, hence

$$Z = \frac{1}{n}$$

Thus:

$$H_i = nE_i \qquad (2.8)$$

and the expression for H_y^i becomes:

$$H_y^i = n_1 E_i \exp\{i\omega [t - n_1(x \sin \vartheta_i + z \cos \vartheta_i)/c]\}$$

Clearly we can construct similar sets of equations for the reflected and refracted waves. Having done this we can impose the boundary conditions to
obtain the required relationships between wave amplitudes. We shall now
derive these relationships, i.e., that between the reflected and incident electric field amplitudes, and that between the refracted and incident electric
field amplitudes for this case.
     We know that the exponential factors are all identical at the boundary
if we are going to be able to satisfy the boundary conditions at all; let us,
therefore, write the universal exponential factor as F.
     For the incident (i) wave, from equations (2.7) we have:

$$E_x^i = -E_i \cos \vartheta_i F$$
$$E_z^i = E_i \sin \vartheta_i F \qquad (i)$$
$$H_y^i = H_i F$$

For the reflected (r) wave:

$$E_x^r = E_r \cos \vartheta_r F$$
$$E_z^r = E_r \sin \vartheta_r F \qquad (r)$$
$$H_y^r = H_r F$$

For the refracted (t) wave:

$$E_x^t = -E_r \cos \vartheta_t F$$
$$E_z^t = E_t \sin \vartheta_t F \qquad\qquad (t)$$
$$H_y^t = H_t F$$

Imposing the condition that the tangential components (i.e., x components) of E must be continuous across the boundary, we have:

$$E_x^i + E_x^r = E_x^t$$

or

$$-E_i \cos \vartheta_i + E_r \cos \vartheta_r = -E_t \cos \vartheta_t \qquad\qquad (2.9)$$

using the appropriate equations from (i), (r) and (t) and cancelling the factor F.

Now doing the same for the tangential H field (y components):

$$H_i + H_r = H_t \qquad\qquad (2.10)$$

We also know, from (2.8), that

$$H_i = n_1 E_i; \;\; H_r = n_1 E_r; \;\; H_t = n_2 E_t$$

hence the H field condition (2.10) becomes

$$n_1 E_i + n_1 E_r = n_2 E_t \qquad\qquad (2.11)$$

We may now eliminate E_t from (2.9) and (2.11) to obtain (remembering, also, that $\vartheta_r = \vartheta_i$):

$$\frac{E_r}{E_i} = \frac{n_2 \cos \vartheta_i - n_1 \cos \vartheta_t}{n_2 \cos \vartheta_i + n_1 \cos \vartheta_t} \qquad\qquad (2.12a)$$

which is the required relationship.

Note also that, since, from Snell's law, $n_1 \sin \theta_i = n_2 \sin \theta_t$, this can be written:

$$\frac{E_r}{E_i} = \frac{\tan (\vartheta_i - \vartheta_t)}{\tan (\vartheta_i + \vartheta_t)}$$

Similarly we may eliminate E_r from (2.9) and (2.11) to obtain:

$$\frac{E_t}{E_i} = \frac{2 n_1 \cos \vartheta_i}{n_2 \cos \vartheta_i + n_1 \cos \vartheta_t} \qquad\qquad (2.12b)$$

We must now consider the wave with the other, orthogonal, polarization: (b) **E** *normal to the plane of incidence;* **H** *in the plane of incidence:* Using the same methods as before we obtain the relations:

$$\frac{E'_r}{E'_i} = \frac{n_1 \cos \vartheta_i - n_2 \cos \vartheta_t}{n_1 \cos \vartheta_i + n_2 \cos \vartheta_t} \qquad (2.12c)$$

$$\frac{E'_t}{E'_i} = \frac{2n_1 \cos \vartheta_i}{n_1 \cos \vartheta_i + n_2 \cos \vartheta_t} \qquad (2.12d)$$

The above four expressions (2.12) are known as *Fresnel's equations*; Fresnel derived them from the elastic-solid theory of light, which prevailed at his time. The equations contain several points worthy of emphasis.

First, we note that there is a possibility of eliminating the reflected wave. For E in the plane of incidence we find from equation (2.12a) that this occurs when:

$$n_1 \cos \vartheta_t = n_2 \cos \vartheta_i$$

But from Snell's law we also have:

$$n_1 \sin \vartheta_i = n_2 \sin \vartheta_t$$

so that, combining the two relations:

$$\sin 2\vartheta_i = \sin 2\vartheta_t$$

Now, of course, this equation has an infinite number of solutions, but the only one of interest is that for which $\vartheta_i \neq \vartheta_t$ ($\vartheta_i = \vartheta_t$ only if $n_1 = n_2$) and for which both ϑ_i and ϑ_t lie in the range $0 \to \pi/2$. The required solution is:

$$\vartheta_i + \vartheta_t = \tfrac{1}{2}\pi$$

and simple geometry then requires that the reflected and refracted rays are normal to each other (Fig. 2.8). Clearly, from Snell's law, this occurs when

$$n_1 \sin \vartheta_i = n_2 \cos \vartheta_i$$

i.e.,

$$\tan \vartheta_i = \frac{n_2}{n_1}$$

This particular value of ϑ_i is known as Brewster's angle (ϑ_B).

For example, for the glass/air boundary we find $\vartheta_B = 56.3°$.

It is instructive to understand the physical reason for the disappearance of the reflected ray at this angle when the electric field lies in the plane of incidence. Referring to Fig. 2.8 we note that the incident wave sets up oscillations of the elementary dipoles in the second medium (see Chapter 4 for details) and, at the Brewster angle, these oscillations take place in the direction of the reflected ray, since the refracted and reflected rays are orthogonal. Hence these oscillations cannot generate any transverse waves

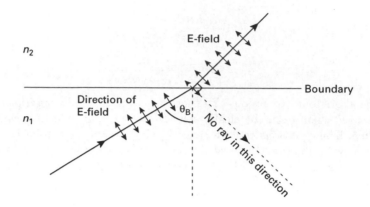

Fig. 2.8 Elimination of the reflected ray at the Brewster angle (ϑ_B).

in the required direction of reflection. Since light waves are, by their very nature, transverse, the reflected ray must be absent. If we ask the same question of the polarization which has E normal to the plane of incidence we find from (2.12c):

$$n_1 \cos \vartheta_i = n_2 \cos \vartheta_t$$

which, with Snell's law, gives:

$$\tan \vartheta_i = \tan \vartheta_t$$

There is no solution of this equation which satisfies the required conditions, so the reflected wave cannot be eliminated in this case. If, then, a wave of arbitrary polarization is incident on the boundary at the Brewster angle, only the polarization with **E** normal to the plane of incidence is reflected. This is a useful way of linearly polarizing a wave.

The second point worthy of emphasis is the condition at normal incidence. Here we have $\vartheta_i = \vartheta_r = \vartheta_t = 0$, hence the relations, identical for both polarizations, become:

$$\frac{E_r}{E_i} = \frac{E_r'}{E_i'} = \frac{n_1 - n_2}{n_1 + n_2} \qquad (2.13a)$$

$$\frac{E_t}{E_i} = \frac{E_t'}{E_i'} = \frac{2n_1}{n_1 + n_2} \qquad (2.13b)$$

Now the wave intensities are proportional to the squares of the electric field amplitudes *but only for a given medium*, since, from equation (2.6c), the intensity is proportional to the refractive index as well as to the square of the field.

Hence, since the incident and reflected waves propagate in the same medium, it is appropriate to write:

$$\frac{I_r}{I_i} = \frac{E_r^2}{E_i^2} = \left(\frac{n_1 - n_2}{n_1 + n_2}\right)^2 \qquad (2.13c)$$

but for the transmitted (refracted) wave, we have:

$$\frac{I_t}{I_i} = \frac{n_2 E_t^2}{n_1 E_i^2} = \frac{4n_1 n_2}{(n_1 + n_2)^2} \qquad (2.13d)$$

Note that now:

$$I_r + I_t = I_i$$

so that energy is conserved, as required.

(2.13c) and (2.13d) are useful expressions, for they tell us how much light is lost by normal reflection when transmitting from one medium (say air) to another (say glass). For example, when passing through a glass lens (air $\rightarrow$ glass $\rightarrow$ air), taking the refractive index of the glass as 1.5 we find from (2.13c) that the fractional loss at the front face of the lens (assumed approximately normal) is:

$$\frac{I_r}{I_i} = \frac{(0.5)^2}{(2.5)^2} = 0.04$$

Another 4% will be lost at the back face, giving a total 'Fresnel' loss of the order of 8%. This figure can be reduced by 'anti-reflection' coatings, of which more later (Chapter 10).

Finally we should notice that all the expressions for the ratios of field amplitudes are mathematically real, and thus any change of phase which occurs at a boundary must be either 0 or π. We shall now look at a rather different type of reflection where this is not the case.

2.6 TOTAL INTERNAL REFLECTION

We return to Snell's law:

$$\boxed{n_1 \sin \vartheta_i = n_2 \sin \vartheta_t}$$

or

$$\sin \vartheta_t = \frac{n_1}{n_2} \sin \vartheta_i \qquad (2.14)$$

The factor $\sin \vartheta_i$ is, of course, always less than unity. However if $n_2 < n_1$ (i.e., the second medium is less optically dense than the first, which contains the incident ray) then it may be that:

$$\sin \vartheta_i > \frac{n_2}{n_1}$$

i.e.,

$$\frac{n_1}{n_2} \sin \vartheta_i > 1$$

If this is so then we have from (2.14):

$$\sin \vartheta_t > 1 \tag{2.15}$$

Equation (2.15) clearly cannot be satisfied for any real value of ϑ_t and there can be no real refracted ray. The explanation of this is that the refracted ray angle (ϑ_t), under these conditions of passage from a less dense to a more dense medium, is always greater than the incident angle (ϑ_i). Consequently ϑ_t will reach a value of 90° (i.e., parallel to the boundary) before ϑ_i, and any greater value of ϑ_i cannot yield a refracted ray (Fig. 2.9). The value of ϑ_i for which (2.15) just becomes true we define as the critical angle, ϑ_c:

$$\sin \vartheta_c = \frac{n_2}{n_1}$$

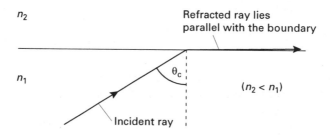

Fig. 2.9 Critical angle (ϑ_c) for total internal reflection (TIR).

For all values of $\vartheta_i > \vartheta_c$ the light is totally reflected at the boundary: the phenomenon is called total internal reflection (TIR). However, Fresnel's equations must still apply, for we made no limitations on the values of the quantities when imposing the boundary conditions. Furthermore, if the fields are to be continuous across the boundary, as required by Maxwell's equations, then there must be a field disturbance of some kind in the second medium. We can use Fresnel's equations to investigate this disturbance.
We write:

$$\cos \vartheta_t = \left(1 - \sin^2 \vartheta_t\right)^{1/2} \tag{2.16}$$

Since $\sin \vartheta_t > 1$ for $\vartheta_t > \vartheta_c$, and since also the function $\cosh \gamma \geq 1$ for all real γ, we may, for convenience, use the substitution:

$$\sin \vartheta_t = \cosh \gamma \quad (\vartheta_i > \vartheta_c)$$

and henceforth, therefore, the TIR condition (2.15) is, implicitly, imposed. We now have, from (2.16):

$$\cos\vartheta_t = i\left(\cosh^2\gamma - 1\right)^{1/2} = \pm i\sinh\gamma$$

Hence we may write the field components in the second medium to vary as:

$$\exp\left\{i\omega\left[t - n_2\left(x\cosh\gamma - iz\sinh\gamma\right)/c\right]\right\}$$

or

$$\exp\left[\left(-\omega n_2 z\sinh\gamma\right)/c\right]\exp\left[i\omega\left(t - n_2 x\cosh\gamma\right)/c\right]$$

where we have used the fact that $\cosh\gamma = \frac{1}{2}\left(e^\gamma + e^{-\gamma}\right)$ which tends to infinity as $\gamma \to \pm\infty$, and has a minimum of 1 at $\gamma = 0$.

This represents a wave travelling in the Ox direction in the second medium (i.e., parallel to the boundary) with amplitude decreasing exponentially in the Oz direction (at right angles to the boundary). The rate at which the amplitude falls with z can be written:

$$\exp\left[\left(-2\pi z\sinh\gamma\right)/\lambda_2\right]$$

or, in terms of the original parameters:

$$\exp\left[-k_2 z\left(n_1^2\sin^2\vartheta_i - n_2^2\right)^{1/2}\right]$$

λ_2 being the wavelength of the light and k_2 the wave number, in the second medium. This shows that the wave is attenuated significantly over distances $\sim \lambda_2$. For example, at the glass/air interface, the critical angle will be $\sim \sin^{-1}(1/1.5)$, i.e., $\sim 42°$. For a wave in the glass incident on the glass/air boundary at $60°$ $(\vartheta_i > \vartheta_c)$ we find that $\sinh\gamma = 1.64$. Hence the amplitude of the wave in the second medium is reduced by a factor of 5.4×10^{-3} in a distance of only one wavelength, the latter being of order $1\,\mu\text{m}$. The wave is called an 'evanescent' wave. Even though the evanescent wave is propagating in the second medium, it transports no light energy in a direction normal to the boundary. All the light is totally internally reflected at the boundary. The fields which exist in the second medium give a Poynting vector which averages to zero in this direction, over one oscillation period of the light wave. All the energy in the evanescent wave is transported parallel to the boundary between the two media. The totally internally reflected wave now suffers a phase change which depends both on the angle of incidence and on the polarization. This can readily be derived from Fresnel's equations. Taking equation (2.12a) we have for the TIR case where E lies in the plane of incidence:

$$\frac{E_r}{E_i} = \frac{n_2\cos\vartheta_i - in_1\sinh\gamma}{n_2\cos\vartheta_i + in_1\sinh\gamma}$$

This complex number provides the phase change on TIR as δ_p where:

$$(E_{para}): \quad \tan\left(\tfrac{1}{2}\delta_p\right) = \frac{n_1\left(n_1^2\sin^2\vartheta_i - n_2^2\right)^{1/2}}{n_2^2\cos\vartheta_i}$$

and for the perpendicular E polarization:

$$(E_{perp}): \quad \tan\left(\tfrac{1}{2}\delta_s\right) = \frac{\left(n_1^2\sin^2\vartheta_1 - n_2^2\right)^{1/2}}{n_1\cos\vartheta_i}$$

We note also that:

$$\tan\left(\tfrac{1}{2}\delta_p\right) = n_1^2\tan\left(\tfrac{1}{2}\delta_s\right)$$

and that:

$$\tan\left[\tfrac{1}{2}\left(\delta_p - \delta_s\right)\right] = \frac{\cos\vartheta_1\left(n_1^2\sin^2\vartheta_i - n_2^2\right)^{1/2}}{n_1\sin^2\vartheta_i}$$

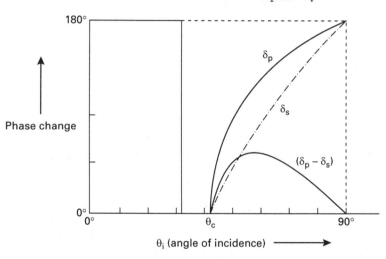

Fig. 2.10 Phase changes on total internal reflection.

The variations δ_p, δ_s and $\delta_p - \delta_s$ are shown in Fig. 2.10 as a function of ϑ_i. It is clear that the polarization state of light undergoing TIR will be changed as a result of the differential phase change $\delta_p - \delta_s$. By choosing ϑ_i appropriately and, perhaps, using two TIRs, it is possible to produce any wanted, final polarization state from any given initial state.

It is interesting to note that the reflected ray in TIR appears to originate from a point which is displaced along the boundary from the point of incidence. This is consistent with the incident ray being reflected from a

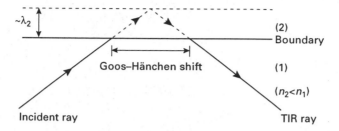

Fig. 2.11 The Goos-Hanchen shift on total internal reflection.

parallel plane which lies a short distance within the second boundary (Fig. 2.11). This view is also consistent with the observed phase shift, which now is regarded as being due to the extra optical path travelled by the ray. The displacement is known as the Goos-Hanchen effect and provides an entirely consistent alternative explanation of TIR. This provides food for further interesting thoughts, which we shall not pursue since they are somewhat beyond the scope of this book.

2.7 INTERFERENCE OF LIGHT

We have seen that light consists of oscillating electric and magnetic fields. We know that these fields are vector fields since they represent forces (on unit charge and unit magnetic pole, respectively). The fields will thus add vectorially. Consequently, when two light waves are superimposed on each other we obtain the resultant by constructing their vector sum at each point in time and space, and this fact has already been used in consideration of the polarization of light (section 2.4.3).

If two sinusoids are added, the result is another sinusoid. Suppose that two light waves given, via their electric fields, as:

$$e_1 = E_1 \cos(\omega t + \varphi_1)$$
$$e_2 = E_2 \cos(\omega t + \varphi_2)$$

have the same polarization and are superimposed at a point in space. We know that the resultant field at the point will be given, using elementary trigonometry or by the complex exponential methods described in section 2.3, by:

$$e_t = E_T \cos(\omega t + \varphi_T)$$

where:

$$E_T^2 = E_1^2 + E_2^2 + 2E_1 E_2 \cos(\varphi_2 - \varphi_1)$$

and

$$\tan \vartheta_T = \frac{E_1 \sin \varphi_1 + E_2 \sin \varphi_2}{E_1 \cos \varphi_1 + E_2 \cos \varphi_2}$$

For the important case where $E_1 = E_2 = E$, say, we have:

$$E_T^2 = 4E^2 \cos^2 \tfrac{1}{2}(\varphi_2 - \varphi_1) \qquad (2.17)$$

and

$$\tan \phi_T = \tan \tfrac{1}{2}(\varphi_2 + \varphi_1)$$

The intensity of the wave will be proportional to E_T^2 so that, from (2.17) it can be seen to vary from $4E^2$ to 0, as $(\varphi_2 - \varphi_1)/2$ varies from 0 to $\pi/2$.

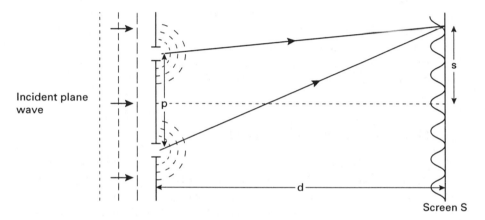

Fig. 2.12 'Young's slits' interference.

Consider now the arrangement shown in Fig. 2.12. Here two slits, separated by a distance p, are illuminated by a plane wave with wavelength λ. The portions of the wave which pass through the slits will interfere on the screen S, a distance d away. Now each of the slits will act as a source of cylindrical waves, from Huygens' principle. Moreover, since they originate from the same plane wave, they will start in phase. On a line displaced a distance s from the line of symmetry on the screen the waves from the two slits will differ in phase by:

$$\delta = \frac{2\pi}{\lambda} \frac{sp}{d} \qquad (d \gg s, p)$$

Thus, as s increases, the intensity will vary between a maximum and zero, in accordance with equation (2.17). These variations will be viewed as fringes, i.e., lines of constant intensity parallel with the slits. They are known as Young's fringes, after their discoverer, and are the simplest example of light interference. We shall now consider some important (and more complex) examples of light interference in action.

2.8 LIGHT WAVEGUIDING

Consider, first, the symmetrical dielectric structure shown in Fig. 2.13. Here we have an infinite (in width and length) dielectric slab of refractive index n_1, sandwiched between two other infinite slabs each of refractive index n_2.

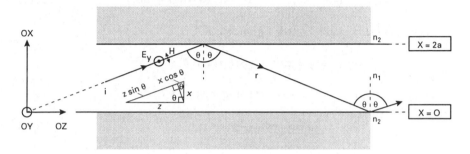

Fig. 2.13 The dielectric-slab waveguide.

Using the Cartesian axes defined in the figure let us consider a light ray starting at the origin of axes and propagating within the first medium at an angle ϑ. If ϑ is greater than the critical angle (ϑ_c) the light will bounce down the first medium by means of a series of total internal reflections at the boundaries with the other media. Since the wave is thus confined to the first medium it is said to be 'guided' by the structure, which is consequently called a 'waveguide'. Let us, firstly, consider guided light which is linearly polarized normal to the plane of incidence. The electric field of the wave represented by ray i (see Fig. 2.13) can be written:

$$E_i = E_0 \exp(i\omega t - kn_1 x \cos \vartheta - ikn_1 z \sin \vartheta)$$

whilst that represented by r, the ray reflected from the first boundary, can be written:

$$E_r = E_0 \exp(i\omega t + kn_1 x \cos \vartheta - ikn_1 z \sin \vartheta + i\delta_s)$$

where δ_s is the phase change at TIR for this polarization. These two waves will be superimposed on each other and will thus interfere. The interference pattern is obtained by adding them:

$$E_T = E_i + E_r = E_0 \exp\left(i\omega t - ikn_1 z \sin \vartheta + i\tfrac{1}{2}\delta_s\right) 2 \cos\left(kn_1 x \cos \vartheta + \tfrac{1}{2}\delta_s\right)$$
$$(2.18)$$

This is a wave propagating in the direction Oz with wave number $kn_1 \sin \vartheta$, and it is amplitude-modulated in the Ox direction according to

$$\cos \left(kn_1 x \cos \vartheta + \tfrac{1}{2}\delta_s \right)$$

Now if the wave propagating in the Oz direction is to be a stable, symmetrical entity resulting from a self-reproducing interference pattern, the intensity of the wave must be the same at each of the two boundaries. This requires that it is the same for $x = 0$ as for $x = 2a$. i.e.,

$$\cos^2 \left(\tfrac{1}{2}\delta_s \right) = \cos^2 \left(kn_1 2a \cos \vartheta + \tfrac{1}{2}\delta_s \right) \tag{2.19}$$

The general solution of this equation is:

$$\tfrac{1}{2}\delta_s = m\pi \pm \left(2akn_1 \cos \vartheta + \tfrac{1}{2}\delta_s \right)$$

where m is any integer (positive or negative).
   Hence either:

$$2akn_1 \cos \vartheta + \delta_s = m\pi \quad (-) \tag{2.20a}$$

or

$$2akn_1 \cos \vartheta = -m\pi \quad (+) \tag{2.20b}$$

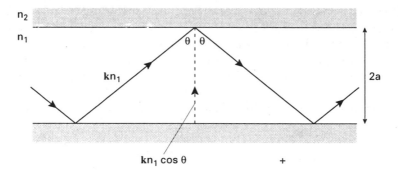

Transverse wave number = $kn_1 \cos \theta$
∴   Phase change across guide, width 2a = $2akn_1 \cos \theta$
     Phase change on reflection = δ_s
∴ $2akn_1 \cos \theta + \delta_s = m\pi$, for reinforcement

Fig. 2.14 Transverse resonance condition.

However, there is another condition to impose. If the interference pattern is to self-reproduce in a consistent way as it propagates down the guide, the phase change experienced by a ray executing one complete 'bounce' down the guide must be an integer times 2π. If this were not so, the waves would not retain mutual phase coherence and the interference pattern would self-destruct. This can be seen from the geometry in Fig. 2.14. The wavefronts resulting from ray reflections at all points along the guide can only be in phase provided that

$$2akn_1 \cos \vartheta + \delta_s = m\pi$$

which corresponds to (2.20a). Equation (2.20b) does not satisfy the condition on wavefronts and is, therefore, invalid. Equation (2.20a) is sometimes known as the 'transverse resonance condition' since it corresponds essentially to the condition that, when resolving the wave vector into directions transverse and parallel to the guide axis, the transverse component has just one half cycle, or an integer multiple thereof $(m\pi)$, fitting into the guide width. This is a 'resonance' in the sense that a string stretched between two points resonates, when plucked, at frequencies which are conditioned in just the same way.

Now since δ_s depends only on ϑ (see Fresnel's equations in section 2.5) it follows that the condition

$$2akn_1 \cos \vartheta + \delta_s = m\pi$$

is a condition on ϑ. The condition tells us that ϑ can have only certain discrete values if the interference pattern is to remain constant in form along the length of the fibre. Each form of interference pattern is, therefore, characterized by a particular value of m which then provides a corresponding value for ϑ. The allowed interference patterns are called the 'modes' of the waveguide, for they are determined by the properties (geometrical and physical) of the guide.

If we now turn to the progression of the wave along the guide (i.e., along the Oz axis) we see from equation (2.18) that this is characterized by a wave number of value:

$$n_1 k \sin \vartheta = \beta (\text{say})$$

Furthermore, since the TIR condition requires that:

$$\sin \vartheta \geq \frac{n_2}{n_1}$$

it follows that:

$$n_1 k \geq \beta \geq n_2 k$$

so that the longitudinal wave number always lies between those of the two guiding media.

Thus we see that waveguiding essentially is a wave interference phenomenon and we shall leave the subject there for the moment. The subject is an extremely important one and there are many other aspects to be considered. Consequently we shall return to it in more detail in Chapter 8.

2.9 INTERFEROMETERS

In section (2.7) the essentials of dual-beam interference were discussed. Although very simple in concept the phenomenon is extremely useful in practice. The reason for this is that the maxima of the resulting fringe pattern appear where the phase difference between the interfering light beams is a multiple of 2π. Any quite small perturbation in the phase of one of the beams will thus cause a transverse shift in the position of the fringe pattern, which, using optoelectronic techniques, is readily observed to about 10^{-4} of the fringe spacing. Such a shift is caused by, for example, an increase in path length of one of the beams by one hundredth of a wavelength, or about 5×10^{-9} m for visible light. This means that differential distances of this order can be measured, leading to obvious applications in, for example, strain monitoring on mechanical structures.

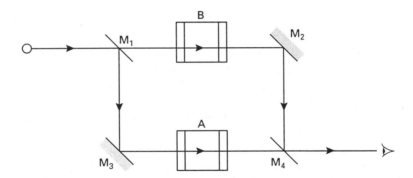

Fig. 2.15 Basic Mach-Zehnder interferometer.

Another example of a dual-beam interferometer is shown in Fig. 2.15. Here the beams are produced from the partial reflection and transmission at a dielectric, or partially silvered, mirror M_1. Another such mirror, M_4, recombines the two beams after their separate passages. Such an arrangement is known as a Mach-Zehnder interferometer and is used extensively to monitor changes in the phase differences between two optical paths. An optical-fibre version of a Mach-Zehnder interferometer is shown in Fig. 2.16. In this case the 'mirrors' are optical couplings between the cores of the two fibres. The 'fringe pattern' consists effectively of just one fringe, since the fibre core acts as an efficient spatial filter. But the light which emerges from the fibre end (E) clearly will depend on the phase relationship between the two optical paths when the light beams recombine at R,

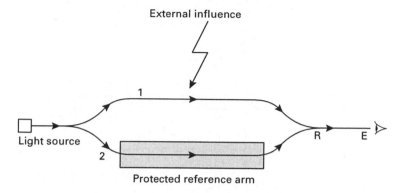

Fig. 2.16 An optical-fibre Mach-Zehnder interferometer.

and thus it will depend critically on propagation conditions within the two arms. If one of the arms varies in temperature, strain, density, etc. compared with the other, then the light output will also vary. Hence the latter can be used as a sensitive measure of any physical parameters which are capable of modifying the phase propagation properties of the fibre.

Finally, Fig. 2.17(a) shows another, rather more sophisticated variation of the Mach-Zehnder idea. In this case the beams are again separated by means of a beam-splitting mirror, but are returned to the same point by fully silvered mirrors placed at the ends of the two respective optical paths. (The plate P is necessary to provide equal optical paths for the two beams in the absence of any perturbation.) This arrangement is called the Michelson interferometer, after the experimenter who in the late 19th century used optical interferometry with great skill to make many physical advances. His interferometer (not to be confused with his 'stellar' interferometer, of which more later) allows for a greater accuracy of fine adjustment via control of the reflecting mirrors, but uses, of course, just the same basic interferometric principles as before. The optical-fibre version of this device is shown in Fig. 2.17(b).

For completeness, and because of its historical importance, mention must be made of the use of Michelson's interferometer in the famous Michelson-Morley experiment of 1887. This demonstrated that light travelled with the same velocity in each of two orthogonal paths, no matter what was the orientation of the interferometer with respect to the earth's 'proper motion' through space. This result was crucial to Einstein's formulation of special relativity in 1905, and thus is certainly one of the most important results in the history of experimental physics.

Valuable as dual-beam interferometry is, it suffers from the limitation that its accuracy depends upon the location of the maxima (or minima) of a sinusoidal variation. For very accurate work, such as precision spec-

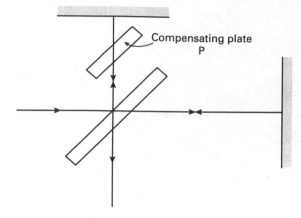

(a) Bulk version

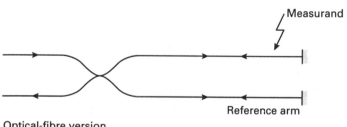

(b) Optical-fibre version

Fig. 2.17 Michelson interferometers.

troscopy, this limitation is severe. By using the interference amongst many beams, rather than just two, we find that we can improve the accuracy very considerably. We can see this by considering the arrangement of Fig. 2.18. Light from a single source gives a large number of phase-related, separate beams by means of multiple reflections and transmissions within a dielectric (e.g. glass) plate. For a given angle of incidence (ϑ) there will be fixed values for the transmission (T, T') and reflection (R) coefficients, as shown. If we start with a wave of amplitude a the waves on successive reflections will suffer attenuation by a constant factor, and will increase in phase by a constant amount. If we consider the transmitted light only, then the total amplitude which arrives at the focus of the lens L is given by the sum:

$$A_T = aTT' \exp(i\omega t) + aTT'R^2 \exp(i\omega t - iks) + aTT'R^4 \exp(i\omega t - 2iks) + \ldots..$$

where s is the optical path difference between successive reflections at the lower surface (including the phase changes on reflection and transmission).

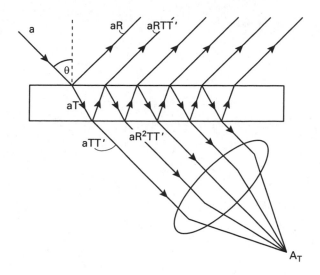

Fig. 2.18 Multiple interference.

The sum can be expressed as:

$$A_T = aTT' \sum_{p=0}^{\infty} R^{2p} \exp(i\omega t - ipks)$$

which is a geometric series whose sum value is:

$$A_T = \frac{aTT' \exp(i\omega t)}{1 - R^2 \exp(-iks)}$$

Hence the intensity I of the light is proportional to $|A_T|^2$, i.e.,

$$I \propto |A_T|^2 = \frac{(aTT')^2}{1 + R^4 - 2R^2 \cos ks} \tag{2.21}$$

We note from this equation that the ratio of maximum and minimum intensities:

$$\frac{I_{\max}}{I_{\min}} = \frac{\left(1 + R^2\right)^2}{\left(1 - R^2\right)^2}$$

so that the fringe contrast increases with R. However, as R increases so does the attenuation between the successive reflections. Hence the total transmitted light power will fall.

    Figure 2.19 shows how I varies with ks for different values of R. We note that the fringes become very sharp for large values of R. Hence the position of the maxima may now be accurately determined. Further, since the spacing of the maxima specifies ks, this information can be used to determine

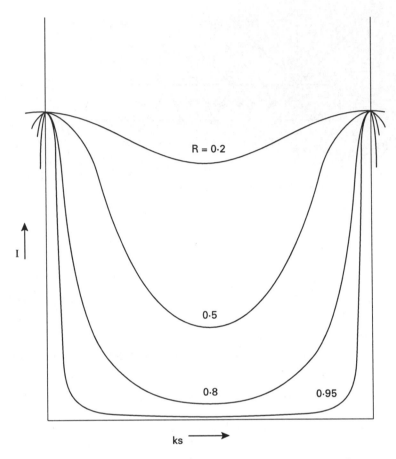

Fig. 2.19 Variation of intensity with optical path, for various reflectivities, in a multiple interference plate.

either k or s, if the other is known. Consequently, multiple interference may be used either to select (or measure) a very specific wavelength, or to measure very small changes in optical path length.

The physical reason for the sharpening of the fringes as the reflectivity increases is indicated in Fig. 2.20. The addition of the multiplicity of waves is equivalent to the addition of vectors with progressively decreasing amplitude, and increasing relative phase. For small reflectivity (Fig. 2.20(a)) the wave amplitudes decrease rapidly, so that the phase increase has a relatively small effect on the resultant wave amplitude. In the case of high reflectivity (Fig. 2.20(b)), the reverse is the case and a small successive phase change rapidly reduces the resultant.

Two important devices based on these ideas of multiple reflection are the Fabry-Perot interferometer and the Fabry-Perot etalon. In the former case

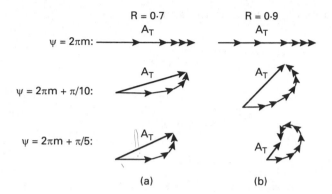

Fig. 2.20 Graphical illustration of the dependence of fringe sharpness on reflectivity (R).

the distance between the two surfaces is finely variable for fringe control; in the case of the etalon the surfaces are fixed. In both cases the flatness and parallelism of the surfaces must be accurate to $\sim \lambda/100$ for good quality fringes. This is difficult to achieve in a variable device, and the etalon is preferred for most practical purposes.

The Fabry-Perot interferometer is extremely important in optoelectronics. We have already noted its wavelength selectivity but we should also note its ability to store optical energy by continually bouncing light between two parallel mirrors. For this reason it is often called a Fabry-Perot 'cavity' and is, roughly speaking, the optical equivalent of an electronic oscillator. The optical term is 'resonator', and it is this property which makes it an integral feature in all lasers.

Because of its importance we must further understand in more detail the parameters which characterize the performance of the Fabry-Perot resonator: there are three main parameters.

These parameters relate, as is to be expected, to the instrument's ability to separate closely-spaced optical wavelengths. The first is a measure of the sharpness of the fringes. This measure is normalized to the separation of the fringes for a single wavelength, since, clearly, there is no advantage in having narrow fringes if they are all crowded together, so that the orders of different wavelengths overlap. We hence define a quantity:

$$\Phi = \frac{\text{separation of successive fringes}}{\text{width at half maximum of a single fringe}}$$

which is called the 'finesse' and is roughly equivalent to the Q ('quality' factor measuring the sharpness of the resonance) of an electronic oscillator.

It is easy to derive an expression for Φ from equation (2.21) as follows.

The equation may be written in the form:

$$I = \frac{I_{\text{max}}}{1 + F \sin^2\left(\frac{1}{2}\Psi\right)} \qquad (2.22)$$

where

$$F = \frac{4R^2}{(1 - R^2)^2}$$

and

$$\Psi = ks$$

From this it is clear that $I = I_{\text{max}}/2$ when:

$$\Psi_h = \frac{2}{\sqrt{F}}$$

Hence the width at half maximum $= 2\Psi_h = 4/\sqrt{F}$

The 'Ψ distance' between successive maxima is just 2π, and thus the finesse is given by:

$$\Phi = \frac{2\pi}{2\Psi_h} = \frac{\pi\sqrt{F}}{2} = \frac{\pi R}{(1 - R^2)}$$

This quantity has a value of 2 for a dual-beam interferometer. For a Fabry-Perot etalon with $R = 0.9$ its value is 15. Clearly, the higher the value the sharper are the fringes for a given fringe separation and the more wavelength selective is the device.

The next quantity we need to look at is the resolving power. This is a measure of the smallest detectable wavelength separation $\delta\lambda$ at a given wavelength λ and is defined as:

$$\rho = \frac{\lambda}{\delta\lambda}$$

If we take λ to be that which corresponds to a Ψ difference equal to the width of the half maximum, we have:

$$\rho = \frac{\lambda}{\delta\lambda} = \frac{\Psi}{2\Psi_h} = \frac{2\pi p}{4/\sqrt{F}} = \frac{\pi p}{2}\sqrt{F} = p \times \text{finesse}$$

i.e.,

$$\rho = p\Phi$$

where p is the 'order' of the maximum (i.e., the number of maxima from the one at $\Psi = 0$). If the etalon is being viewed close to normal incidence, then p will be effectively just the number of wavelengths in a double passage across the etalon. If the etalon has optical thickness t we have $p = 2t/\lambda$ and thus:

$$\rho = \pi t \frac{\sqrt{F}}{\lambda}$$

This is typically of the order of 10^6, compared with a figure $\sim 10^4$ for a dual-beam interferometer such as the Michelson. The ratio of these figures thus represents the improvement in accuracy afforded by multiple-beam interferometry over dual-beam techniques.

Finally, we define a quantity concerned with the overlapping of orders. If the range of wavelengths $(\Delta\lambda)$ under investigation is such that the $(p+1)$th maximum of λ is to coincide with the pth maximum of $(\lambda + \Delta\lambda)$, then, clearly, there is an unresolvable confusion. For this just to be so:

$$(p+1)k = p(k + \Delta k)$$

so that

$$\frac{\Delta k}{k} = \frac{\Delta\lambda}{\lambda} = \frac{1}{p}$$

Again, close to normal incidence we may write, with $p = 2t/\lambda$:

$$\Delta\lambda = \frac{\lambda}{p} = \frac{\lambda^2}{2t}$$

$\Delta\lambda$ is called the 'free spectral range'of the etalon and represents the maximum usable wavelength range without recourse to prior separation of the confusable wavelengths.

We shall need to return to the Fabry-Perot interferometer later on.

2.10 DIFFRACTION

In section (2.5) it was noted that each point on a wavefront could be regarded formally and rigorously as a source of spherical waves. In section (2.7) it was noted that any two waves, when superimposed, will interfere. Consequently wavefronts can interfere with themselves and with other, separate, wavefronts. To the former usually is attached the name 'diffraction' and to the latter 'interference', but the distinction is somewhat arbitrary and, in several cases, far from clear cut.

Diffraction of light may be regarded as the limiting case of multiple interference as the source spacings become infinitesimally small. Consider the slit aperture in Fig. 2.21. This slit is illuminated with a uniform plane wave and the light which passes through the slit is observed on a screen which is sufficiently distant from the slit for the light which falls upon it to be effectively, again, a plane wave. These are the conditions for

Fraunhofer diffraction. If source and screen are close enough to the slit for the waves not to be plane we have a more complex situation, known as Fresnel diffraction. Fraunhofer diffraction is by far the more important of the two, and is the only form of diffraction we shall deal with here. Fresnel diffraction usually can be transformed into Fraunhofer diffraction, in any case, by the use of lenses which render the waves effectively plane, even over short distances.

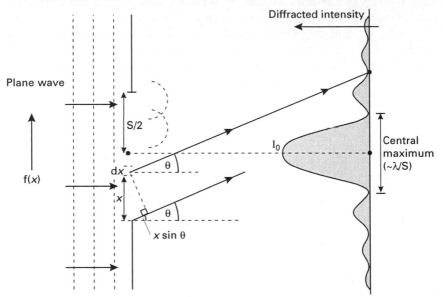

Fig. 2.21 Diffraction at a slit.

Suppose that in Fig. 2.21 the amplitude of the wave at distances between x and $x + dx$ along the slit is given by the complex quantity $f(x)dx$, and consider the effect of this at angle ϑ, as shown. (Since each point on the wavefront acts as a source of spherical waves, all angles will, of course, be illuminated by the strip). The screen, being effectively infinitely distant from the slit, will be illuminated at one point by the light leaving the slit at angles between ϑ and $\vartheta + d\vartheta$. Taking the bottom of the slit as the phase reference, the light, on arriving at the screen, will lead by a phase:

$$\Phi = kx \sin \vartheta$$

and hence the total amplitude in directions ϑ to $\vartheta + d\vartheta$ will be given by:

$$A(\vartheta) = \int_{-\infty}^{\infty} f(x) \exp(-ikx \sin \vartheta) dx$$

We can also write:

$$A(\alpha) = \int\limits_{-\infty}^{\infty} f(x) \exp(-i\alpha x) dx$$

with

$$\alpha = k \sin \vartheta$$

Hence $A(\alpha)$ and $f(x)$ constitute a reciprocal Fourier transform pair (see Appendix II), i.e. each is the Fourier transform of the other. This is an important result. For small values of ϑ it implies that the angular distribution of the diffracted light is the Fourier transform of the aperture's amplitude distribution.

Let us see how this works for some simple cases. Take first a uniformly illuminated slit of width s. The angular distribution of the diffracted light will now be:

$$A(k \sin \vartheta) = \int\limits_{-s/2}^{s/2} a \exp(-ikx \sin \vartheta) dx$$

where a is the (uniform) amplitude at the slit per unit of slit width.
Hence:

$$A(k \sin \vartheta) = a \frac{\sin \left(\frac{1}{2} ks \sin \vartheta\right)}{\left(\frac{1}{2} k \sin \vartheta\right)}$$

Writing, for convenience, $\beta = \frac{1}{2}ks \sin \vartheta$, we find that the intensity in a direction ϑ is given by:

$$I(\vartheta) = (as)^2 \frac{\sin^2 \beta}{\beta^2} = I_0 \frac{\sin^2 \beta}{\beta^2} \tag{2.23}$$

where I_0 is intensity at the centre of the diffraction pattern. This variation is shown in Fig. 2.21 and, as in the case of multiple interference between discrete sources, its shape is a result of the addition of wave vectors with phase increasing steadily with ϑ.

This form of variation occurs frequently in physics across a broad range of applications and it is instructive to understand why. The function appropriate to the variation is given the name 'sinc'(pronounced 'sink'), i.e.,

$$\text{sinc } \beta = \frac{\sin \beta}{\beta}$$

$$\text{sinc}^2 \beta = \frac{\sin^2 \beta}{\beta^2}$$

Let us examine the physical reason for the sinc function in the case we have been considering, i.e. a uniformly illuminated slit. In this case each infinitesimal element of the slit provides a wave amplitude $a\,dx$ and at the centre of the screen all of these elements are in phase, producing a total amplitude, as. Hence it is possible to represent all these elementary vectors as a straight line (since they are all in phase) of length as [Fig. 2.22(a)]. Now consider the situation at angle ϑ to the axis. As already shown the ray from the bottom of the slit lags that from the top by a phase:

$$\varphi_T = ks\sin\vartheta = 2\beta$$

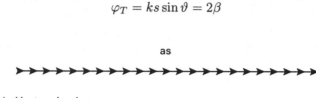

(a) Vectors in phase

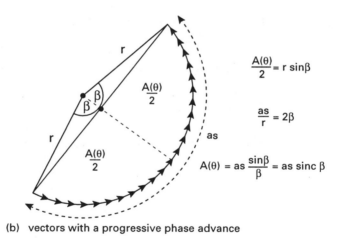

$$\frac{A(\theta)}{2} = r\sin\beta$$

$$\frac{as}{r} = 2\beta$$

$$A(\theta) = as\,\frac{\sin\beta}{\beta} = as\,\text{sinc}\,\beta$$

(b) vectors with a progressive phase advance

Fig. 2.22 Graphical explanation of the 'sinc' function.

The result can, therefore, be depicted as in Fig. 2.22(b). The first and last infinitesimal vectors are inclined at 2β to each other and the intervening vectors form an arc of a circle which subtends 2β at the circle's centre. The vector addition of all the vectors thus leads to a resultant which is the chord across the arc in Fig. 2.22(b). Simple geometry gives the length of this chord as:

$$A(\theta) = 2r\sin\beta \tag{2.24}$$

where r is the radius of the circle. Now the total length of the arc is the same as that of the straight line when all vectors were in phase (i.e., as)

hence

$$\frac{as}{r} = 2\beta$$

and thus, substituting for r in (2.24) we have:

$$A(\vartheta) = as\frac{\sin\beta}{\beta}$$

Hence the resultant intensity at angle ϑ will be:

$$I(\vartheta) = (as)^2\frac{\sin^2\beta}{\beta^2} = I_0\frac{\sin^2\beta}{\beta^2}$$

as in equation (2.23).

The reason for the ubiquity of this variation in physics can now be seen to be due to the fact that one very often encounters situations where there is a systematically increasing phase difference amongst a large number of infinitesimal vector quantities: optical interference, electron interference, mass spectrometer energies, particle scattering, etc., etc.

The principles which lead to the sinc function are all exactly the same, and are those which have just been described.

Let us return now to the intensity diffraction pattern for a slit.

$$I(\vartheta) = I_0\frac{\sin^2\beta}{\beta^2}$$

An important feature of this variation is the scale of the angular divergence. The two minima immediately on either side of the principal maximum (at $\vartheta = 0$) occur when:

$$\beta = \tfrac{1}{2}(ks)\sin\vartheta = \pm\pi$$

giving:

$$\sin\vartheta = \pm\frac{\lambda}{s}$$

so that, if ϑ is small, the width of the central maximum is given by:

$$\theta_w = 2\vartheta = \pm\frac{2\lambda}{s}$$

Thus the smaller is s for a given wavelength the more quickly does the light energy diverge, and vice versa. This is an important determinant of general behaviour in optical systems.

As a second example consider a sinusoidal variation of amplitude over the aperture. The Fourier transform of a sinusoid consists of one positive and one negative 'frequency' equally spaced around the origin. Thus

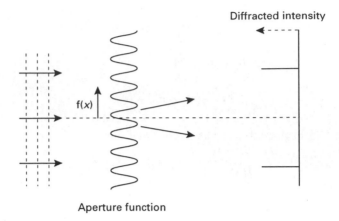

Aperture function

Fig. 2.23 Sinusoidal diffracting aperture.

the diffraction pattern consists of just two lines of intensity equally spaced about the centre position of the observing screen (Fig. 2.23). Those two lines of intensity could themselves be photographed to provide a 'two-slit' aperture plate which would then provide a sinusoidal diffraction (interference?) pattern. This latter pattern will be viewed as an 'intensity' pattern, however, not an 'amplitude' pattern. Consequently it will not comprise the original aperture, which must have positive and negative amplitude in order to yield just two lines in its diffraction pattern. Thus, whilst this example illustrates well the strong relationship which exists between the two functions, it also serves to emphasize that the relationship is between the *amplitude* functions, while the observed diffraction pattern is (in the absence of special arrangements) the *intensity* function.

Finally, we consider one of the most important examples of all: a rectangular-wave aperture amplitude function. The function is shown in Fig. 2.24. This is equivalent to a set of narrow slits, i.e. to a diffraction grating. The Fourier transform (and hence the Fraunhofer diffraction pattern) will be a set of discrete lines of intensity, spaced uniformly to accord with the 'fundamental' frequency of the aperture function, and enveloped by the Fourier transform of one slit. If the aperture function extended to infinity in each direction then the individual lines would be infinitely narrow (delta functions), but, since it cannot do so in practice, their width is inversely proportional to the total width of the grating (i.e. the intensity distribution across one line is essentially the Fourier transform of the envelope function for the rectangular wave).

To fix these ideas, consider a grating of N slits, each of width d, and separated by distance s. The diffracted intensity pattern is now given by:

$$I\left(\vartheta\right) = I_0 \frac{\sin^2 \beta}{\beta^2} \cdot \frac{\sin^2 N\gamma}{\sin^2 \gamma}$$

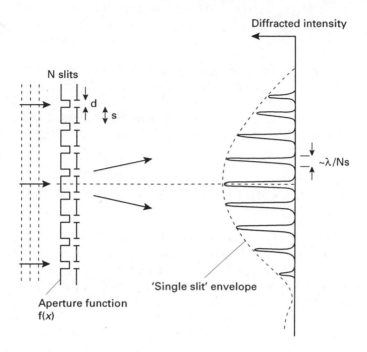

Fig. 2.24 Diffraction grating.

where

$$\beta = \tfrac{1}{2}(kd)\sin\vartheta$$
$$\gamma = \tfrac{1}{2}(ks)\sin\vartheta$$

The pattern is shown in Fig. 2.24. It is similar in many ways to that of the Fabry-Perot etalon, as we would expect, since it is a case of multiple interference. Clearly each wavelength present in the light incident on a diffraction grating will produce its own separate diffraction pattern. This fact is used to analyse the spectrum of incident light, and also to select and measure specific component wavelengths. Its ability to perform these tasks is most readily characterized by means of its resolving power, which is defined as it was for the Fabry-Perot etalon, i.e.,

$$\rho = \frac{\lambda}{\delta\lambda}$$

where $\delta\lambda$ is the smallest resolvable wavelength difference. If we take λ to be that wavelength difference which causes the pattern from $\lambda + d\lambda$ to produce a maximum, of order p, which falls on the first minimum of λ at that same order then we have:

$$pN\lambda + \lambda = pN(\lambda + \delta\lambda)$$

and thus:

$$\rho = \frac{\lambda}{\delta\lambda} = pN$$

Gratings are ruled either on glass (transmission) or on mirrors (reflection) with $\sim 10^5$ 'lines' (slits) in a distance $\sim 150\,\text{mm}$. The first-order resolving power is thus $\sim 10^5$, which is an order down on that for a Fabry-Perot etalon. However, the grating is less demanding of optical and mechanical tolerances in production and use, and is thus cheaper, and less prone to degradation with time.

2.11 GAUSSIAN BEAMS AND STABLE OPTICAL RESONATORS

In the discussions of Fabry-Perot etalons the reflecting surfaces were parallel and plane. The more recent discussions on diffraction provide further insights into the detailed behaviour of such an arrangement, for we have assumed that the light incident on the mirrors is a plane wave, with uniform amplitude and phase across its aperture. For a circular mirror of diameter d our considerations of Fraunhofer diffraction have indicated that such an aperture will yield a reflected beam which diverges at angle $\sim \lambda/d$. Hence, if the mirrors are a distance D apart, and $D >> d$, only a fraction $\sim d^4/\lambda^2 D^2$ of the light power will be interrupted by the second mirror, and this loss will be sustained for each mirror-to-mirror passage. How can this loss be reduced? To answer this question it is reasonable first to look for a *stable* solution to the problem, i.e. one which does not involve an additional loss for each pass between mirrors. To find this we may employ our knowledge of diffraction theory to ask the subsidiary question: what aperture amplitude distribution is stable in the face of aperture diffraction effects? Since we know that the far field diffraction pattern is the Fourier transform of the aperture distribution function, it is clear that we are asking, effectively, which function Fourier transforms into itself, or in more mathematical language: which function is invariant under Fourier transformation? There is only one such function: the Gaussian function, of the form:

$$f(r) = A \exp\left(-\frac{r^2}{\sigma^2}\right)$$

where r is the radial dimension in the aperture plane and σ is a constant known, in this context, as the 'spot size'.

Suppose, then, that we consider a wave with uniform phase in the plane $(x_0 y_0)$ at P_0 as shown in Fig. 2.25, and that this wave has a Gaussian amplitude distribution

$$f(r) = A \exp\left(-\frac{x^2 + y^2}{\sigma^2}\right)$$

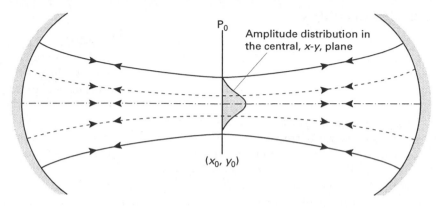

Fig. 2.25 Gaussian stable resonator.

In the far field this will diffract into a spherical wave with the same form of amplitude distribution and thus, if we place in that field a perfectly reflecting spherical mirror whose diameter is much greater than the spot size at that distance from P_0, essentially all the light will be returned along its incident path (Fig. 2.25) (99% of the light will be reflected for a mirror diameter three times the spot size). If another such mirror is also placed on the opposite side of P_0, the light continues to bounce between the spherical mirrors with very little loss. Such an arrangement is known as a 'stable resonator', and it is clear that the light within it is in the form of a 'Gaussian beam'. Such arrangements are the preferred ones for laser structures, since the losses are minimized and the laser becomes more efficient. It also follows that the light which emerges from the partially-silvered mirror of the laser source will possess a Gaussian intensity distribution. The condition on the size of the mirror will be satisfied automatically since the settling position for the resonance will be that which minimizes the losses. One may readily see, also, that if a plane mirror is placed at the central plane which contains $(x_0 y_0)$ the optical situation is essentially unchanged, so that a spherical mirror can also be used with a plane mirror to create a stable resonator. This configuration is, indeed, sometimes used in laser design.

As the radii of curvature of the mirrors increase, so their diameters must also increase, for a given spacing, in order to obtain a stable Gaussian resonator mode. In the limit as the radius tends to infinity, and the two mirrors thus become plane, the configuration is right on the limit of stability. The diffraction approximations, in fact, break down, and other methods must be used to obtain the aperture intensity distribution, which is now critically dependent on mirror alignment and surface finish.

2.12 CONCLUSION

The wave description of light provides a valuable and powerful analytical tool for the understanding and manipulation of many of its properties. Maxwell's wave equation was an important advance which established the electromagnetic nature of light and pointed the way towards an understanding of many of its interactions with matter.

Using the wave description, we have seen in this chapter how it is possible to explain satisfactorily the phenomena of reflection, refraction, interference and diffraction. With the understanding acquired we have seen also how to design useful devices such as interferometers and gratings for the analysis and control of light. We noted that the light wave is comprised of field vibrations which take place transversely to the propagation direction; we have touched only briefly, however, on the effects which depend upon the particular transverse direction in which this takes place, i.e. upon the polarization state of the light. A more detailed look at this is the subject of our next chapter.

PROBLEMS

2.1 The wavelength of visible light usually is taken to extend from 400 nm to 700 nm. To what frequency range does this correspond? To what range of wavenumbers (k) does it correspond?

2.2 A complex wave amplitude is given by:

$$A = a \cos \varphi + ib \sin \varphi$$

What is its modulus and argument?

If $\mathbf{A}^*$ is the complex conjugate of $\mathbf{A}$ what is the value of $\mathbf{A}.\mathbf{A}^*$? How does $\mathbf{A}.\mathbf{A}^*$ relate to the modulus of $\mathbf{A}$?

2.3 A linearly-polarized sinusoidal plane wave has (scalar) electric field amplitude $10\,\mathrm{Vm}^{-1}$ and is propagating along a line in the xy plane at 45° to the Ox axis, with field vibrations occurring in the xy plane. Write a vector expression describing the wave. What is the intensity of the wave if it is propagating in free space?

What is its intensity if it is propagating in a medium of relative permittivity 2.3? (Look up any fundamental constants you need.)

2.4 Light is incident upon an air-glass interface at 30° to the normal. If the refractive indices are 1 and 1.5 respectively, what are the amplitudes of the reflected light for the two polarization components, compared with that of the incident wave?

2.5 Light is incident upon a medium of refractive index 1.68, from air. What is Brewster's angle for this case? What effects could be observed if the light were incident at Brewster's angle? Of what use might these be?

2.6 Explain what is meant by an 'evanescent' optical wave.

Two optical media, 1 and 2, with refractive indices n_1 and n_2, respectively, are separated by a plane boundary. A plane optical wave within medium 1 strikes the boundary between the media at angle ϑ_1. Show that, under certain conditions, which you should define, the electric field of the optical wave in the second medium has the magnitude:

$$E_2(x) = E_1 \exp\left(\frac{-2\pi x \sinh \gamma}{\lambda}\right)$$

where E_1 is the amplitude of the wave at the boundary, λ is the wavelength of the light, x is the distance from the boundary in the normal direction into medium 2, and:

$$\sinh \gamma = \frac{\left(n_1^2 \sin^2 \vartheta_1 - n_2^2\right)^{1/2}}{n_2}$$

If $n_1 = 1.47, n_2 = 1.35$ and the second medium is in the form of a film $0.2\,\mu$m thick, what fraction of the light power, incident on the boundary from within the first medium at an angle of 75° and with a wavelength of 514 nm, penetrates through the film into the air beyond?

2.7 Two narrow parallel slits are illuminated with sodium light (wavelength 589.29 nm) and the resulting fringe pattern consists of fringes with a separation of 0.5 mm on a screen 2.25 m away. What is the slit separation?

2.8 Seven sinusoidal waves of the same frequency are superimposed. The waves are of equal amplitude but each differs in phase from the next by an equal amount. For what value of this phase difference does the resultant wave have zero amplitude? What is the resultant amplitude if this phase difference is 2π?

2.9 The mirrors of a Fabry-Perot interferometer have an amplitude reflection coefficient of 0.895; the separation of the mirrors is 10 mm. Calculate, for a wavelength of 500 nm:
  a) the finesse
  b) the half-width of a fringe (in radians)
  c) the free spectral range
  d) the resolving power.

2.10 What is meant by a stable optical resonator?

If one of the mirrors in such a resonator is partially silvered so that some light escapes from the resonator, why does this light have a Gaussian transverse distribution of light intensity?

REFERENCES

[1] Bleaney, B. I. and Bleaney, B. (1975), *Electricity and Magnetism*, Clarendon Press, Oxford.

[2] Born, M. and Wolf, E., (1975), *Principles of Optics*, Pergamon Press, 5th edn., section 8.3.2.

FURTHER READING

Guenther, R. (1990), *Modern Optics*, John Wilcy and Sons (for general wave optics).

Hecht, E. (1987), *Optics*, Addison-Wesley, 2nd edn., chaps 9 and 10, (for particularly good treatments of wave interference and diffraction, respectively).

Lipson, S.G. and Lipson, H. (1969), *Optical Physics*, Cambridge University Press (for physical insight into most important wave-optical processes).

And, for an excellent rigorous mathematical treatment of classical optics, reference [2] above.

3
Polarization optics

3.1 INTRODUCTION

The essential idea of optical polarization was introduced in section (2.4.3) but we must now consider this important topic in more detail. We know that the electric and magnetic fields, for a freely propagating light wave, lie transversely to the propagation direction and orthogonally to each other.

Normally, when discussing polarization phenomena, we fix our attention on the electric field, since it is this which has the most direct effect when the wave interacts with matter.

In saying that an optical wave is polarized we are implying that the direction of the optical field is either constant or is changing in an ordered, prescribable manner. In general, the tip of the electric vector circumscribes an ellipse, performing a complete circuit in a time equal to the period of the wave, or in a distance of one wavelength. Clearly, the two parameters are equivalent in this respect.

As is well known, linearly polarized light can conveniently be produced by passing any light beam through a sheet of polarizing film. This is a material which absorbs light of one linear polarization (the 'acceptance' direction) to a much greater extent ($\sim$ 1000 times) than the orthogonal polarization, thus, effectively, allowing just one linear polarization state to pass. The material's properties result from the fact that it consists of long-chain polymeric molecules aligned in one direction (the acceptance direction) by stretching a plastic, and then stabilizing it. Electrons can move more easily along the chains than transversely to them, and thus the optical wave transmits easily only when its electric field lies along this acceptance direction. The material is cheap and allows the use of large optical apertures. It thus provides a convenient means whereby, for example, a specific linear polarization state can be defined; this state then provides a ready polarization reference which can be used as a starting point for other manipulations. In order to study these manipulations and other aspects of polarization optics, we shall begin by looking more closely at the polarization ellipse.

3.2 THE POLARIZATION ELLIPSE

In section (2.4) the most general form of polarized light wave propagating in the Oz direction was derived from the two linearly polarized components in the Ox and Oy directions (Fig. 3.1):

$$E_x = e_x \cos(\omega t - kz + \delta_x)$$
$$E_y = e_y \cos(\omega t - kz + \delta_y)$$

(3.1a)

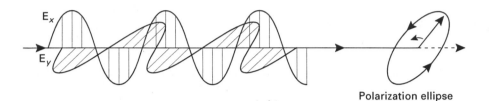

Polarization ellipse

Fig. 3.1 Components for an elliptically-polarized wave.

If we eliminate $(\omega t - kz)$ from these equations we obtain the expression:

$$\frac{E_x^2}{e_x^2} + \frac{E_y^2}{e_y^2} + \frac{2 E_x E_y}{e_x e_y} \cos(\delta_y - \delta_x) = \sin^2(\delta_y - \delta_x)$$

(3.1b)

which is the ellipse (in the variables E_x, E_y) circumscribed by the tip of the resultant electric vector at any one point in space over one period of the combined wave. This can only be true, however, if the phase difference $(\delta_y - \delta_x)$ is constant in time, or, at least, changes only slowly when compared with the speed of response of the detector. In other words, we say that the two waves must have a large mutual 'coherence'. If this were not so then relative phases and hence resultant field vectors would vary randomly within the detector response time, giving no ordered pattern to the behaviour of the resultant field and thus presenting to the detector what would be, essentially, unpolarized light.

Assuming that the mutual coherence is good, we may investigate further the properties of the polarization ellipse.

Note, firstly, that the ellipse always lies in the rectangle shown in Fig. 3.2, but that the axes of the ellipse are not parallel with the original x, y directions.

The ellipse is specified as follows: with $e_x, e_y, \delta (= \delta_y - \delta_x)$ known, then we define $\tan \beta = e_y / e_x$.

The orientation of the ellipse, α, is given by (see Appendix IV):

$$\tan 2\alpha = \tan 2\beta \cos \delta$$

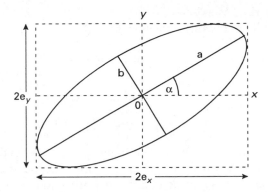

Fig. 3.2 The polarization ellipse.

Semi-major and semi-minor axes a, b are given by:

$$e_x^2 + e_y^2 = a^2 + b^2 \sim I$$

$e = \tan \chi = \pm b/a$ (the sign determines the sense of the rotation) where:

$$\sin 2\chi = -\sin 2\beta \sin \delta$$

We should note also that the electric field components along the major and minor axes are always in quadrature (i.e., $\pi/2$ phase difference, the sign of the difference depending on the sense of the rotation).

Linear and circular states of polarization may be regarded as special cases where the polarization ellipse degenerates into a straight line or a circle, respectively.

A linear state is obtained with the components in equations (3.1a) when either:

$$\left.\begin{array}{l} e_x = 0 \\ e_y \neq 0 \end{array}\right\} \qquad \text{linearly polarized in } Oy \text{ direction}$$

$$\left.\begin{array}{l} e_x \neq 0 \\ e_y = 0 \end{array}\right\} \qquad \text{linearly polarized in } Ox \text{ direction}$$

or,

$$\delta_y - \delta_x = m\pi$$

where m is an integer. In this latter case the direction of polarization will be at an angle:

$$+ \tan^{-1}\left(\frac{e_y}{e_x}\right) \qquad m \quad \text{even}$$

$$- \tan^{-1}\left(\frac{e_y}{e_x}\right) \qquad m \quad \text{odd}$$

with respect to the Ox axis.

A circular state is obtained when

$$e_x = e_y$$

and

$$(\delta_y - \delta_x) = (2m + 1)\pi/2$$

i.e., in this case the two waves have equal amplitudes and are in phase quadrature. The waves will be right-hand circularly polarized when m is even and left-hand circularly polarized when m is odd.

Light can become polarized as a result of the intrinsic directional properties of matter: either the matter which is the original source of the light, or the matter through which the light passes. These intrinsic material directional properties are the result of directionality in the bonding which holds together the atoms of which the material is made. This directionality leads to variations in the response of the material according to the direction of an imposed force, be it electric, magnetic or mechanical. The best known manifestation of directionality in solid materials is the crystal, with the large variety of crystallographic forms, some symmetrical, some asymmetrical. The characteristic shapes which we associate with certain crystals result from the fact that they tend to break preferentially along certain planes known as cleavage planes, which are those planes between which atomic forces are weakest.

It is not surprising, then, to find that directionality in a crystalline material is also evident in the light which it produces, or is impressed upon the light which passes through it.

In order to understand the ways in which we may produce polarized light, control it and use it, we must make a gentle incursion into the subject of crystal optics.

3.3 CRYSTAL OPTICS

Light propagates through a material by stimulating the elementary atomic dipoles to oscillate and thus to radiate. In our previous discussions the forced oscillation was assumed to take place in the direction of the driving electric field, but in the case of a medium whose physical properties vary with direction, an anisotropic medium, this is not necessarily the case. If an electron in an atom or molecule can move more easily in one direction than another, then an electric field at some arbitrary angle to the preferred direction will move the electron in a direction which is not parallel with the field direction (Fig. 3.3). As a result, the direction in which the oscillating

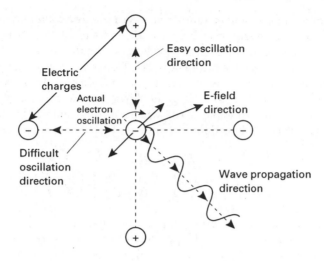

Fig. 3.3 Electron response to electric field in an anisotropic medium.

dipole's radiation is maximized (i.e., normal to its oscillation direction) is not the same as that of the driving wave.

The consequences, for the optics of anisotropic media, of this simple piece of physics are complex.

Immediately we can see that the already-discussed relationship between the electric displacement **D** and the electric field **E**, for an *isotropic* (i.e., no directionality) medium:

$$\mathbf{D} = \varepsilon_R \varepsilon_o \mathbf{E}$$

must be more complex for an anisotropic medium; in fact the relation must now be written in the form (for any, arbitrary three orthogonal directions Ox, Oy, Oz):

$$D_x = \varepsilon_o \left(\varepsilon_{xx} E_x + \varepsilon_{xy} E_y + \varepsilon_{xz} E_z \right)$$
$$D_y = \varepsilon_o \left(\varepsilon_{yx} E_x + \varepsilon_{yy} E_y + \varepsilon_{yz} E_z \right)$$
$$D_z = \varepsilon_o \left(\varepsilon_{zx} E_x + \varepsilon_{zy} E_y + \varepsilon_{zz} E_z \right)$$

Clearly what is depicted here is an array which describes the various electric field susceptibilities in the various directions within the crystal: ε_{ij} (a scalar quantity) is a measure of the effect which an electric field in direction j has in direction i within the crystal, i.e., the ease with which it can move electrons in that direction and thus create a dipole moment.

The array can be written in the abbreviated form

$$D_i = \varepsilon_o \varepsilon_{ij} E_j \quad (i, j = x, y, z)$$

and ε_{ij} is now a *tensor* known, in this case, as the permittivity tensor. A tensor is a physical quantity which characterizes a particular physical property of an anisotropic medium, and takes the form of a matrix. Clearly **D** is not now (in general) parallel with **E**, and the angle between the two also will depend upon the direction of **E** in the material.

Now it can be shown (see Appendix III) from energy considerations that the permittivity tensor is symmetrical, i.e., $\varepsilon_{ij} = \varepsilon_{ji}$. Also, symmetrical tensors can be cast into their diagonal form by referring them to a special set of axes (the principal axes) which are determined by the crystal structure [1]. When this is done, we have:

$$\begin{pmatrix} D_x \\ D_y \\ D_z \end{pmatrix} = \varepsilon_0 \begin{pmatrix} \varepsilon_{xx} & 0 & 0 \\ 0 & \varepsilon_{yy} & 0 \\ 0 & 0 & \varepsilon_{zz} \end{pmatrix} \begin{pmatrix} E_x \\ E_y \\ E_z \end{pmatrix}$$

The new set of axes, Ox, Oy, Oz, is now this special set.

Suppose now that $\mathbf{E} = E_x \mathbf{i}$, i.e., we have, entering the crystal, an optical wave whose **E** field lies in one of these special crystal directions.

In this case we simply have:

$$D_x = \epsilon_0 \epsilon_{xx} E_x$$

as our tensor relation and e_{xx} is, of course, a scalar quantity. In other words, we have **D** parallel with **E**, just as for an isotropic material, and the light will propagate, with refractive index $e_{xx}^{1/2}$, perfectly normally. Furthermore, the same will be true for:

$$\mathbf{E} = E_y \mathbf{j}, \quad \text{(refractive index } e_{yy}^{1/2})$$
$$\mathbf{E} = E_z \mathbf{k}, \quad \text{(refractive index } e_{zz}^{1/2})$$

Before going further we should note an important consequence of all this: the refractive index varies with the direction of **E**. If we have a wave travelling in direction Oz, its velocity now will depend upon its polarization state: if the wave is linearly polarized in the Ox direction it will travel with velocity $c_0/e_{xx}^{1/2}$ while if it is linearly polarized in the Oy direction its velocity will be $c_0/e_{yy}^{1/2}$. Hence the medium is offering two refractive indices to the wave travelling in this direction: we have the phenomenon known as double refraction or 'birefringence'. A wave which is linearly polarized in a direction at 45° to Ox will split into two components, linearly polarized in directions Ox and Oy, the two components travelling at different velocities. Hence the phase difference between the two components will steadily increase and the composite polarization state of the wave will vary progressively from linear to circular and back to linear again.

This behaviour is, of course, a direct consequence of the basic physics which was discussed earlier: it is easier, in the anisotropic crystal, for the electric field to move the atomic electrons in one direction than in another.

Hence, for the direction of easy movement, the light polarized in this direction can travel faster than when it is polarized in the direction for which the movement is more sluggish. Birefringence is long word, but the physical principles which underlie it really are very simple. It follows from these discussions that an anisotropic medium may be characterized by means of three refractive indices, corresponding to polarization directions along Ox, Oy, Oz, and that these will have values $e_{xx}^{1/2}, e_{yy}^{1/2}, e_{zz}^{1/2}$, respectively. We can use this information to determine the refractive index (and thus the velocity) for a wave in any direction with any given linear polarization state.

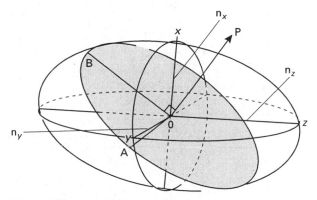

OA and OB represent the linearly polarized eigenstates for the direction OP

Fig. 3.4 The index ellipsoid.

To do this we construct an 'index ellipsoid' or 'indicatrix', as it is sometimes called (see Fig. 3.4), from the form of the permittivity tensor for any given crystal. This ellipsoid has the following important properties.

Suppose that we wish to investigate the propagation of light, at an arbitrary angle to the crystal axes (polarization as yet unspecified). We draw a line, OP, corresponding to this direction within the index ellipsoid, passing through its centre O (Fig. 3.4). Now we construct the plane, also passing through O, which lies at right angles to the line. This plane will cut the ellipsoid in an ellipse. This ellipse has the property that the directions of its major and minor axes define the directions of linear polarization for which **D** and **E** are parallel for this propagation direction, and the lengths of these axes OA and OB are equal to the refractive indices for these polarizations. Since these two linear polarization states are the only ones which propagate without change of polarization form for this crystal direction, they are sometimes referred to as the 'eigenstates' or 'polarization eigenmodes' for this direction, conforming to the matrix terminology of eigenvectors and eigenvalues.

The propagation direction we first considered, along Oz, corresponds, of course, to one of the axes of the ellipsoid, and the two refractive indices $e_{xx}^{1/2}$ and $e_{yy}^{1/2}$ are the lengths of the other two axes in the central plane normal to Oz.

The refractive indices $e_{xx}^{1/2}, e_{yy}^{1/2}, e_{zz}^{1/2}$ are referred to as the *principal* refractive indices and we shall henceforth denote them n_x, n_y, n_z.

Several other points are very well worth noting. Suppose, firstly, that

$$n_x > n_y > n_z$$

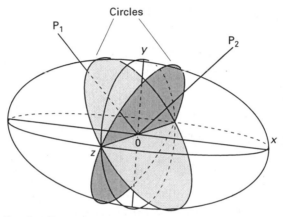

$n_x > n_y > n_z$
P_1 and P_2 are the optic axes of the crystal

Fig. 3.5 Ellipsoid for a biaxial crystal.

It follows that there will be a plane which contains Oz for which the two axes of interception with the ellipsoid are equal (Fig. 3.5). This plane will be at some angle to the yz plane and will thus intersect the ellipsoid in a circle. This means, of course, that, for the light propagation direction corresponding to the normal to this plane, all polarization directions have the same velocity; there is no double refraction for this direction. This direction is an *optic axis* of the crystal and there will, in general, be two such axes, since there must also be such a plane at an equal angle to the yz plane on the other side (see Fig. 3.5). Such a crystal with two optic axes is said to be biaxial.

Suppose now that:

$$n_x = n_y = n_o(say), \text{ the 'ordinary' index}$$

and

$$n_z = n_e(say), \text{ the 'extraordinary' index}$$

In this case one of the principal planes is a circle and it is the only circular section (containing the origin) which exists. Hence, in this case there is only one optic axis, along the Oz direction. Such crystals are said to be uniaxial (Fig. 3.6). The crystal is said to be positive when $n_e > n_o$ and negative when $n_e < n_o$. For example, quartz is a positive uniaxial crystal, and calcite a negative uniaxial crystal. These features are, of course, determined by the crystal class to which these materials belong.

It is clear that the index ellipsoid is a very useful device for determining the polarization behaviour of anisotropic media.

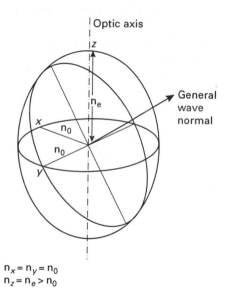

$$n_x = n_y = n_0$$
$$n_z = n_e > n_0$$

Fig. 3.6 Ellipsoid for a (positive) uniaxial crystal.

Let us now consider some practical consequences of all that we have just learned.

3.4 RETARDING WAVE-PLATES

Consider a positive uniaxial crystal plate (e.g., quartz) cut in such a way (Fig. 3.7) as to set the optic axis parallel with one of the faces. Suppose a wave is incident normally on to this face. If the wave is linearly polarized with its **E** field parallel with the optic axis, it will travel with refractive index n_e as we have described; if it has the orthogonal polarization, normal to the optic axis, it will travel with refractive index n_o.

The two waves travel in the same direction through the crystal but with different velocities. For a positive uniaxial crystal $n_e > n_o$, and thus the

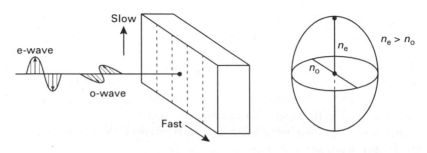

Fig. 3.7 Plate with face parallel with the optic axis in quartz.

light linearly polarized parallel with the optic axis will be a 'slow' wave, whilst the one at right angles to the axis will be 'fast'. For this reason the two crystal directions are often referred to as the 'slow' and 'fast' axes.

Suppose that the wave is linearly polarized at 45° to the optic axis. The phase difference between the components parallel with and orthogonal to the optic axis will now increase with distance l into the crystal according to:

$$\varphi = \frac{2\pi}{\lambda}(n_e - n_o)l$$

Hence, if, for a given wavelength λ

$$l = \frac{\lambda}{4(n_e - n_o)}$$

then

$$\varphi = \frac{\pi}{2}$$

and the light emerges from the plate circularly polarized. We have inserted a phase difference of $\pi/2$ between the components, equivalent to a distance shift of $\lambda/4$, and the crystal plate, when of this thickness, is called a 'quarter-wave' plate. It will (for an input polarization direction at 45° to the axes) convert linearly polarized light into circularly polarized light or vice versa. If the input linear polarization direction lies at some arbitrary angle α to the optic axis then the two components

$$E \cos \alpha$$
$$E \sin \alpha$$

will emerge with a phase difference of $\pi/2$. We noted in section (3.2) that the electric field components along the two axes of a polarization ellipse were always in phase quadrature. It follows that these two components are now the major and minor axes of the elliptical polarization state which emerges from the plate. Thus the ellipticity of the ellipse (i.e., the ratio of the major and minor axis) is just $\tan \alpha$ and by varying the input polarization direction α we have a means by which we can generate an ellipse of any ellipticity. The orientation of the ellipse will be defined by the direction of the optic axis of the waveplate (Fig. 3.8(a)).

Suppose now that the crystal plate has twice the previous thickness and is used at the same wavelength. It becomes a 'half-wave' plate. A phase difference of π is inserted between the components (linear eigenstates). The result of this is that an input wave which is linearly polarized at angle α to the optic axis will emerge still linearly polarized but with its direction now at $-\alpha$ to the axis. The plate has rotated the polarization direction through an angle -2α. And, indeed, any input polarization ellipse will emerge with the same ellipticity but with its orientation rotated through -2α (Fig. 3.8(b)).

It follows that, with the aid of these two simple plates, we can generate elliptical polarization of any prescribed ellipticity and orientation from linearly polarized light, which can itself be generated from any light source plus a simple polarizing sheet.

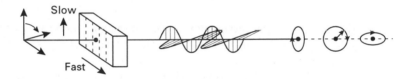

(a) Quarter-wave plate

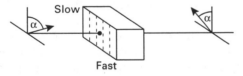

(b) Half-wave plate

Fig. 3.8 Polarization control with waveplates.

Equally valuable is the reverse process: that of the analysis of an arbitrary elliptical polarization state or its conversion to a linear state.

Suppose we have light of unknown elliptical polarization. By inserting a polarizing sheet and rotating it around the axis parallel to the propagation direction (Fig. 3.9) we shall find a position of maximum transmission and an orthogonal position of minimum transmission. These are the major and minor axes of the ellipse (respectively) and the ratio of the two intensities

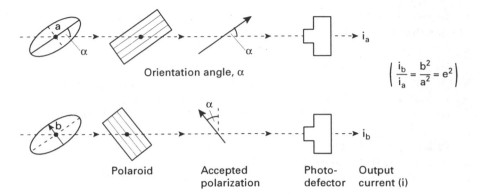

Orientation angle, α

$$\left(\frac{i_b}{i_a} = \frac{b^2}{a^2} = e^2\right)$$

| Polaroid | Accepted polarization | Photo-defector | Output current (i) |

Fig. 3.9 Determination of the polarization ellipse.

at these positions will give the square of the ellipticity of the ellipse, i.e.,

$$e = \frac{b}{a} = \frac{E_b}{E_a} = \left(\frac{I_b}{I_a}\right)^{1/2}$$

Clearly, the orientation of the ellipse also is known since this is, by definition, just the direction of the major axis, and is given by the position at which the maximum occurs. In order to convert the elliptical state into a linear one, all we need is a quarter-wave plate (appropriate to the wavelength of the light used, of course). Since the components of the electric field along the major and minor axes of the ellipse are always in phase quadrature (see section 3.2), the insertion of a quarter-wave plate with its axes aligned with the axes of the polarization ellipse will bring the components into phase or into antiphase, and the light will thus become linearly polarized. The quarter-wave plate is used in conjunction with a following polaroid sheet (or prism polarizer) and the two are rotated (independently) about the propagation axis until the light is extinguished. The quarter-wave plate must then have the required orientation in line with the ellipse axes, since only when the light has become linearly polarized can the polarizer extinguish it completely. (If there are no positions for which the light is extinguished, then it is not fully polarized.)

Such are the quite powerful manipulations and analyses which can be performed with very simple devices. However, manual human intervention via rotation of plates is not always convenient or even possible.

In many cases polarization analysis and control must be done very quickly (perhaps in nanoseconds) and automatically, using electronic processing. For these cases more advanced polarization devices must be used and, in order to understand and use these, a more advanced theoretical framework is necessary. We shall introduce this in section (3.8).

3.5 A VARIABLE WAVEPLATE - THE SOLEIL-BABINET COMPENSATOR

Consider the structure shown in Fig. 3.10. A pair of wedges, cut from a crystal (e.g., quartz) so that its optic axis lies parallel with the front faces, rests on a rectangular block of the same crystal with its optic axis orthogonal to that of the wedges. The wedges may be moved laterally, as shown in the diagram, so that the total thickness of the upper slab, which the wedges comprise, is variable.

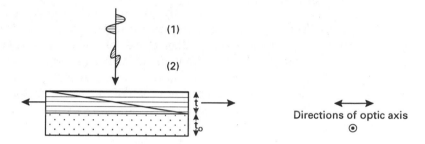

Fig. 3.10 The Soleil-Babinet compensator.

Consider now the incidence of a plane wave (1), normal to the upper surface and linearly polarized in a direction parallel with the optic axis of the wedges. Clearly it will travel through the wedges seeing a refractive index of n_e, and through the lower block with refractive index n_o. For a wave (2) with the orthogonal direction of polarization, the order of the refractive indices is reversed.

Suppose that the 'wedge' block thickness (variable) is t, whilst the lower block thickness (fixed) is t_0. Then it is clear that the phase delay suffered by the first wave will be:

$$\varphi_1 = \frac{2\pi}{\lambda}\left(n_e t + n_o t_0\right)l$$

and by the second:

$$\varphi_2 = \frac{2\pi}{\lambda}\left(n_o t + n_e t_0\right)l$$

giving a phase difference between the two:

$$\varphi_2 - \varphi_1 = \frac{2\pi}{\lambda}\left(n_e - n_o\right)\left(t_0 - t\right)$$

This phase difference will be constant across that part of the aperture of the device which includes both wedges, and will be continuously variable from 0 to 2π, for any given wavelength, by sliding the wedges apart, and thus varying t. The device is known as a Soleil-Babinet (pronounced 'Sollay-Babbinay') compensator (sometimes, but less commonly, a Babinet-Soleil compensator) and is very useful in both the control and analysis of optical polarization states. Clearly, the Soleil-Babinet compensator can be adjusted to form either a quarter-wave plate or a half-wave plate for any optical wavelength, if that is what is desired.

3.6 POLARIZING PRISMS

The same ideas as those just described are also useful in devices which produce linearly polarized light with a higher degree of polarization than a polarizing sheet is capable of, and without its intrinsic loss (even for the 'acceptance' direction there is a significant loss). We shall look at just two of these devices, in order to illustrate the application of the ideas, but there are several others (these are described in most standard optics texts).

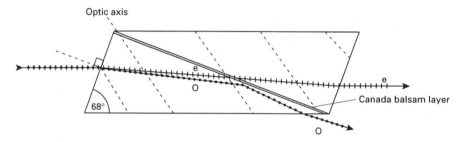

Fig. 3.11 Action of the Nicol prism.

The first device is the Nicol prism, illustrated in Fig. 3.11. Two wedges of calcite crystal are cut as shown, with their optic axes in the same direction (in the plane of the page) and cemented together with 'Canada balsam', a material whose refractive index at visible wavelengths lies midway between n_e and n_o. When unpolarized light enters parallel to the axis of the prism (as shown) and at an angle to the front face, it will split, as always, into the e and o components, each with its own refractive index, and thus each with its own refractive angle according to Snell's law. (Calcite is a negative uniaxial crystal so $n_o > n_e$.) When the light reaches the Canada balsam interface between the two wedges, it finds that the geometry and refractive indices have been arranged such that the ordinary (o) ray, with the larger deflection angle, strikes this interface at an angle greater than the total internal reflection (TIR) angle and is thus not passed into the second wedge, whereas the extraordinary (e) ray is so passed. Hence only the e ray emerges from the prism and this is linearly polarized. Thus we have an effective

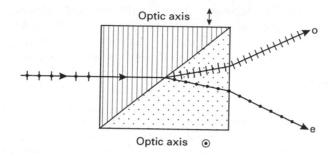

Fig. 3.12 Action of the Wollaston prism.

prism polarizer, albeit one of limited angular acceptance ($\sim 14°$) since the TIR condition is quite critical in respect of angle of incidence.

The second prism we shall discuss is widely used in practical polarization optics: it is called the Wollaston prism, and is shown in Fig. 3.12.

Again we have two wedges of positive (say) uniaxial crystal. They are equal in size, placed together to form a rectangular block (sometimes a cube) and have their optic axes orthogonal, as shown.

Consider a wave entering normally from the left. The e and o waves travel with differing velocities and strike the boundary between the wedges at the same angle. On striking the boundary one of the waves sees a positive change in refractive index ($n_e - n_o$), the other a negative change ($n_o - n_e$), so that they are deflected, respectively, up and down (Fig. 3.12) through equal angles. The e and o rays thus diverge as they emerge from the prism, allowing either to be isolated, or the two to be observed (or detected) simultaneously but separately. Also it is clear that, by rotating this prism around the propagation axis, we may reverse the positions of the two components.

It is extremely useful to be able to separate the two orthogonally polarized components in this controllable way.

3.7 CIRCULAR BIREFRINGENCE

So far we have considered only linear birefringence, where two orthogonal linear polarization eigenstates propagate, each remaining linear, but with different velocities. Some crystals also exhibit circular birefringence. Quartz (again) is one such crystal and its circular birefringence derives from the fact that the crystal structure spirals around the optic axis in a right-handed (dextro-rotatory) or left-handed (laevo-rotatory) sense depending on the crystal specimen: both forms exist in nature.

It is not surprising to find, in view of this knowledge, and our understanding of the easy motions of electrons, that light which is right-hand circularly polarized (clockwise rotation of the tip of the electric vector as viewed by a receiver of the light) will travel faster down the axis of a matching right-hand spiralled crystal structure than left-hand circularly polarized light. We now have circular birefringence: the two circular polarization components propagate without change of form (i.e., they remain circularly polarized) but at different velocities. They are the circular polarization eigenstates for this case.

The term 'optical activity' has been traditionally applied to this phenomenon, and it is usually described in terms of the rotation of the polarization direction of a linearly polarized wave as it passes down the optic axis of an 'optically active' crystal. This fact is exactly equivalent to the interpretation in terms of circular birefringence, since a linear polarization state can be resolved into two oppositely rotating circular components (Fig. 3.13). If these travel at different velocities, a phase difference is inserted between them. As a result of this, when recombined, they again form a resultant which is linearly polarized but rotated with respect to the original direction (Fig. 3.13). Hence 'optical activity' is equivalent to circular birefringence. In general, both linear and circular birefringence might be present simultaneously in a material (such as quartz). In this case the polarization eigenstates which propagate without change of form (and at different velocities) will be elliptical states, the ellipticity and orientation depending upon the ratio of the magnitudes of the linear and circular birefringences, and on the direction of the linear birefringence eigen-axes within the crystal.

It should, again, be emphasised that only the polarization eigenstates propagate without change of form. All other polarization states will be changed into different polarization states by the action of the polarization element (e.g., a crystal component). These changes of polarization state are very useful in opto-electronics. They allow us to control, analyse, modulate and demodulate polarization information impressed upon a light beam, and to measure important directional properties relating to the medium through which the light has passed. We must now develop a rigorous formalism to handle these more general polarization processes.

3.8 POLARIZATION ANALYSIS

As has been stated, with both linear and circular birefringence present, the polarization eigenstates (i.e., the states which propagate without change of form) for a given optical element are elliptical states, and the element is said to exhibit elliptical birefringence, since these eigenstates propagate with different velocities.

In general, if we have, as an input to a polarization-optical element, light of one elliptical polarization state, it will be converted, on emergence, into a

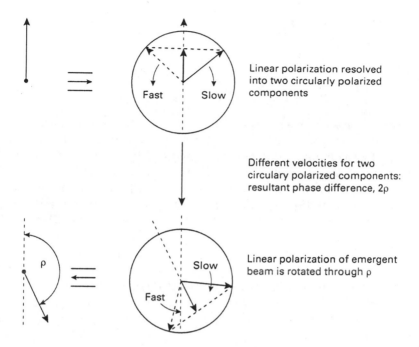

Linear polarization resolved into two circularly polarized components

Different velocities for two circulary polarized components: resultant phase difference, 2ρ

Linear polarization of emergent beam is rotated through ρ

Fig. 3.13 Resolution of linear polarization into circularly polarized components in circular birefringence (2ρ).

different elliptical polarization state (the only exceptions being, of course, when the input state is itself an eigenstate). We know that any elliptical polarization state can always be expressed in terms of two orthogonal electric field components defined with respect to chosen axes Ox, Oy, i.e.,

$$E_x = e_x \cos(\omega t - kz + \delta_x)$$
$$E_x = e_y \cos(\omega t - kz + \delta_y)$$

or, in complex exponential notation:

$$E_x = |E_x| \exp(i\varphi_x); \quad \varphi_x = \omega t - kz + \delta_x$$
$$E_y = |E_y| \exp(i\varphi_y); \quad \varphi_y = \omega t - kz + \delta_y$$

When this ellipse is converted into another by the action of a lossless polarization element, the new ellipse will be formed from components which are linear combinations of the old, since it results from directional resolutions and rotations of the original fields. Thus these new components can be written:

$$E'_x = m_1 E_x + m_4 E_y$$
$$E'_y = m_3 E_y + m_2 E_y$$

or, in matrix notation

$$E' = M.E$$

where

$$M = \begin{pmatrix} m_1 & m_4 \\ m_3 & m_2 \end{pmatrix} \tag{3.2}$$

and the m_n are, in general, complex numbers. M is known as a 'Jones' matrix after the mathematician who developed an extremely useful 'Jones calculus' for manipulations in polarization optics [2]. Now in order to make measurements of the input and output states in practice we need a quick and convenient experimental method. In section (3.4) there was described a method for doing this which involved the manual rotation of a quarter-wave plate and/or a polarizer, but the method we seek now must lend itself to automatic operation.

A convenient method for this practical determination is again to use the linear polarizer and the quarter-wave plate, but to measure the light intensities for a series of fixed orientations of these elements.

Suppose that $I(\vartheta, \varepsilon)$ denotes the intensity of the incident light passed by the linear polarizer set at angle ϑ to Ox, after the Oy component has been retarded by angle ε as a result of the insertion of the quarter-wave plate with its axes parallel with O_x, O_y. We measure what are called the four Stokes parameters, as follows:

$$S_0 = I(0°, 0) + I(90°, 0) = e_x^2 + e_y^2$$
$$S_1 = I(0°, 0) - I(90°, 0) = e_x^2 - e_y^2$$
$$S_2 = I(45°, 0) - I(135°, 0) = 2e_x e_y \cos \delta$$
$$S_3 = I(45°, \frac{\pi}{2}) - I(135°, \frac{\pi}{2}) = 2e_x e_y \sin \delta$$
$$\delta = \delta_y - \delta_x$$

If the light is 100% polarized, only three of these parameters are independent, since:

$$S_0^2 = S_1^2 + S_2^2 + S_3^2$$

S_0 being the total light intensity.

If the light is only partially polarized, the fraction

$$\eta = \frac{S_1^2 + S_2^2 + S_3^2}{S_0^2}$$

defines the degree of polarization. In what follows we shall assume that the light is fully polarized ($\eta = 1$). It is easy to show (see Appendix IV) that measurement of the S_n provides the ellipticity, e, and the orientation α of the polarization ellipse according to the relations:

$$e = \tan\chi$$

$$\sin 2\chi = \frac{S_3}{S_0}$$

$$\tan 2\alpha = \frac{S_2}{S_1}$$

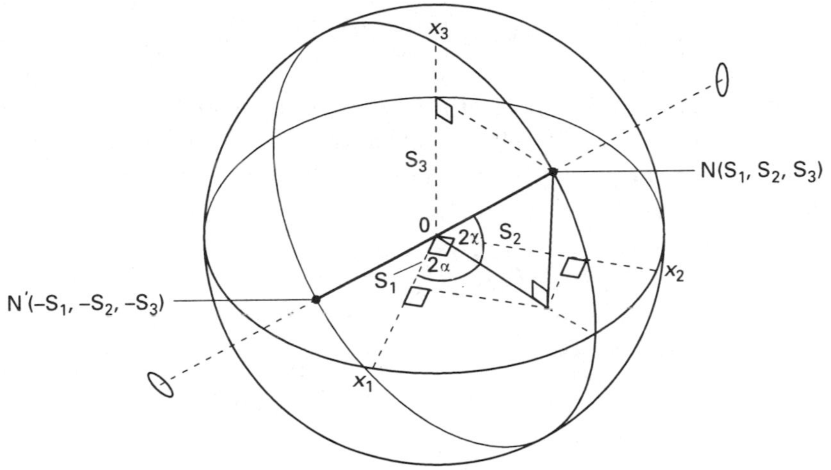

Fig. 3.14 The Poincaré sphere: the eigen-mode diameter (NN').

Now, the above relations suggest a geometrical construction which provides a powerful and elegant means for description and analysis of polarization-optical phenomena. The Stokes parameters S_1, S_2, S_3 may be regarded as the Cartesian co-ordinates of a point referred to axes Ox_1, Ox_2, Ox_3. Thus every elliptical polarization state corresponds to a unique point in three-dimensional space. For a constant S_0 (lossless medium) it follows that all such points lie on a sphere of radius S_0 - the Poincaré sphere (Fig. 3.14). The properties of the sphere are quite well known (see, for example [3]). We can see that the equator will comprise the continuum of linearly polarized states, while the two poles will correspond to the two oppositely-handed states of circular polarization.

It is clear that any change, resulting from the passage of light through a lossless element, from one polarization state to another, corresponds to a rotation of the sphere about a diameter. Now any such rotation of the sphere may be expressed as a unitary 2×2 matrix **M**. Thus, the conversion from one polarization state **E** to another **E'** may also be expressed in the form:

$$\mathbf{E'} = \mathbf{M}\mathbf{E}$$

or

$$\begin{pmatrix} E'_x \\ E'_y \end{pmatrix} = \begin{pmatrix} m_1 & m_4 \\ m_3 & m_2 \end{pmatrix} \begin{pmatrix} E_x \\ E_y \end{pmatrix}$$

i.e.,

$$E'_x = m_1 E_x + m_4 E_y$$
$$E'_y = m_3 E_x + m_2 E_y$$

where

$$\mathbf{M} = \begin{pmatrix} m_1 & m_4 \\ m_3 & m_2 \end{pmatrix}$$

and **M** may be immediately identified with our previous **M** (equation 3.2).

M is a Jones matrix [2] which completely characterizes the polarization action of the element and is also equivalent to a rotation of the Poincaré sphere.

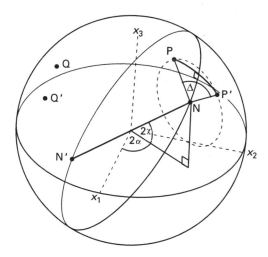

Fig. 3.15 Rotation of the Poincaré sphere about the eigenmode diameter NN'.

The two eigenvectors of the matrix correspond to the eigenmodes (or eigenstates) of the element (i.e., those polarization states which can propagate through the element without change of form). These two polarization eigenstates lie at opposite ends of a diameter (NN') of the Poincaré sphere and the polarization effect of the element is to rotate the sphere about this diameter (Fig. 3.15) through an angle Δ which is equal to the phase which the polarization element inserts between its eigenstates.

The polarization action of the element may thus be regarded as that of resolving the input polarization state into the two eigenstates with appropriate amplitudes, and then inserting a phase difference between them before recombining to obtain the emergent state. Thus, a pure rotator (e.g., optically active crystal) is equivalent to a rotation about the polar axis, with the two oppositely handed circular polarizations as eigenstates. The phase velocity difference between these two eigenstates is a measure of the circular birefringence. Analogously, a pure linear retarder (such as a wave plate) inserts a phase difference between orthogonal linear polarizations which measures the linear birefringence. The linear retarder's eigenstates lie at opposite ends of an equatorial diameter.

It is useful for many purposes to resolve the polarization action of any given element into its linear and circular birefringence components. The Poincaré sphere makes it clear that this may always be done since any rotation of the sphere can always be resolved into two sub-rotations, one about the polar diameter and the other about an equatorial diameter.

From this brief discussion we can begin to understand the importance of the Poincaré sphere.

It is a construction which converts all polarization actions into visualizable relationships in three-dimensional space.

To illustrate this point graphically let us consider a particular problem. Suppose that we ask what is the smallest number of measurements necessary to define completely the polarization properties of a given lossless polarization element, about which we know nothing in advance. Clearly, we must provide known polarization input states and measure their corresponding output states, but how many input/output pairs are necessary: one, two, more?

The Poincaré sphere answers this question easily. The element in question will possess two polarization eigenmodes and these will be at opposite ends of a diameter. We need to identify this diameter. We know that the action of the element is equivalent to a rotation of the sphere about this diameter, and through an angle equal to the phase difference which the element inserts between its eigenmodes. Hence, if we know one input/output pair of polarization states (NN'), we know that the rotation from the input to the output state must have taken place about a diameter which lies in the plane which perpendicularly bisects the line joining the two states (see Fig. 3.15). Two other input/output states (QQ') will similarly define another such plane, and thus the required diameter is clearly seen as the common line of intersection of these planes.

Further, the phase difference Δ inserted between the eigenstates (i.e., the sphere's rotation angle) is easily calculated from either pair of states, once the diameter is known.

Hence the answer is that *two* pairs of input/output states will define completely the polarization properties of the element. Simple geometry has provided the answer. A good general approach is to use the Poincaré sphere to determine (visualise?) the nature of the solution to a problem, and then to revert to the Jones matrices to perform the precise calculations. Alternatively, some simple results in spherical trigonometry will usually suffice.

3.9 APPLICATIONS OF POLARIZATION OPTICS

In practice, polarization effects may arise naturally, or may be induced deliberately.

Of those which occur naturally the most common are the ones which are a consequence of an anisotropic material, an asymmetrical material strain or asymmetrical waveguide geometries.

If an optical medium is compressed in a particular direction, there results the same kind of directional restriction on the atomic or molecular electrons as in the case of crystals, and hence the optical polarization directions parallel and orthogonal to these imposed forces (for isotropic materials) will encounter different refractive indices.

Somewhat similarly, if an optical wave is being guided in a channel, or other type of guide, with a refractive index greater than its surroundings, we have to be aware of the effect of any asymmetry in the geometry of the guide's cross-section. Clearly, if the cross-section is a perfect circle, as in the case of an ideal optical fibre, all linear polarization directions must propagate with the same velocity. If, however, the cross-section were elliptical, then it is not difficult to appreciate that a linear polarization direction parallel with the minor axis will propagate at a different velocity from that parallel with the major axis. This is indeed the case, and we shall be looking more closely at this, and other aspects of optical waveguiding, later on (Chapter 8).

The optical fibre is, in fact, a good medium for illustrating these passive polarization effects, since all real fibres possess some directional asymmetry due to one or more of the following: non-circularity of core cross-section; linear strain in the core; twist strain in the core. Bending will introduce linear strain, and twisting will introduce circular strain (Fig. 3.16). Linear strain leads to linear birefringence, circular (twist) strain to circular birefringence.

The linear birefringence in 'standard' telecommunications optical fibre can be quite troublesome for high performance links since it introduces velocity differences between the two orthogonal linear polarization states, which lead to relative time lags of the order of $1 - 10 \, \mathrm{ps \, km^{-1}}$. Clearly, this distorts the modulating signal: a pulse in a digital system, for example, will be broadened, and thus degraded, by this amount. This so called 'polarization dispersion' can be reduced by spinning the preform from which

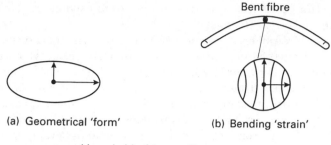

(a) Geometrical 'form'   (b) Bending 'strain'

Linearly birefringent fibres

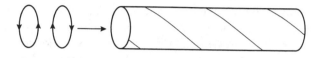

(c) Twist-strain circularly birefringent fibre

Fig. 3.16 Birefringence in optical fibres.

the fibre is being drawn, whilst it is being drawn, so as to average out the cross-sectional anisotropies. This 'spun preform' technique [4] reduces this form of dispersion to $\sim 0.01 \, \text{ps km}^{-1}$, i.e., by two orders of magnitude.

Fig. 3.17 Asymmetrically-doped linearly-birefringent optical fibre ('bow-tie'). [5]

It is sometimes valuable deliberately to introduce linear or circular bire-fringence into a fibre (we shall deal with such fibres in more detail in Chapter 8).

In order to introduce linear birefringence the fibre core may be made elliptical (with the consequences previously discussed) or stress may be introduced by asymmetric doping of the cladding material which surrounds the core (Fig. 3.17) [5]. The stress results from asymmetric contraction as the fibre cools from the melt.

Circular birefringence may be introduced by twisting and then clamping the fibre or by spinning an àsymmetric preform (from which the fibre is being pulled). One important application of fibre with a high value of linear birefringence ('hi-bi' fibre) is that linearly polarized light launched into one of the two linear eigenmodes will tend to remain in that state, thus providing a convenient means for conveying linearly polarized light between two points. The reason for this 'polarization holding' property is that light, when coupled (i.e., transferred) to the other eigenmode, will be coupled to a mode with a different velocity and will not, in general, be in phase with other previous light couplings into the mode; thus the various couplings will interfere destructively overall and only a small amplitude will result. There is said to be a 'phase mismatch'. (This is yet another example of wave interference!). Clearly, however, if a deliberate attempt is made to couple light only at those points where the two modes are in phase, then constructive interference can occur and the coupling will be strong. This is known as 'resonant' coupling and has a number of important applications (see section 9.4 and Appendix VIII).

An extremely convenient way of inducing polarization anisotropies into materials is by subjecting them to electric and/or magnetic fields. As we know very well, these fields can exert forces on electrons, so it is not surprising to learn that, via their effects on atomic electrons, the fields can influence the polarization properties of media, just as the chemical-bond restrictions on these electrons in crystals were able to do. The use of electric and magnetic fields thus allows us to build convenient polarization controllers and modulators. Some examples of the effects which can be used will help to establish these ideas. The effects themselves will be treated in more detail in Chapter 7.

3.9.1 *The electro-optic effect*

When an electric field is applied to an optical medium the electrons suffer restricted motion in the direction of the field, when compared with that orthogonal to it. Thus the material becomes linearly birefringent in response to the field. This is known as the electro-optic effect.

Consider the arrangement of Fig. 3.18. Here we have incident light which is linearly polarized at 45° to an electric field and the field acts on a medium transversely to the propagation direction of the light. The field-induced linear birefringence will cause a phase displacement between components of the incident light which lie, respectively, parallel and orthogonal to the field; hence the light will emerge elliptically polarized.

A (perfect) polarizer placed with its acceptance direction parallel with the input polarization direction will of course, pass all the light in the

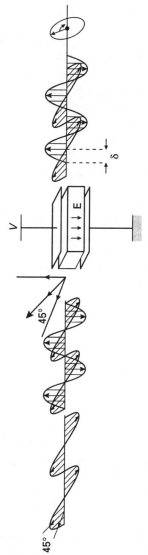

Linear polarization becomes elliptical by passing
through an electro-optic medium with applied field E

Fig. 3.18 The electro-optic effect.

absence of a field. When the field is applied, the fraction of light power which is passed will depend upon the form of the ellipse, which in turn depends upon the phase delay introduced by the field. Consequently, the field can be used to modulate the intensity of the light, and the electro-optic effect is, indeed, very useful for the modulation of light (see section 7.3.1).

The phase delay introduced may be proportional either to the field (Pockels effect) or to the square of the field (Kerr effect). All materials manifest a transverse Kerr effect. Only crystalline materials can manifest any kind of Pockels effect, or longitudinal (E field parallel with propagation direction) Kerr effect. The reason for this is physically quite clear. If a material is to respond linearly to an electric field the effect of the field must change sign when the field changes sign. This means that the medium must be able to distinguish (for example) between 'up' (positive field) and 'down' (negative field). But it can only do this if it possesses some kind of directionality in itself, otherwise all field directions must be equivalent in their physical effects. Hence, in order to make the necessary distinction between up and down, the material must possess an intrinsic asymmetry, and hence must be crystalline. By a similar argument a longitudinal E field can only produce a directional effect orthogonally to itself (i.e., in the direction of the optical electric field) if the medium is anisotropic (i.e., crystalline) for otherwise all transverse directions will be equivalent. In addition to the modulation of light (phase or intensity/power) it is clear that the electro-optic effect could be used to measure an electric field and/or the voltage which gives rise to it. Modulation and sensors based on this idea will be discussed in Chapters 7 and 10, respectively.

3.9.2 *The magneto-optic effect*

If a magnetic field is applied to a medium in a direction parallel with the direction in which light is passing through the medium, the result is a rotation of the polarization direction of whatever is the light's polarization state: in general, the polarization ellipse is rotated. The phenomenon, known as the Faraday (after its discover, in 1845) magneto-optic effect, normally is used with a linearly polarized input, so that there is a straightforward rotation of a single polarization direction (Fig. 3.19(a)). The magnitude of the rotation due to a field H, over a path length L, is given by:

$$\rho = V \int_0^L H \, dl$$

where V is a constant known as the Verdet constant: V is a constant for any given material, but is wavelength dependent. Clearly, if H is constant over the optical path, we have:

$$\rho = VHL$$

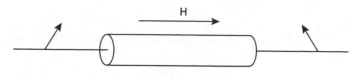

(a)

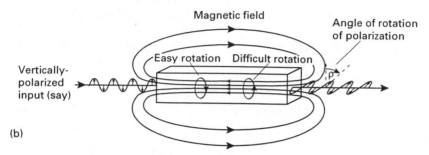

(b)

Fig. 3.19 The Faraday magneto-optic effect.

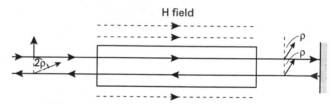

(a) Non-reciprocal rotation (Faraday effect). Rotation in
same direction in relation to the magnetic field.

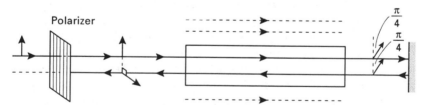

(b) Optical isolator action. Total rotation
of π/2 for polarization blocking.

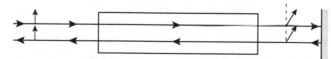

(c) Reciprocal rotation (optical activity). Rotation in
same direction in relation to propagation direction.

Fig. 3.20 Reciprocal and non-reciprocal polarization rotation.

From the discussion in section (3.7), we see that this is a magnetic-field-induced circular birefringence.

The physical reason for the effect is easy to understand in qualitative terms. When a magnetic field is applied to a medium the atomic electrons find it easier to rotate in one direction around the field than in the other: the Lorentz force acts on a moving charge in a magnetic field, and this will act radially on the electron as it circles the field. The force will be outward for one direction of rotation and inward for the other. The consequent electron displacement will lead to two different radii of rotation and thus two different rotational frequencies and electric permittivities. Hence the field will result in two different refractive indices, and thus to circular bire-fringence. Light which is circularly polarized in the 'easy' (say clockwise) direction will travel faster than that polarized in the 'hard' direction (anti-clockwise), leading to the observed effect (Fig. 3.19(b)). Another important aspect of the Faraday magneto-optic effect is that it is 'non-reciprocal'. This means that linearly polarized light (for example) is always rotated in the same absolute direction in space, independently of the direction of propagation of the light (Fig. 3.20(a)). For an optically active crystal this is not the case: if the polarization direction is rotated from right to left (say) on forward passage (as viewed by a fixed observer) it will be rotated from left to right on backward passage (as viewed by the same observer), so that back-reflection of light through an optically active crystal will result in light with zero final rotation, the two rotations having cancelled out (Fig. 3.20(c)). This is called a reciprocal rotation because the rotation looks the same for an observer who always looks in the direction of propagation of the light (Fig. 3.20(c)).

For the Faraday magneto-optic case, however, the rotation always takes place in the same direction *with respect to the magnetic field* (not the prop-agation direction) since it is this which determines the 'easy' and 'hard' directions. Hence, an observer always looking in the direction of light prop-agation will see different directions of rotation since he/she is, in one case, looking along the field and, in the other, against it. It is a non-reciprocal effect. The Faraday effect has a number of practical applications. It can be used to modulate light, although it is less convenient for this than the electro-optic effect, owing to the greater difficulty of producing and manip-ulating large, and rapidly varying (for high modulation bandwidth) mag-netic fields when compared with electric fields (large solenoids have large inductance!).

It can very effectively be used in optical isolators, however. In these de-vices light from a source passes through a linear polarizer and then through a magneto-optic element which rotates the polarization direction through 45°. Any light which is back-reflected by the ensuing optical system suffers a further 45° rotation during the backward passage, and in the same rota-tional direction, thus arriving back at the polarizer rotated through 90°; it is thus blocked by the polarizer (Fig. 3.20(b)). Hence the source is isolated from back-reflections by the magneto-optic element/polarizer combination which is thus known as a Faraday magneto-optic isolator. This is very valuable for use with devices whose stability is sensitive to back-reflection, such as lasers and optical amplifiers, and it effectively protects them from feedback effects. The Faraday magneto-optic effect also can be used to measure magnetic fields, and the electric currents which give rise to them. This topic comes up again in Chapter 10.

There are other magneto-optic effects (e.g., Kerr, Cotton-Mouton, Voigt) but the Faraday effect is by far the most important for optoelectronics.

3.9.3 The electrogyration effect

For some materials (e.g., quartz, bismuth germanium oxide) an *electric* field will induce a circular birefringence. This effect is thus the electric field analogue of the Faraday magneto-optic effect. This effect occurs only when the material possesses a spirality in the crystal structure, and thus an intrinsic circular birefringence (optical activity). The electric field effectively alters the pitch of the spiral and hence the magnitude of the circular birefringence. The effect is known as the electrogyration effect [6] and can be used to measure electric field/voltage, and also as a source of reciprocal (rather than non-reciprocal, as in the Faraday effect) field-induced polarization rotation.

This is a small effect and, since the electro-optic effect also always occurs in the materials which exhibit it, the electrogyration effect is often swamped by it. It is thus of only limited practical usefulness; but it has diagnostic value for crystal optics.

3.10 CONCLUSIONS

In this chapter we have looked closely at the directionality possessed by the optical transverse electric field, i.e., we have looked at optical polarization. We have seen how to describe it, to characterize it, to control it, to analyse it, and how, in some ways, to use it.

We have also looked at the ways in which the transverse electric and magnetic fields interact with directionalities (anisotropies) in material media through which the light propagates. In particular, we firstly looked at ways in which the interactions allow us to probe the nature and extent of the material directionalities, and thus to understand better the materials themselves. We found also that we could use our knowledge of the interactions to make measurements of the external fields acting on the media during the light propagation.

Secondly, we looked briefly at the ways in which these material interactions allow us to control light: to modulate it, and perhaps to analyse it.

We shall find later that the knowledge we have gained bears upon more advanced phenomena, such as those which allow light to switch light and to process light, opening up a new range of possibilities in the world of very fast (femtosecond: 10^{-15} s) phenomena.

PROBLEMS

3.1 Discuss the concept of optical polarization. How is partial polarization quantified?

An elliptically polarized beam of light is incident upon a linear polarizer. The ellipticity of the ellipse is 0.25, and the (non-attenuating) acceptance direction of the polarizer lies at an angle of 30° to its major axis. What fraction of the light intensity is passed by the analyser? For what angle of the polarizer would this fraction be a maximum?

How could the original light be converted to a linear polarization state of the same intensity?

3.2 An optical wave has X and Y linearly polarized components given by:

$$E_x = 5\sin(\omega t + \tfrac{1}{3}\pi)$$
$$E_y = 7\sin(\omega t - \tfrac{1}{8}\pi)$$

What is the ellipticity and the orientation of the resultant polarization ellipse? Plot its position on the Poincaré sphere.

3.3 You are presented with two plates and are told that one is a linear polarizer and the other is a quarter-wave plate. What experiments would you perform to determine which was which?

3.4 An optical polarization state is described by its Jones vector:

$$E = \begin{pmatrix} 1 \\ -2i \end{pmatrix}$$

What is the Jones vector of the state which is diametrically opposite to it on the Poincaré sphere?

3.5 What is the index ellipsoid? Show how it can be used to determine the linear eigenmodes for any given propagation direction in a crystal. How does the index ellipsoid for a uniaxial crystal differ from that for a biaxial crystal?

3.6 An ideal linear polarizer is placed between a pair of crossed linear polarizers. The acceptance direction of this central polarizer is rotated at a uniform rate. Show that the intensity of the light emerging from the system will be modulated at a rate equal to four times the rotation rate of the polarizer.

3.7 Describe the Faraday magneto-optic effect. How can it be used for impressing information on a beam of light?

A cylindrical rod of magneto-optic material, of length L and with Verdet constant V at the wavelength of light to be used, is placed wholly within an N-turn solenoid, also of length L, so that the rod and solenoid axes are parallel. A beam of laser light, with power P watts, passes through the rod along its axis. The light entering the rod is linearly polarized in the vertical direction. The light emerging from the rod is passed through a polarization analyser whose acceptance direction is vertical, and then on to a photodiode which has a sensitivity of S amps watt^{-1}.

A current ramp is applied to the solenoid, of the form:

$$i = kt$$

where i is the current at time t, and k is constant.

Assuming that optical losses are negligible for the arrangement, derive expressions for the amplitude and frequency of the a.c. component of the signal delivered by the photodiode.

3.8 Explain why the Kerr electro-optic effect is reciprocal and the Faraday magneto-optic effect is non-reciprocal.

REFERENCES

[1] Nye, J.F. (1976) *Physical Properties of Crystals*, Clarendon Press, Ox- ford, chap. 2.
[2] Jones, R.C. (1941-1956) 'A new calculus for the treatment of optical systems', J. Opt. Soc. Am. **31**; (1941) to (1956) **46**, 234-241.
[3] Jerrard, H.G. (1954) 'Transmission of light through optically active media', J. Opt. Soc. Am. **44**(8), 634-64.
[4] Barlow, A.J., Payne, D.N., Hadley, M.R. and Mansfield, R.J. (1981) 'Production of single-mode filres with negligible intrinsic birefringence and polarization mode dispersion', Elect. Lett. **17**, 725-726.
[5] Varnham, P., et al. (1983) 'Single polarization operation of highly-birefringent bow-tie optical filters', Elect. Lett. **19**, 246-247.
[6] Rogers, A.J. (1977), 'The electrogyration effect in crystalline quartz', Proc. Roy. Soc. (Series A), 353 177-192.

FURTHER READING

Born, M. and Wolf, E. (1975), *Principles of Optics*, 5th edn, Pergamon Press, section 1.4.
Collett, E. (1993), *Polarised Light: Fundamentals and Applications*, Marcel Dekker.
Kliger, D.S., Lewis, J.W. and Randall, C.E. (1990), *Polarised Light in Optics and Spectroscopy*, Academic Press.
Shurchiff, W.A. (1962), *Polarised Light: Production and Use*, Harvard University Press (an excellent introduction).

Plus reference [1], above, for an excellent account of crystal optics.

4

Light and matter: emission, propagation and absorption processes

4.1 INTRODUCTION

In this chapter we shall deal with the various processes by which light and matter interact. It is impossible to overestimate the importance of this subject, since it is only via this interaction that we can even become aware of the existence of light, and certainly we need to understand well the processes of interaction in order to study and to use light.

The detailed understanding of the interactive processes requires a deep knowledge of quantum theory, which is beyond the purpose of this book. Much insight can be gained, however, from a combination of classical (i.e., quasi-intuitive) ideas and elementary quantum physics. This is the approach which will be adopted.

A familiarity with the ideas in this chapter will ease the path for appreciation of most of the later chapters in the book, but especially Chapters 7 and 9.

We shall begin by considering the nature of light propagation in an optical medium. This uses the classical wave theory of Chapter 2, which provides a useful picture of the processes involved. The following sections, on optical dispersion and the emission and absorption processes, develop further the ideas introduced in the first section and also provide the groundwork for the more comprehensive treatments in later chapters.

4.2 CLASSICAL THEORY OF LIGHT PROPAGATION IN UNIFORM DIELECTRIC MEDIA

Consider the standard expression for the electric field component of an electromagnetic wave (of arbitrary polarization) propagating in the Oz direction in an optical medium of refractive index n:

$$E = E_0 \exp\left[i\left(\omega t - kz\right)\right]$$

We know that:

$$\frac{\omega}{k} = c = \frac{c_0}{n}$$

and hence may write:

$$E = E_0 \exp\left[i\omega\left(t - \frac{nz}{c_0}\right)\right]$$

We may conveniently include both the amplitude attenuation and the phase behaviour of the wave in this expression by defining a complex refractive index:

$$n = n' - in'' \tag{4.1}$$

so that

$$E = E_0 \exp\left(\frac{-\omega n'' z}{c_0}\right) \exp\left[i\omega\left(t - \frac{n'z}{c_0}\right)\right]$$

The first exponential clearly represents an attenuation factor (real exponent) while the second represents the propagating wave (imaginary exponent).

When electromagnetic radiation propagates through a material medium it stimulates the atomic electrons to oscillate, and thus to radiate, in the manner of elementary electric dipoles (Appendix V). The resulting amplitude distribution of the radiation will depend upon the interference between the original wave and the scattered radiations from these elementary dipoles; it will thus depend upon the distribution of the dipoles (Fig. 4.1).

As is well known (see Appendix V) the radiation pattern for an oscillating dipole follows a squared cosine law (Lambert's Law) (Fig. 4.1(b)) with a maximum normal to the line of oscillating charge, and zero radiation along the line of the oscillation. Clearly, in the forward direction (i.e., parallel with the driving wave) all secondary radiations will bear a constant phase relationship with the driving wave (the actual phase difference depending on the relationship between the driving frequency and their natural resonant frequency) and will all be in phase with each other since, as they progress along the Oz direction, their phases will advance in the same way as that of the driving wave. Thus there is strong reinforcement in the forward direction, resulting in a forward wave of large amplitude but of different phase from the primary, owing to the aforementioned phase difference between primary and secondaries. Now a phase change of the resultant wave is equivalent to a refractive index other than unity since it has the same effect as a change of velocity and this is, in fact, the origin of refractive index in material media.

The elementary dipole oscillators leading to the secondary radiation will also be damped, to some extent, by radiation loss and by atomic collisions. Hence some absorption of the primary wave also will occur, leading to attenuation (and giving rise to scattering, and heating of the medium).

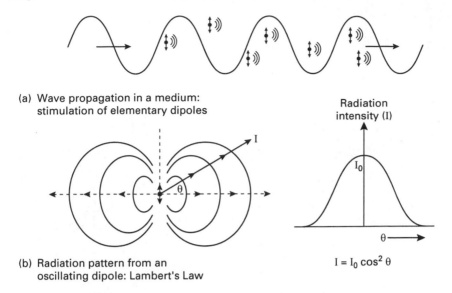

(a) Wave propagation in a medium:
stimulation of elementary dipoles

Radiation
intensity (I)

(b) Radiation pattern from an
oscillating dipole: Lambert's Law

$I = I_0 \cos^2 \theta$

Fig. 4.1 Stimulation of oscillators in a material medium.

Hence, we can best represent simultaneously both the change of phase and
the loss by using a complex refractive index for this forward-scatter process.

For radiation scattered in directions other than forward, the resultant
amplitude will depend upon the distribution of the radiating dipoles, i.e.,
the distribution of the atoms or molecules of the medium. If the scatterers are regularly arranged, as in a crystal, then there can be maxima in
directions other than forward provided that (see Fig. 4.2(a)):

$$\left. \begin{aligned} a \sin \vartheta \\ b \cos \vartheta \end{aligned} \right\} = m\lambda \qquad (m \text{ integer}) \tag{4.2}$$

where a and b are spacings in two (say) planes and ϑ is the scatter angle
w.r.t. the forward direction. It is clear, however, that no secondary maxima
are possible if either $m\lambda/a > 1$ or $m\lambda/b > 1$ for there are then no real
values of ϑ which satisfy (4.2). Hence if $\lambda > a, b$ there are no secondary
maxima, and this is the condition which obtains for most crystals, except
when λ becomes smaller than the interatomic spacing. This happens first
at X-ray wavelengths ($< 10\,\text{nm}$) and leads to the subject known as X-ray
crystallography, wherein X-rays are used very successfully to probe crystal
structures. But for optical wavelengths there are no maxima in crystals
other than that in the forward direction, so that, in a pure crystal, all the
light passes straight through without sideways scatter: a laser beam passing
through a pure crystal will not be visible from the direction orthogonal to
the line of propagation.

Suppose, however, that the scatterers are randomly distributed, as in
an amorphous (i.e., non-crystalline) medium. What happens in this case?

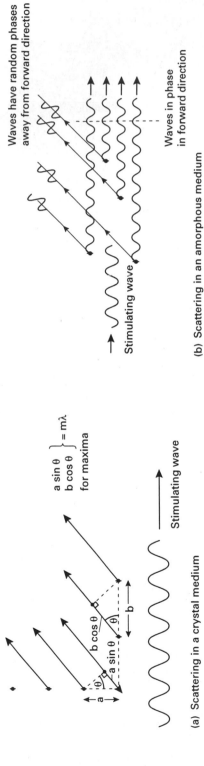

Waves have random phases away from forward direction

Waves in phase in forward direction

Stimulating wave

(b) Scattering in an amorphous medium

Blue light scattered preferentially $(\sim \frac{1}{\lambda^4})$ from setting Sun's rays

Red light persists for observer

Atmosphere

Sun

Earth

$$\left.\begin{array}{l} a\sin\theta \\ b\cos\theta \end{array}\right\} = m\lambda$$
for maxima

Stimulating wave

(a) Scattering in a crystal medium

$b\cos\theta$

$a\sin\theta$

θ

a

b

No wave

No wave

(All scattered waves in the horizontal plane are vertically polarized)

Vertically-polarized driving wave

(c) Rayleigh scattering (and the red setting sun)

Fig. 4.2 Light scattering processes.

Well here, although there will still be reinforcement in the forward direction since all scatterers remain in phase in that direction, in any other direction the phases of the scattered waves will be randomly distributed along with the atomic spacings (Fig. 4.2(b)), so there are no fixed phase relationships. (It is also necessary for the scatterers to be small in size compared with a wavelength otherwise different parts of the scatterer could give rise to coherent scatter and thus to peaks in the scattered radiation). The consequence of the random scatter is that (in any direction other than forward!) the resultant scattered amplitude is a summation of waves of the form:

$$\sum a \sin(\omega t - kz + \delta_i)$$

where δ_i is a randomly distributed phase. Hence the optical *intensity* of the light in that direction for a total of N_T scatterers will be given by:

$$I = \left(\sum_{i=1}^{N_T} a \sin(\omega t - kz + \delta_i) \right)^2$$

giving

$$I = \sum_{i=1}^{N_T} a^2 \sin^2(\omega t - kz) \cos^2 \delta_i$$

$$+ \sum_{i=1}^{N_T} a^2 \cos^2(\omega t - kz) \sin^2 \delta_i$$

$$+ \text{ terms in } \sum_{i=1}^{N_T} q \sin \delta_i \cos \delta_i$$

Since δ_i is as likely to be positive as it is to be negative (if randomly distributed) we are left only with the first two terms in this expression for I, and, since the average value of $\cos^2 \delta_i$ and $\sin^2 \delta_i$ is 1/2, the result is:

$$I = \tfrac{1}{2} N_T a^2$$

Now $\tfrac{1}{2} a^2$ is the mean square value of the intensity from each dipole, hence the result in this case is an intensity which is just the sum of the separate dipole intensities. This is a result which is intuitively correct for random scatterers. One final point is worth making before moving on to a more formal treatment of these ideas.

The power radiated by an oscillating dipole is proportional to the fourth power of the frequency and thus inversely to the fourth power of the wavelength (Appendix V). Hence, for sideways scatter in an amorphous medium the light intensity will be inversely proportional to the fourth power of the wavelength. This is a feature of what is known as Rayleigh scattering. The best known manifestation of this is the blue colour of a clear sky. When viewed away from the sun the light observed will be that scattered sideways from the sun's rays (Fig. 4.2(c)) by the atmosphere, which consists of randomly arranged molecules. Since blue light has a smaller wavelength than red light it will be scattered much more effectively than red ($\sim \lambda^4$) and hence the sky looks blue. For the same reason the sun itself looks red (more blue than red has been scattered away from the line of sight) and even redder at sunrise and sunset, when the direct rays have to pass through a greater thickness of atmosphere (Fig. 4.2(c)).

Having now acquired a qualitative 'feel' for these ideas we may move on to a more quantitative treatment of them.

We have seen that the refractive index of a medium is due to its atomic/molecular scatterers. We shall now derive quantitative relationships between the atomic scatterers and the refractive index of the medium (remembering that this latter is a complex quantity) to allow for both velocity changes and attenuation. Our problem then is to derive an expression for the complex refractive index:

$$n = n' - in''$$

in terms of the elementary atomic radiators, in order to be able to use our understanding of the processes for design and control. How should this be done?

We can start by remembering that the refractive index is related to the dielectric constant by (equation 2.4):

$$n = \varepsilon_R^{1/2}$$

Remember also that ε_R is the result of the electric polarization (separation of positive and negative charges) of the medium produced by an applied electric field. In fact, if the electric dipole moment (charge multiplied by displacement) per unit volume is P, elementary electrostatics tells us that

$$D = \varepsilon_0 E_0 + P = \varepsilon_R \varepsilon_0 E_0 \tag{4.3a}$$

where D is the electric 'displacement' and E_0 the applied field.

Now it is just this displacement of charge, in sympathy with the oscillating optical electric field, which gives rise to the secondary radiation; so immediately the required connection is made. The approach to be adopted, then, is to relate the electric polarization of the medium to the electronic displacements. We may note, at this point, that the ease with which these displacements can be made, for a given medium, is characterized by what is called the volume susceptibility, χ, which is defined by:

$$\chi = \frac{P}{\varepsilon_0 E_0}$$

i.e., the electric dipole moment per unit volume produced for a given free space displacement field.

Note also from equation (4.3a) that this means that:

$$\varepsilon_R = 1 + \chi \qquad (4.3b)$$

which is a well known relation in electrostatics.

Consider, then, a material through which is passing an electromagnetic wave with electric field given by:

$$E = E_0 \exp[i(\omega t - kz)]$$

and let us assume that the wavelength $(2\pi/k)$ of this propagating radiation is large compared with the interatomic spacing (i.e., we are not dealing with X- or γ-rays!).

An electron within an atom or molecule of the medium will be displaced, as a result of the action of the field, by an amount s, say, and the bonding in the atom will then exert a 'restoring force'. This restoring force, for moderate values of s compared with atomic dimensions, will be proportional to s, and we readily recognize this state of affairs as the prescription for simple harmonic motion (SHM) where the electron will oscillate with a frequency ω_r, say, with the atomic restoring force given by $m\omega_r^2 s$, m being the electron mass. Additionally, there will be a damping of the oscillation, owing to various dissipative forces such as radiation from the oscillating electron, collisions with other atoms/molecules, and so on; these will be proportional to the instantaneous velocity of the electron, ds/dt. Hence we may represent these losses by a term of the form: $-m\gamma ds/dt$.

The parameter γ will always be positive for a passive medium such as is presently being considered.

Now the force acting to displace the electrons will be that due to the electric field of the original wave plus that due to the electric polarization of the surrounding medium.

For an electron displacement s the electric dipole moment of an elementary dipole is: $p = es$ where e is the electronic charge. For N dipoles *per unit volume* the total electric polarization of the medium is thus:

$$P = Nes \qquad (4.4)$$

The effect of this local polarization on a charge embedded within it is equivalent to that on a charge sitting at the centre of a sphere which has surface density of charge equal to P (see any book covering elementary electrostatics, e.g., [1]) and this effect can readily be calculated to be equivalent to a field of $P/3\varepsilon_0$. Hence the total field acting on a given electron can be written:

$$E_e = E + \frac{P}{3\varepsilon_0}$$

The term $P/3\varepsilon_0$, representing the effect of the polarization of the surrounding molecules, is known as the Lorentz correction (see, for example, [1]).

We may now write the equation of motion of the displaced electron as a forced simple harmonic oscillator:

$$\frac{d^2 s}{dt^2} + \gamma \frac{ds}{dt} + \omega_r^2 s = \frac{e}{m} \left(E + \frac{P}{3\varepsilon_0} \right)$$

Substituting for s from (4.4) we obtain:

$$\frac{d^2 P}{dt^2} + \gamma \frac{dP}{dt} + P \left(\omega_r^2 - \frac{Ne^2}{3\varepsilon_0 m} \right) = \frac{Ne^2}{m} E \qquad (4.5)$$

Note that the 'restoring force' term has been modified to

$$\left(\omega_r^2 - \frac{Ne^2}{3\varepsilon_0 m} \right)$$

reflecting the fact that the electron displacements in the surrounding atoms themselves alter the force on any given electron. This has the effect, of course, of altering the natural oscillation frequency of the electron to ω_e where

$$\omega_e = \omega_r - \frac{Ne^2}{3\varepsilon_0 m}$$

Remembering that the driving field is given by

$$E = E_0 \exp[i(\omega t - kz)]$$

we try a solution for equation (4.5) of the form:

$$P = P_0 \exp[i(\omega t - kz)]$$

and, ignoring transients, obtain an expression for the polarizability (volume susceptibility) of the medium

$$\chi = \frac{P_0}{\varepsilon_0 E_0} = \frac{Ne^2}{\varepsilon_0 m} \cdot \frac{1}{(\omega_e^2 - \omega^2) + i\gamma\omega}$$

It is important to note here that ω_e^2 will always be positive for, if it were negative, the electron's restoring force would also be negative, and the atom would then ionize under the influence of the electric field; normal linear propagation could not occur under these conditions.

Finally, then, it is possible to generalize the expression for χ to include electrons with various natural frequencies of oscillation; this will be the case in any real atomic system. We obtain a total effective volume susceptibility:

$$\chi = \sum_j \frac{C_j}{(\omega_j^2 - \omega^2) + i\gamma_j \omega} \tag{4.6}$$

and all of the C_j, ω_j^2 and γ_j will always be positive.

Now we have already noted that:

$$\varepsilon_R = 1 + \chi$$

and

$$n^2 = \varepsilon_R$$

Hence we have:

$$n^2 = 1 + \chi$$

so that, from (4.6):

$$n^2 = 1 + \chi_e = 1 + \sum_j \frac{C_j}{(\omega_j^2 - \omega^2) + i\gamma_j \omega}$$

For media of low density, such as gases, χ_e is very small, and we can use the binomial expansion to obtain:

$$n \approx 1 + \tfrac{1}{2} \sum_j \frac{C_j}{(\omega_j^2 - \omega^2) + i\gamma_j \omega}$$

However, for solid media this expansion is invalid, and the full expression for n must be used. Of course, we can now identify n with the complex refractive index defined by equation (4.1), i.e., :

$$n = \left(1 + \sum_j \frac{C_j}{(\omega_j^2 - \omega^2) + i\gamma_j \omega}\right)^{1/2} = n' - in'' \tag{4.7}$$

The imaginary component of this (i.e., $-in''$) is, as already noted, simply a convenient way of expressing the loss term in the exponential representation. Of course this loss is due to the γ_j, without which n would be wholly real. γ_j, again as already noted, is due to radiation other than in the forward direction, and to atomic collisions, which heat the material. The collisions depend upon structural kinetics and are beyond this present discussion. However, because the radiation loss also is the result of just those interactive processes we have been considering, this can be readily calculated.

To find the non-forward scatter component we first note that for an oscillating electron with an electric dipole moment of amplitude p_0, the total radiated power (Appendix V) is:

$$W_s = \frac{\mu_0 \omega^4 p_0^2}{12\pi c_0}$$

where μ_0 is the magnetic permeability and c_0 the velocity of light in free space. If there are N dipoles per unit volume then:

$$P_0 = N p_0$$

where P_0 is the electric polarization of the medium. We have, by definition, that the volume susceptibility is:

$$\chi_e = \frac{P_0}{\varepsilon_0 E_0}$$

and thus

$$p_0 = \varepsilon_0 \chi_e \frac{E_0}{N}$$

giving

$$W_s = K \frac{E_0^2}{N^2} \omega^4 \chi_e^2 \tag{4.8}$$

with

$$K = \frac{\mu_0 \varepsilon_0^2}{12\pi c_0}$$

Now from (4.6) it is noted that if $\omega_j \gg \omega$ and the damping is small, then χ_e is approximately independent of ω. Under this condition W_s is proportional to ω^4 and thus to $1/\lambda^4$. This is the Rayleigh scattering condition and it occurs for frequencies well removed from any molecular resonances. In the region of the resonances the dependence of the scattering on frequency is more complex; the scattering itself is much stronger (resonance scattering).

Schematic variations with frequency for the real and imaginary components of the refractive index of doped silica (of which optical fibres are made) are shown in Fig. 4.3. The effects of molecular resonances are clearly seen on both components. In fact, it is possible to relate the real and imaginary components mathematically. This is due to the fact that they each rely on the same physical phenomenon of resonant absorption. The relevant mathematical expressions are known as the Kramers-Kronig relationships [2], and can be used, in principle, to derive one of the two variations when the other is known over a broad frequency range.

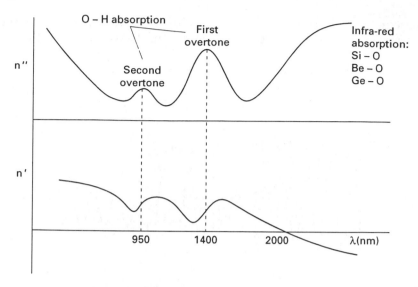

Fig. 4.3 Refractive index components for doped silica.

4.3 OPTICAL DISPERSION

The fact that the real part of the refractive index varies with frequency has some important implications for the propagation of the wave, for it means that the wave velocity varies with frequency. All real sources of light provide their radiation over a range of frequencies. This range is large for an incandescent radiator such as a light bulb, and very small for a gas laser; but it can never be zero. Consequently, in the cases of a medium whose refractive index varies with frequency, different portions of the source spectrum will travel at different velocities and thus will experience different refractive indices. This causes 'dispersion' of the light energy, and the medium is thus said to be 'optically dispersive'.

The phenomenon has a number of manifestations and practical consequences. One of the best known manifestations is that of the rainbow, where the variation of the refractive index with wavelength in water causes raindrops in the atmosphere to refract the sun's rays through different angles, according to the colour of the light, and thus to provide for us a wonderful technicolour display. Another well known example of dispersion is the experiment performed by Isaac Newton with a glass prism, allowing him to demonstrate quantitatively the different angles of refraction in glass for the spectral colours of which the sun's light is composed.

In the modern idiom of present-day optoelectronics we are rather more concerned with the effect that dispersion has on the information carried by a light beam, especially a guided one; so it is useful to quantify the dispersion effect with this in mind. In order to understand some of these consequences of dispersion, suppose that just two closely-spaced frequency components, of equal amplitude, are present in the source spectrum, i.e.,

$$E = E_0 \cos(\omega t - kz) + E_0 \cos(\overline{\omega + \delta\omega}\, t - \overline{k + \delta k}\, z)$$

where $\delta\omega, \delta k$ are small compared with ω and k, respectively.

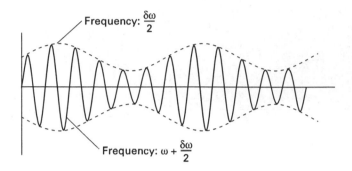

Fig. 4.4 Amplitude-modulated wave: sum of two waves of different frequencies.

Using elementary trigonometry we have:

$$E = 2E_0 \cos \tfrac{1}{2}\left(\delta\omega\, t - \delta k\, z\right) \cos\left(\overline{\omega + \tfrac{1}{2}\delta\omega}\, t - \overline{k + \tfrac{1}{2}\delta k}\, z\right)$$

This represents a sinusoidal wave (second factor) whose amplitude is modulated by another sinusoid (first factor) of lower frequency (Fig. 4.4). The wave itself travels at a velocity:

$$\frac{\omega + \tfrac{1}{2}\delta\omega}{k + \tfrac{1}{2}\delta k} \approx \frac{\omega}{k} = c$$

which is the mean velocity of the two waves. However, the point of maximum amplitude of the wave will always occur when the amplitude modulation has maximum value, i.e., when:

$$\tfrac{1}{2}\delta\omega\, t - \tfrac{1}{2}\delta k\, z = 0$$

so that:

$$\frac{\delta\omega}{\delta k} = \frac{z}{t} = c_g$$

and hence, in the limit as $\delta\omega,\ \delta k \to 0$:

$$c_g = \frac{d\omega}{dk} \tag{4.9}$$

where c_g is called the *group* velocity, and is the velocity (in this case) with which any given wave maximum progresses.

Now we also know that:

$$\frac{\omega}{k} = c = \frac{c_0}{n}$$

and hence $\omega = (c_0/n)k$ where n is the refractive index of the medium.

In general, n will vary with optical frequency and thus will be a function of k, so that we can differentiate this expression for ω to obtain:

$$\frac{d\omega}{dk} = \frac{c_0}{n} \left(1 - \frac{k}{n} \frac{dn}{dk} \right)$$

or, in terms of the wavelength λ:

$$c_g = \frac{d\omega}{dk} = \frac{c_0}{n} \left(1 - \frac{\lambda}{n} \frac{dn}{d\lambda} \right) \tag{4.10}$$

If n does not vary with wavelength, then

$$\frac{dn}{d\lambda} = \frac{dn}{dk} = 0$$

and then

$$\frac{d\omega}{dk} = c_g = \frac{c_0}{n} = c$$

However, if $dn/d\lambda \neq 0$ (i.e., the medium is dispersive) then $c_g \neq c$ and the maximum of the disturbance travels at a different velocity from the 'carrier' optical wave. These ideas may readily be generalized to include the complete spectrum of a practical source. Provided that $dn/d\lambda$ is sensibly constant over the spectrum of wavelengths, it follows that a pulse of light from the source will effectively travel undistorted at a velocity of c_g rather than c. The quantity c_g is called the group velocity of the pulse. For convenience, a 'group refractive index' is defined by:

$$N_g = \frac{c_0}{c_g}$$

and, from (4.10), if $dn/d\lambda \ll n/\lambda$ (i.e., dispersion is small)

$$N_g \approx n - \lambda \frac{dn}{d\lambda}$$

Suppose, however, that $dn/d\lambda$ is not constant over the source spectrum. In this case different portions of the pulse travel with different *group* velocities, so that the pulse's width increases as it propagates. This can be seen clearly by examining the expression for the propagation time over a distance L for a given group velocity c_g:

$$\tau = \frac{L}{c_g}$$

The variation of this over a spread of frequencies $\delta\omega$ will be given by:

$$\Delta\tau = L\frac{\mathrm{d}\left(\frac{1}{c_g}\right)}{\mathrm{d}\omega}\delta\omega$$

and, substituting for c_g from (4.9):

$$\Delta\tau = L\frac{\mathrm{d}^2 k}{\mathrm{d}\omega^2}\delta\omega \tag{4.11}$$

This can also be expressed, conveniently, in terms of n and λ, by using (4.10) and assuming that the dispersion is small

$$\tau = \frac{L}{c_g} = \frac{L}{c_0}\left(n - \lambda\frac{\mathrm{d}n}{\mathrm{d}\lambda}\right)$$

Hence

$$\Delta\tau = \frac{\mathrm{d}\tau}{\mathrm{d}\lambda}\delta\lambda = -\frac{L}{c_0}\lambda\frac{\mathrm{d}^2 n}{\mathrm{d}\lambda^2}\delta\lambda \tag{4.12}$$

Thus it is noted that the spread of arrival times depends upon $\mathrm{d}^2 n/\mathrm{d}\lambda^2$ (or $\mathrm{d}^2 k/\mathrm{d}\omega^2$) and thus on a non-linearity in the relationship between n and λ (or ω and k) over the source spectrum. This phenomenon is known as *group velocity dispersion* (GVD). Clearly in order to minimize GVD, and thus minimize pulse spreading, it is necessary to minimize $\mathrm{d}^2 n/\mathrm{d}\lambda^2$ and $\delta\lambda$.

To set these ideas into a practical context we may note that, in silica, the material from which optical fibres are made, the pulse spreading at 850 nm for a light-emitting diode (~ 30 nm spectral width) is ~ 2.5 ns, whereas for a semi-conductor laser (width ~ 3 nm) the spread is only ~ 0.25 ns. The material condition for minimum GVD (i.e., $\mathrm{d}^2 n/\mathrm{d}\lambda^2 = 0$) occurs at a wavelength of 1.28 nm, where the spreading for each of the above sources is an order of magnitude lower. The presently preferred wavelength for optical-fibre communications systems is, therefore, 1.28 nm. We shall be returning to this important topic in Chapters 8 and 10.

4.4 EMISSION AND ABSORPTION OF LIGHT

4.4.1 The elementary processes

In considering the processes by which light is emitted and absorbed by atoms we must again quickly come to terms with the corpuscular or, to use the more modern term, the particulate nature of light.

In classical (i.e., pre-quantum theory) physics, the atom was held to possess natural resonant frequencies. These corresponded to the electromagnetic wave frequencies which the atom was able to emit when excited into oscillation. Conversely, when light radiation at any of these frequencies fell upon the atom, the atom was able to absorb energy from the radiation in the way of all classical resonant system driving-force interactions.

However, these ideas are incapable of explaining why, in a gas discharge, some frequencies which are emitted by the gas are not also absorbed by it under quiescent conditions; neither can it explain why, in the photoelectric effect (where electrons are ejected from atoms by the interaction with light radiation), the energy with which the electrons are ejected depends not on the intensity of the light, but only on its frequency.

We know that the explanation of those facts is that atoms and molecules can exist only at discrete energy levels. These energy levels may be listed in order of ascending magnitude, $E_0, E_1, E_2, \ldots, E_n$. Under conditions of thermal equilibrium the number of atoms having energy E_i is related to that having energy E_j by the Boltzmann relation:

$$\frac{n_i}{n_j} = \exp\left(-\frac{E_i - E_j}{kT}\right) \qquad (4.13)$$

where k is Boltzmann's constant and T is the absolute temperature. Light can only be absorbed by the atomic system when its frequency ν corresponds to at least one of the values ν_{ji} where:

$$\nu_{ji} = E_j - E_i. \quad (j > i) \qquad (4.14)$$

(The symbol ν is used now for the frequency rather than $\omega/2\pi$, to emphasize that the light is exhibiting its particulate character.) Here, h is Planck's quantum constant, with value 6.626×10^{-34} joule.seconds.

In this case the interpretation is that one quantum of light, or photon, with energy $h\nu_{ji}$, has been absorbed by the atom, which in consequence has increased in energy from one of its allowed values E_i, to another, E_j. Correspondingly, a photon will be emitted when a downward transition occurs from E_j to E_i, this photon having the same frequency ν_{ji}.

In this context we must think of the light radiation as a stream of photons. If there is a flux of p photons across unit area per unit time then we may write:

$$I = ph\nu$$

where I is the light intensity defined in equation (1.7). Similarly, any other quantity defined within the wave context also has its counterpart in the particulate context.

In attempting to reconcile the two views, the electromagnetic wave should be regarded as a probability function whose intensity at any point in space defines the probability of finding a photon there. But only in the specialized study of quantum optics are such concepts of real practical significance. For almost all other purposes (including the present one) either the wave representation or the particle representation is appropriate in any given practical situation, without any mutual contradiction.

Each atom or molecule has a characteristic set of energy levels, so that the light frequencies emitted or absorbed by atoms and molecules are themselves characteristic of the material concerned. When an excited system returns to its lowest state, some return pathways are more probable than others, and these probabilities are also characteristic of the particular atoms or molecules in question. (They can be calculated from quantum principles.) Consequently, the emission and/or the absorption spectrum of a material can be used to identify it, and to determine its concentration. These ideas form the substance of the subject known as analytical spectroscopy, which is a very extensive and powerful tool in materials analysis. But it is a highly specialized topic and we shall not deal with it further here.

4.4.2 Elements of laser action

The laser is a very special source of light. Modern optics, of which the subject of optoelectronics forms a part, effectively dates from the invention of the laser in 1960 [3]. The word 'laser' is an acronym for Light Amplification by Stimulated Emission of Radiation, and we will now proceed to determine the processes on which it depends, although these will be considered in more detail in Chapter 7.

It was noted in the previous section that a photon could cause an atomic system to change from one of its allowed states to another, provided that equation (4.14) was obeyed. This equation related to the action of the photon in raising the system from a lower to a higher energy state. However, if the system were already in the higher of the two states when the photon acted, then it is also true that its action would be to cause the transition down to the lower state, still in accordance with equation (4.14) (but now with $j < i$). This process is called 'stimulated emission' since the effect is to cause the system to emit a photon, with energy $h\nu_{ij}$, corresponding to that lost by the system; so now we have two photons - the 'driving' photon and the emitted one. This process is crucial to laser action. (A rough classical analogy is that where an a.c. driving force is 'anti-resonant' with a naturally oscillating system, i.e., in negative phase quadrature. In this case the driving force will receive energy from the system.)

We must also be aware of the fact that a system which is not in its lowest energy state is not in stable equilibrium. If it has any interaction with the outside world it will eventually fall to its lowest state. Thus an atomic system in a state E_i will fall spontaneously to the lower state E_j, even without the stimulus of $h\nu$, in a time which depends on the exact nature of the equilibrium conditions; these, broadly, may be classed as unstable, or metastable (a long-lived non-stable state). The photon which results from this type of transition is thus said to be due to spontaneous emission.

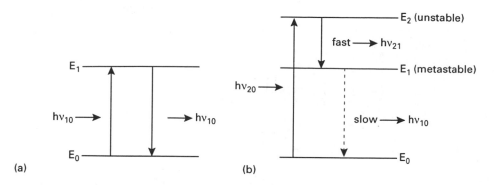

Fig. 4.5 Energy level diagram for laser action.

Let us now consider a two-level atomic system with the energy levels E_0 and E_1 (Fig. 4.5(a)). Suppose we illuminate this system with electromagnetic radiation of frequency

$$\nu_{10} = \frac{E_1 - E_0}{h}$$

Initially, if the system is in thermal equilibrium at temperature T, the numbers of atoms in the two levels will be related, according to equation (4.13), by:

$$\frac{n_1}{n_0} = \exp\left[-\frac{E_1 - E_0}{kT}\right] \qquad (4.15)$$

so that if $E_1 > E_0, n_1 < n_0$. Suppose now that the intensity of the radiation at frequency ν_{10} is steadily increased from zero. At low levels, assuming that the probability of transition is the same for the two transition directions, more atoms will be raised from the lower to the higher state than vice versa, since there are more atoms in the lower state, according to (4.13). As the intensity is increased the number of downward transitions (stimulated and spontaneous) will increase as the occupancy of the upper state rises, tending towards the saturation condition where the (dynamic) occupancies of the two states, and the rates of transition in the two directions, are equal.

Consider now the three-level system shown in Fig. 4.5(b). Here we have the lowest level E_0, a metastable level E_1 and an unstable level E_2. If this system (initially in thermal equilibrium) is irradiated with light of frequency $\nu_{20} = (E_2 - E_0)/h$, then the effect is to raise a large number

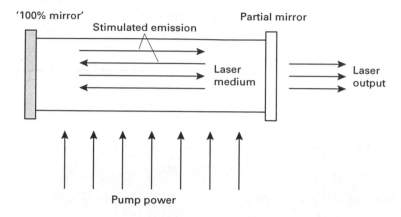

Fig. 4.6 Schematic laser construction.

of atoms from E_0 to E_2. These then decay quickly to the state E_1 by spontaneous emission only (since the input light frequency does not correspond to this transition), and thence only slowly from this metastable (i.e long-lived) state back to the ground state. The consequence of this is that, under these circumstances, there can be a larger number of atoms in state E_1 than in state E_0. Since this does not correspond to a Boltzmann distribution (which requires fewer atoms in a state of higher energy), it is known as an 'inverted' population. Suppose that a second beam of light is incident on this inverted population at frequency

$$\nu_{10} = \frac{E_1 - E_0}{h}$$

This light encounters a situation where it can more frequently produce downward transitions by stimulated emission from E_1 to E_0 than it can excite atoms from E_0 to E_1. Thus more stimulated photons are produced than are absorbed by excitation, and this beam receives 'gain' from the medium, i.e., it is amplified. The medium is said to be 'pumped' by the first beam to provide gain for the second. We have 'Light Amplification by Stimulated Emission of Radiation', i.e., a laser. If, now, the medium is enclosed in a tube with parallel mirrors at the ends (Fig. 4.6), then the stimulated photons can be made to bounce back and forth between the mirrors and themselves act to stimulate even more photons. We have provided the 'amplifier' with positive feedback and have produced an 'oscillator'. If one of the two mirrors is only partially reflecting, some of the oscillator energy can emerge from the tube. This energy will be in the form of a light wave of frequency

$$\nu_{10} = \frac{E_1 - E_0}{h}$$

which is accurately defined if the energy levels are sharp; of relatively large intensity if the volume of the tube is large, the pump power is large and the cross-sectional area of the tube is small; well-collimated, since the light will only receive amplification within the tube if it is able to bounce between the two parallel mirrors; and has an accurately defined phase since the phase of the stimulated photon is locked to that of the photon which stimulates it. Thus we have monochromatic (narrow frequency range), coherent (well-defined phase), well-collimated light: we have laser light.

The simple picture that has been painted of laser action illustrates the main ideas involved in laser action. However, the excitation and de-excitation pathways for most of the commonly used lasers are quite complex, and it would be as well for the reader to be aware of this. In pursuit of this aim we may consider the action of what, presently, is probably the most commonly used visible-light laser: the helium-neon (He-Ne) laser. The energy level structure for this laser system is shown in Fig. 4.7(a), and the basic physical construction of the laser is shown in Fig. 4.7(b).

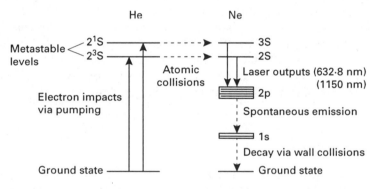

(a) Energy-level pathways

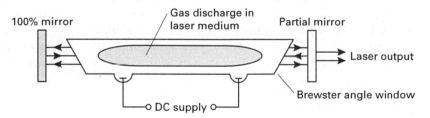

(b) Basic construction

Fig. 4.7 The helium-neon laser.

An electrical discharge is struck in a mixture of, typically, He at 1 mmHg pressure, and Ne at 0.1 mmHg. The discharge electrons excite the atoms into a variety of excited states resulting, in equilibrium, in a relatively large number resting in the metastable states 2^1S and 2^3S, shown in Fig. 4.7(a). (These designations derive from a spectroscopic notation which is detailed in many books on atomic physics (e.g., [4]), and which it is not necessary to understand fully for our present purposes.) These metastable levels happen to correspond to S levels, in Ne atoms, which are not readily excited by the discharge (owing to the unfavourably acting quantum selection rules). There is a high probability of transfer of energy from the metastable He atoms to these Ne atoms via direct atomic collisional exchange. The excited Ne atoms clearly will now constitute an inverted population since there is no corresponding collisional tendency to depopulate them. The two excited Ne levels decay quickly to the sparsely populated 2p levels shown, emitting optical wavelengths of 632.8 nm and 1150 nm, respectively. Optical feedback, and thus simulated emission, is arranged for either one or the other of these wavelengths via the wavelength selectivity of the end mirrors in the structure. Atoms in the 2p levels then decay spontaneously to 1s levels and subsequently, primarily via tube-wall collisions, to the ground state. (This last feature introduces a geometrical factor into the design of the laser, and illustrates one of the many such considerations which must be taken into account in the optimization of laser design.) Typically, a He-Ne laser at 632.8 nm will provide $0.5 - 50$ mW of laser light for $5 - 10$ W of electrical excitation (pump) power. Thus it is not especially efficient (0.05%).

Having understood these basic ideas of laser mechanism and design, the more detailed treatment of lasers given in Chapter 7 should now become much more comfortable.

4.4.3 Luminescence

In section 4.4.1 quantum emission processes from atoms were discussed, and it was noted that an atom, when raised to an excited state by an incoming photon, could return to the ground state via various intermediate levels. Indeed, this was seen to be the basis for laser action. Whenever energy is absorbed by a substance, a fraction of the energy may be re-emitted in the form of visible or near-visible light radiation via such processes. This phenomenon is given the name 'luminescence'. Thus we may have photoluminescence (excitation by photons), cathodoluminescence (excitation by electrons), thermoluminescence (by heat), chemiluminescence (by chemical reaction), electroluminescence (by electric field), and so on.

If the light is emitted during the excitation process the phenomenon is often called 'fluorescence', while in the case of emission after the excitation has ceased it is called 'phosphorescence'. Clearly both are aspects of luminescence, and the distinction between them must be arbitrary, since there must always be *some* delay between emission and absorption. A delay of 10^{-8} s is usually taken as the upper limit for fluorescence; beyond lies phosphorescence.

Phosphorescence usually is due to metastable (i.e., long-lived) excited states. These, in turn, are usually due to impurity 'activators' in solids, which, by distorting the host lattice in some way, lead to 'traps' in which excited atomic electrons can 'rest' for a relatively long time before returning to their ground state.

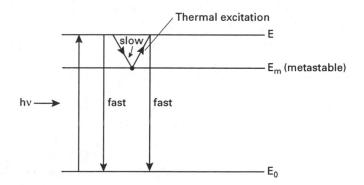

Fig. 4.8 Temperature-dependent phosphorescence.

One consequence of this is that the phosphorescence is often temperature sensitive. This can readily be seen by reference to Fig. 4.8. Here we have the situation where a substance absorbs sufficient energy to excite atoms to an excited state E, from its ground state E_0. The atoms find that they can then do one of two things: either return to the ground state directly or drop into the metastable state E_m. Let us suppose that the first probability is very much greater than the second, so that most of the light at frequency $(E - E_0)/h$ is emitted rapidly, as fluorescence. There will then follow a much longer period of phosphorescence which is a result of the atoms being excited thermally from E_m to E, and then quickly decaying again to E_0 (see Fig. 4.8). This latter process is thus controlled by the thermal $E_m \to E$ excitation, which has probability of the form $p_0 \exp[-(E - E_m)/kT]$, and is thus strongly temperature dependent. The decay time can thus be used to measure temperature, using purely optical means (both excitation and detection) and this has already proved useful in some application areas.

Another device in which luminescence features strongly is the light-emitting diode (LED). In this case (Fig. 4.9) a junction is constructed from a p-type and an n-type semiconductor. The p-n junction comes to equilibrium by diffusion of majority carriers (i.e., holes and electrons) across the physical junction, until an equilibrium is established between the force exerted by the field (which results from the separation of charge), and the tendency to diffuse. If now an external electric field is imposed on the junction, in opposition to the equilibrium field ('forward bias'), the result is to cause electrons and holes, with differing energy levels, to annihilate each

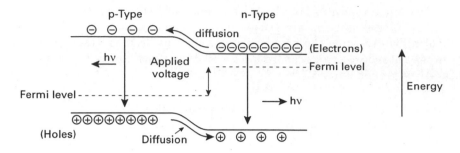

Fig. 4.9 Energy diagram for light-emitting diode (LED).

other and thus give rise to photons equal in energy to the band gap of the material. This is thus an example of electroluminescence. The efficiency with which this process occurs (i.e., fraction of injected electrons which gives rise to photons) depends significantly on the difference in *momentum* between the combining electrons and holes. Any momentum difference must be lost or gained, in order to conserve momentum overall. The greater is this momentum difference the more difficult is the compensation process (it is usually lost or gained via 'phonons' i.e., lattice vibrational energy) and the less efficient will be the conversion process. Substances for which the two momenta are the same are called 'direct band-gap' semiconductors, and two of the best known of these are gallium arsenide (GaAs) and gallium phosphide (GaP). Doping these substances with impurities can give some control over the emission wavelengths, which lie in the near infrared for GaAs, and in the visible range for GaP. The two substances can also be combined into a ternary alloy with general formula $GaAs_{1-x}P_x$ to give a useful range of devices, by varying x. These materials are widely used in optical devices for their conveniences of robustness, ease of activation, compactness and range of possible emitted wavelengths. If a crystal of, say, GaAs has two opposite facets polished accurately parallel to each other, then these form a Fabry-Perot laser cavity which allows laser action when 'pumped' by injecting electrons from a current source. This is the basis for the semiconductor laser diode (SLD) which is very widely used both in optical-fibre sensors and in optical-fibre telecommunications.

Once again an understanding of all these ideas will be useful when optical sources are dealt with in more detail in Chapter 7.

4.4.4 Photodetection

The processes which enable light powers to be measured accurately depend directly upon the effects which occur when photons strike matter. In most cases of quantitative measurement the processes rely on the photon to raise an electron to a state where it can be observed directly as an electric current.

It should also be noted, in this context, that photons arrive at a detector as a stream of particles obeying (normally) Poisson statistics in their arrival rate. This topic was dealt with quantitatively in section 1.7, and it is worth emphasizing one of the important results from that section, that the signal-to-noise ratio in a photodetection process is given by:

$$\text{SNR} = \left(\frac{P_m}{Bh\nu}\right)^{1/2} \tag{4.16}$$

where P_m is the mean optical power, B is the bandwidth of the photode-tector, h Planck's constant and ν the optical frequency. The importance resides in the fact that the accuracy of measurement (SNR) improves as the square root of the mean power and inversely as the square root of the optical frequency. In both cases this is because the 'granular', particulate nature of the photon arrival process is more noticeable when the power (P_m) is low, since there will then be fewer photons arriving in a given time; and also when ν is large, since this means more energy per photon and thus, again, fewer photons in a given time, for a given power. For detectors which are 'quantum noise limited' (i.e., limited by the statistical nature of this arrival process) the expression (4.16) for SNR is a vital design feature.

Consider, again, the $p-n$ junction of Fig. 4.9. When considering luminescence it was noted that the physical contact between these two types of semiconductor (i.e., p and n) led to a diffusion of majority carriers across the junction in an attempt to equalize their concentrations on either side. The result, however, was to establish an electric field across the junction as a consequence of the charge polarization. Suppose now that a photon is incident upon the region of the semiconductor exposed to this field. If this photon has sufficient energy to create an electron-hole pair, these two new charge carriers will be swept quickly in opposite directions across the junction to give rise to an electric current which can then be measured. The process is assisted by application of an external 'reverse bias' electric field. This simple picture of the process enables us to establish two important relationships appropriate to such devices (which are called photodiodes).

First, for the photon to yield an electron-hole pair its optical frequency (ν) must satisfy $h\nu > E_g$, where E_g is the bandgap energy of the material. If ν is too high, however, all the photons will be absorbed in a thin surface layer and the charge pairs will not be collected efficiently by the junction. Thus there is a frequency 'responsivity' spectrum for each type of photo-diode, which, con-sequently, must be matched to the spectrum of the light which is to be detected.

Secondly, suppose that we are seeking to detect a light power of P at an optical frequency ν. This means that $P/h\nu$ photons are arriving every second. Suppose now that a fraction η of these produce electron-hole pairs. Then there are $\eta P/h\nu$ charge carriers of each sign produced every second, so, if all are collected, the observed electric current is given by $i = e\eta P/h\nu$.

Thus the current is proportional to the optical power. This means that the electrical power is proportional to the square of the optical power. It is important, therefore, when specifying the signal-to-noise ratio for a detection process, to be sure about whether the ratio is stated in terms of electrical or optical power. (This is a fairly common source of confusion in the specification of detector noise performance.)

Photodetectors are covered more comprehensively in Chapter 7.

4.4.5 Photo-emission

This chapter on light-matter interactions would be incomplete without at least a mention of the topic of photo-emission, the process whereby a photon can eject an electron from an atom or molecule, so that it becomes freed from its bound state. In a gas, the process is relatively straightforward, for, in this case, there is a calculable (and measurable) energy which is required to allow the electron to become free of the coulomb forces in the molecule, and the energy of the photon clearly must exceed that value. Hence we can impose the simple condition: $h\nu > E_i$ where E_i is the 'ionization' energy. A molecule which has lost at least one electron must be left with a residual positive charge (since it was originally electrically neutral) and thus it becomes a positive 'ion'. Furthermore the kinetic energy with which the electron leaves its binding will be given by:

$$E_k = h\nu - E_i$$

since energy must be conserved. Thus, when we expose a mass of gas to radiation of sufficiently high frequency (usually uv or X-radiation) we create a mass of separate positive and negative charges, which behaves very differently from the original neutral gas, especially, of course, under the action of electric and magnetic fields. Such a state of matter is called a 'plasma' and it has some very interesting and important properties and applications, not least because it often appears naturally in the physical world (e.g., the ionosphere and stellar atmospheres). However, this represents a subject all of its own and is somewhat removed from mainstream optoelectronics, so we cannot pursue it.

If we turn our attention now to photo-emission processes in liquids and solids, the basics become much more complex. This is because the electrons are not bound to individual atoms or molecules but to the material as a whole, in energy bands. However, it is still possible to eject electrons, completely, from solids and liquids, and there are devices which depend on this phenomenon (the photoelectric effect). To study this properly we need to understand something about the band structure of solids, so it will be deferred until we have had a chance to deal with this (Chapter 6). A very sensitive photoetector, the photomultiplier, depends upon photo-emission from solid materials, and this will be treated in Chapter 7.

4.5 CONCLUSIONS

In this chapter we have gained some familiarity with the way in which light and matter interact.

We have seen that when the light wave allows the atomic electrons to remain bound in their atoms we may readily understand the phenomena of wave propagation through the medium as an interaction between a driving force (the wave) and a natural oscillator (the atom or molecule). The phenomena related to scattering by the medium can also be explained in the same way, using classical ideas. When we need to deal with the processes which involve the generation of frequencies other than that of the driving wave, quantum ideas are required. Hence laser action and luminescence can only be explained satisfactorily with their assistance.

We also need these ideas when dealing with the ejection of electrons from their bound states. As has been stated before, there is no easy way round this mix of ideas: some processes are easier to understand in one picture, others in the other picture. The ultimate test for optoelectronic technology is the extent to which our understanding allows us to make advances in the provision of devices and systems.

PROBLEMS

4.1 Show that for a medium where electric polarization as a result of the action of an external electric field is P, the electric field acting on any given electron in the medium is given by:

$$E_e = E + \frac{P}{3\epsilon_0}$$

where ϵ_0 is the permittivity of free space. (Consider the action on an electron sitting at the centre of a sphere within the medium.) If the electric susceptibility is given by

$$\chi = \frac{P}{\epsilon_0 E}$$

derive an expression for the dielectric constant ϵ_R of the medium. What, then, does ϵ_R actually mean?

4.2 What is meant by the terms 'optical dispersion' and 'group velocity dispersion'? The refractive index n of a medium varies with wavelength λ in accordance with the Cauchy formula:

$$n = A + \frac{B}{\lambda^2} + \frac{C}{\lambda^4}$$

A narrow pulse of light, with mean wavelength λ contains a range of wavelengths $\delta\lambda$, with $\delta\lambda \ll \lambda$. What is the increase in the pulse width after it has travelled a distance L in the medium?

At what wavelength would the group velocity dispersion be zero in such a medium? What is the condition that it will never be zero?

4.3 Derive the relationship between light intensity and photon flux. What light power is required to pass 10^{25} photons per second through an area, normal to the propagation direction, of $100 \, \mu m^2$ at a wavelength of 633 nm?

4.4 Why does the signal-to-noise ratio at the output from an optical detector depend upon the mean optical power incident upon it? Derive an expression for this dependence.

The output power level from an optical source at a wavelength of $10.6 \, \mu m$ must be measured to an accuracy of 0.1% in a time of 1 ms. What is the minimum output level which would allow this to be done, assuming that the detector is quantum-noise limited?

4.5 Discuss what is meant by a 'plasma'. What kinds of effects would you expect to occur if an intense laser beam were sent into a highly-ionized plasma?

4.6 What is meant by Rayleigh scattering? Why should it be proportional to the fourth power of the optical frequency?

An optical wave travelling in a medium suffers both scattering and absorption. The Rayleigh scatter coefficient in the medium is given by:

$$S = \sigma/\lambda^4 \text{ m}^{-1}$$

and the absorption coefficient by:

$$\alpha = \left(a + \frac{b}{\lambda} \right) \text{ m}^{-1}$$

where λ is the wavelength of light.

Design and describe in detail an experiment to measure σ, a and b.

If two optical waves with equal intensities but different wavelengths λ_1 and λ_2 are co-propagated in the medium, derive an expression for the ratio of the two intensities, scattered at right angles to the direction of propagation, at distance l into the medium.

REFERENCES

[1] Bleaney, B. I. and Bleaney, B. (1959) *Electricity and Magnetism*, Oxford, chap. 18.
[2] Lipson, S. G. and Lipson, H. (1969) *Optical Physics*, Cambridge University Press, chap. 12.

[3] Maiman, T. H. (1960) 'Stimulated optical radiation in ruby masers', Nature (Lond), 187, 493.
[4] Richtmeyer, F. K., Kennard, E. H. and Lauritsen, T. (1955) *An Introduction to Modern Physics*, McGraw-Hill, chap. 7.

FURTHER READING

Ditchburn, R. W. (1976) *Light*, Academic Press, chaps 15 and 17.
Yariv, A. (1976) *Introduction to Optical Electronics*, Holt, Reinhart and Winston, chaps 5 and 6.

5

Optical coherence and correlation

5.1 INTRODUCTION

In dealing with the subjects of interference and diffraction in Chapter 2 the assumption was made that each of the interfering waves bore a constant phase relationship to the others in both time and space. Such an assumption cannot be valid for all time and space intervals since, as was seen in section 4.4, the atomic emission processes which give rise to light are largely uncorrelated, except for the special case of laser emission. In this section the topic of 'coherence' will be dealt with. Clearly it will have a bearing on interference phenomena.

The coherence of a wave describes the extent to which it can be represented by a pure sine wave. A pure sine wave has infinite extension in space and time, and hence cannot exist in reality. Perfect coherence is thus unachievable, but it is nevertheless a valuable concept.

Coherence, in general, is a valuable concept because it is a measure of the constancy of the relationships between one part of a wave (in time and/or space) and another; and between one wave and another. This is why it is so important from the point of view of interference: a wave can only interfere with itself or with another wave (of the same polarization) to produce a sensible interference pattern if the phases and amplitudes remain in constant relationship whilst the pattern is being sensed. Additionally, it is clear that if we wish to impose information on an optical wave by modulating one of its defining parameters (i.e. amplitude, phase, polarization or frequency) then the extent to which that information remains intact is mirrored by the extent to which the modulated parameter remains intact on the wave itself; thus coherence is an important parameter in respect of the optical wave's information-carrying capacity, and our ability generally to control and manipulate it.

Waves which quite accurately can be represented by sine waves for a limited period of time or in a limited region of space are called partially coherent waves. Figure 5.1 shows examples of time-like and space-like partial coherence.

A normal light source emits quanta of energy at random. Each quantum conveniently can be regarded as a finite wave train having angular frequency ω_0, say, and duration $2\tau_c$ (Fig. 5.2).

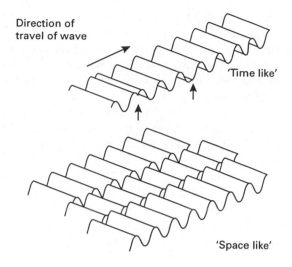

Fig. 5.1 Illustrations for the idea of partial coherence.

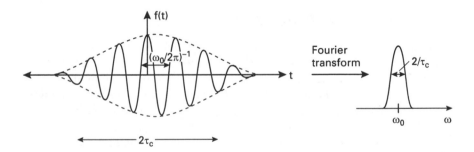

Fig. 5.2 The optical wave packet.

Fourier theory tells us that this wave train can be described in the frequency domain as a set of waves lying in the frequency range $\omega_0/2\pi \pm 1/\tau_c$. For a large number of randomly-emitted wave packets all the components at any given frequency will possess random relative phases. Spatial and temporal (time-like) coherence will thus only exist for, respectively, a distance of the order of the length of one packet $(2c\tau_c)$ and a time of the order of its duration $(2\tau_c)$. If the wave packets were of infinitely short duration (δ-function pulses, see Appendix VI) then Fourier theory tells us that all frequencies would be present in equal amounts and they would be completely uncorrelated in relative phase. This is the condition we call white light. Its spatial and temporal coherence are zero (i.e. there is perfect *incoherence*) and, again, it is an unachievable fiction. Between the two fictions of perfect coherence and perfect incoherence there lies the real world.

In this real world we have to deal with real sources of light. As has been noted before (Section 4.3) real sources always have a non-zero spectral width, i.e. the light power is spread over a range of frequencies, and the result of superimposing all the frequency components, as again we know from Fourier theory, is to produce a disturbance which is not a pure sine wave: an example is the wave packet we have just considered. Another is the infinitesimally narrow δ-function.

Hence there is seen to be a clear connection between spectral width and coherence. The two are inversely proportional. The narrower is a pulse of light (or any other waveform) the less it is like a pure sine wave and, by Fourier theory, the greater is its spectral width.

In the practical case of a two-slit interference pattern, the pattern will be sharp and clear if the light used is of narrow linewidth (a laser perhaps); but if a broad-linewidth source, such as a tungsten filament lamp, is used, the pattern is multicoloured, messy and confused.

Of course, we need to quantify these ideas properly if we are to use them effectively.

Let us begin.

5.2 MEASURE OF COHERENCE

If we are to determine the measurable effect which the degree of coherence is to have on optical systems, especially those which involve interference and diffraction phenomena, we need a quantitative measure of coherence. This must measure the extent to which we can, knowing the (complex) amplitude of a periodic disturbance at one place and/or time, predict its magnitude at another place and/or time. We know that this measure can be expected to have its maximum value for a pure sine wave, and its minimum value for white light. A convenient definition will render these values 1 and 0, respectively.

To fix ideas, and to simplify matters, let us consider first just temporal coherence.

We may sensibly postulate that for a time function $f(t)$ a knowledge of its value at t will provide us with some knowledge of its value up to a later time, say t', when it becomes completely independent of its value at t. If two time functions are completely independent then we expect the average value of their product over a time which is long compared with the characteristic time constant for their variations (i.e. the reciprocal of bandwidth) to be equal to the product of their individual average values, i.e.,

$$\langle f(t)f(t')\rangle = \langle f(t)\rangle\langle f(t')\rangle \tag{5.1}$$

and if, as is the case for the vast majority of optical disturbances, the functions oscillate about a zero mean, i.e.,

$$\langle f(t)\rangle = \langle f(t')\rangle = 0$$

then it follows from (5.1) that:

$$\langle f(t)f(t')\rangle = 0$$

In words: the product of two independent functions, when averaged over a term which is long compared with the time over which each changes significantly, is zero.

On the other hand, if we set $t = t'$, our 'delay average' above must have its maximum possible value, since a knowledge of the value of $f(t)$ at t enables us to predict its value at that time t with absolute certainty! Hence we have

$$\langle f(t)f(t)\rangle = \langle f^2(t)\rangle$$

and this must be the maximum value of the 'delay-average' function.

Clearly, then, the value of this product (for all real-world functions) will fall off from a value of unity when $t' = t$, to a value of zero when the two variations are completely independent, at some other value of t'; and it is also clear that the larger the value of t' for which this occurs, the stronger is the dependence and thus the greater is the coherence. It might well be, then, that the quantity which we seek in order to measure the coherence is a quantity which characterizes the speed at which this product function decays to zero.

Suppose, for example, we consider a pure, temporal sine wave in this context. We know in advance that this is a perfectly coherent disturbance, and conveniently we would thus require our coherence measure to be unity.

Let us write the wave as:

$$f(t) = a\sin\omega t$$

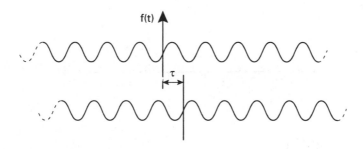

Fig. 5.3 Self-correlation delay.

To obtain the 'delay-average' function we first multiply this by a replica of itself, displaced by time τ (i.e., $t' = t + \tau$) (Fig. 5.3). We have:

$$f(t)f(t+\tau) = a\sin\omega t \; a\sin\omega(t+\tau)$$
$$= \tfrac{1}{2}a^2[\cos\omega\tau - \cos(2\omega t + \omega\tau)]$$

We now average this over all time (zero bandwidth gives a characteristic time constant of infinity!) and, since $\langle \cos(2\omega t + \omega\tau)\rangle = 0$, we have:

$$\langle f(t)f(t+\tau)\rangle = \tfrac{1}{2}a^2 \cos\omega\tau$$

This quantity we call the self-correlation function (sometimes the auto-correlation function), $c(\tau)$, of the disturbance. In more formal mathematical terms we would calculate it according to:

$$c(\tau) = \lim_{T\to\infty} \frac{1}{T}\int_0^T f(t)f(t+\tau)dt \tag{5.2}$$

i.e.,

$$c(\tau) = \lim_{T\to\infty} \frac{1}{T}\int_0^T \tfrac{1}{2}a^2[\cos\omega\tau - \cos(2\omega t + \tau)]dt$$

and thus, again:

$$c(\tau) = \tfrac{1}{2}a^2 \cos\omega\tau$$

This function does not decay, but oscillates with frequency ω and constant *amplitude* $a^2/2$. It is this latter amplitude which we take as our measure of coherence of the light wave, since the sinusoidal term in $\omega\tau$ will always be present for an oscillatory field such as an optical wave, and provides no useful information on the coherence. The mathematical form of $c(\tau)$ is known as a 'convolution' integral.

Since we require, for convenience, that this measure be unity for the sine wave, we choose to normalize it to its value at $\tau = 0$ [i.e., $c(0) = a^2/2$, in this case].

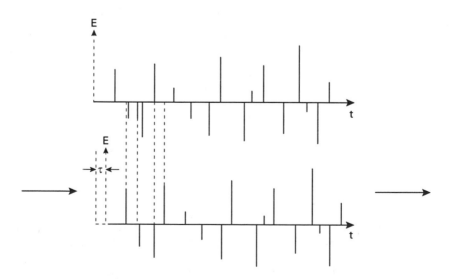

Fig. 5.4 Auto-correlation for randomly-spaced δ-functions.

This normalized function we call the coherence function $\gamma(\tau)$, and hence for the case we have considered, $\gamma(\tau) = 1$ (perfect coherence).

Consider now white light. We have noted that this is equivalent to a series of randomly spaced δ-functions which are also randomly positive or negative (Fig. 5.4). Clearly this series must have a mean amplitude of zero if it is to represent a spread of optical sinusoids over an infinite frequency range, each with a mean amplitude of zero. If now to obtain the 'delay-average' function we multiply this set of δ-functions by a displaced replica of itself, then only a fraction of the total will overlap, and an overlap between two δ-functions of the same sign will have equal probability with that between two of opposite signs. Consequently, the mean value of the overlap function will also be zero, always, regardless of the time delay. Hence for this case, $c(\tau) = 0$ for all τ and $\gamma(\tau) = 0$, also (i.e., perfect incoherence).

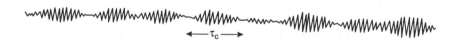

Fig. 5.5 A random stream of wave packets.

Consider, finally, a random stream of quanta, or wave packets (Fig. 5.5).

The packets run into each other, but each packet is largely coherent within itself. If this stream waveform is multiplied by a displaced replica of itself the result will be of the form shown in Fig. 5.6(a). Only when the displacement exceeds the duration of one packet does the correlation fall essentially to zero. Thus, in this case, we have a decaying sine wave, and the quantity which characterizes the decay rate of its amplitude will be our

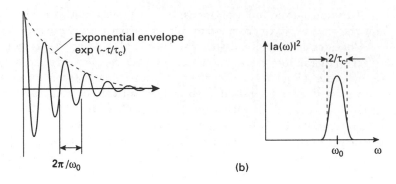

(a) $2\pi/\omega_0$

(b)

Fig. 5.6 Stream correlation.

measure of coherence (for example, time to $1/e$ point for an exponential decay).

All of the above requirements are taken care of in the general mathematical expression for the coherence function:

$$\gamma(\tau) = \frac{\left| \int\limits_0^\infty f(t) f^*(t + \tau) dt \right|}{\int\limits_0^\infty f(t) f^*(t) dt} = \frac{|c(\tau)|}{|c(0)|}$$

The integration performs the time averaging, the use of the complex form allows the complex conjugate in one of the functions to remove the oscillatory term in the complex exponential representation, the use of the modulus operation returns the complex value to a real value, and the division effects the required normalization. The function $c(\tau)$, as defined in equation (5.2), is called the correlation coefficient and is sometimes separately useful. Note that it is, in general, a complex quantity.

To cement ideas let us just see how these functions work, again for the pure sine wave.

We must express the sine wave as a complex exponential, so that we write:

$$f(t) = a \exp(i\omega t)$$

Then we have:

$$c(\tau) = \int_0^\infty a \exp(i\omega t) \, a \exp[-i\omega(t + \tau)] dt$$
$$= a^2 \exp(-i\omega\tau)[T]_0^\infty \to \infty$$

and

$$c(0) = \int_0^\infty a \exp(i\omega t) \, a \exp(-i\omega t) dt$$
$$= a^2 [T]_0^\infty \to \infty$$

Hence:

$$\gamma(\tau) = \frac{|c(\tau)|}{|c(0)|} = |\exp(i\omega\tau)| = 1$$

as required.

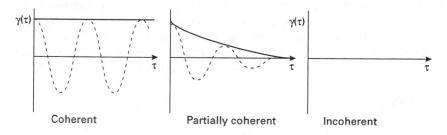

Fig. 5.7 Coherence functions.

    The coherence functions for the three cases we have been considering are shown in Fig. 5.7. The sine wave's function does not decay and therefore its coherence time is infinite; the white light's function decays in zero time and thus its coherence time is zero; the 'stream of wave packets' function decays in time τ_c, and thus it is partially coherent with coherence time τ_c. All other temporal functions may have their coherence time quantified in this way. It is clear, also, that the quantity $c\tau_c$ (where c is the velocity of light), will specify a coherence length.

    The same ideas, fairly obviously, can also be used for spatial coherence, with τ replaced by σ, the spatial delay. Specifically, in this case:

$$\gamma(\sigma) = \frac{\left| \int\limits_0^\infty f(s)f^*(s+\sigma)ds \right|}{\int\limits_0^\infty f(s)f^*(s)ds} = \frac{|c(\tau)|}{|c(0)|}$$

and a 'decay' parameter σ_c will define the coherence length.

    Finally, the mutual coherence of two separate functions $f_1(t)$ and $f_2(t)$ may be characterized by a closely similar *mutual coherence* function:

$$\gamma_{12}(\tau) = \frac{\left| \int\limits_0^\infty f_1(t)f_2^*(t+\tau)dt \right|}{\int\limits_0^\infty f_1(t)f_2^*(t)dt} = \frac{|c_{12}(\tau)|}{|c_{12}(0)|}$$

with t, τ again replaceable by s, σ respectively for the mutual spatial coherence case.

    γ_{12} is sometimes called the 'degree of coherence' between the two functions.

5.3 WIENER-KHINCHIN THEOREM

The Wiener-Khinchin theorem is one which has considerable practical value.

This theorem states that the Fourier transform (FT) of the self-correlation function, $c(\tau)$, for a function $f(t)$ provides the power spectrum of $f(t)$. This follows directly from the convolution theorem in Fourier analysis, which states that the Fourier transform of the convolution of two functions is the product of their individual Fourier transforms (see any text on Fourier theory, e.g., [1]). 'Convolution' is essentially just the 'delayed average' process we have been considering.

Thus we have:

$$\text{FT} \int_0^\infty f(t)f^*(t+\tau)\mathrm{d}t \; [= \text{FT}\, c(\tau)] = \text{FT}\, f(t)\; \text{FT}\, f^*(t)$$

where FT now denotes the Fourier transform operator. If we now write $f(t)$ in the form:

$$f(t) = \int_0^\infty a(\omega)\exp(i\omega t)\mathrm{d}\omega$$

we have

$$\text{FT}\, f(t)\; \text{FT}\, f^*(t) = |a(\omega)|^2$$

which is the 'power' spectrum of $f(t)$. Hence the power spectrum of a light source may be determined from its self-correlation function. This is not too much of a surprise since it was noted in section 5.1 that the spectral width was intimately related to the coherence, and self-correlation is, of course, a measure of coherence. The power spectrum of the random stream of Fig. 5.5 is shown in Fig. 5.6(b).

The Wiener-Khinchin theorem is a useful practical result since, as we are about to discover, we can determine (as we should expect) the self-correlation function from the light's interference properties. This means that interference patterns can determine for us the frequency components present in a source output and their relative amplitudes, i.e. the complete power spectrum.

In order to appreciate more clearly how all of these ideas help us in practical situations, we shall now look at some particular examples of coherence theory in action.

We begin by considering a simple arrangement for measuring the coherence of a light source.

5.4 DUAL-BEAM INTERFERENCE

We shall first consider in more detail the conditions for interference between two light beams (Fig. 5.8). It is clear, from our previous look at this topic, that interference fringes will be formed if the two waves bear a constant phase relationship to each other, but we must now consider the form of the interference pattern for varying degrees of mutual coherence. In particular, we must consider the 'visibility' of the pattern; in other words the extent to which it contains measurable structure and contrast.

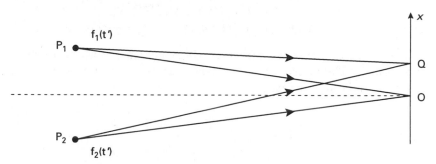

Fig. 5.8 Two-source interference.

At the point O (in Fig. 4.8) the (complex) amplitude resulting from the two sources P_1 and P_2 is given by:

$$A = f_1(t'') + f_2(t'')$$

where $t'' = t' + \tau_0$, τ_0 being the time taken for light to travel from P_1 or P_2 to O. If f_1, f_2 represent the electric field amplitudes of the waves, the observed intensity at O will be given by the square of the modulus of this complex number. Remember that the modulus of the complex number: $A = a + ib$ is written

$$|a + ib| = (a^2 + b^2)^{1/2}$$

and a convenient stratagem for obtaining the square of the modulus of a complex number is to multiply it by its complex conjugate, i.e.:

$$AA^* = (a + ib)(a - ib) = (a^2 + b^2)$$

Hence in this case the optical intensity is given by

$$I_0 = \langle AA^* \rangle = \langle [f_1(t'') + f_2(t'')][f_1^*(t'') + f_2^*(t'')] \rangle$$

where the triangular brackets indicate an average taken over the response time of the detector (e.g., the human eye) and we assume that f_1 and f_2 contain the required constant of proportionality ($K^{1/2}$) to relate optical intensity with electric field strength, i.e., $I = KE^2$.

At point Q the amplitudes will be:

$$f_1(t'' - \tfrac{1}{2}\tau), f_2(t'' + \tfrac{1}{2}\tau)$$

τ being the time difference between paths P_2Q and P_1Q.

Writing $t = t' - \frac{1}{2}\tau$ we have the intensity at Q:

$$I_Q = \langle [f_1(t) + f_2(t + \tau)][f_1^*(t) + f_2^*(t + \tau)] \rangle$$

i.e..

$$I_Q = \langle f_1(t)f_1^*(t) \rangle + \langle f_2(t)f_2^*(t) \rangle$$
$$+ \langle f_2(t + \tau)f_1^*(t) \rangle + \langle f_1(t)f_2^*(t + \tau) \rangle$$

The first two terms are clearly the independent intensities of the two sources at Q. The second two terms have the form of our previously defined mutual correlation function, in fact:

$$\langle f_1(t)f_2^*(t + \tau) \rangle = c_{12}(\tau)$$

$$\langle f_1^*(t)f_2(t + \tau) \rangle = c_{12}^*(\tau)$$

We may note, in passing, that each of these terms will be zero if f_1 and f_2 have orthogonal polarizations, since in that case neither field amplitude has a component in the direction of the other, there can be no superposition, and the two cannot interfere. Hence the average value of their product is again just the product of their averages, each of which is zero, being a sinusoid.

If $c_{12}(t)$ is now written in the form:

$$c_{12}(\tau) = |c_{12}(\tau)| \exp(i\omega\tau)$$

(which is valid provided that f_1 and f_2 are sinusoids in ωt) we have:

$$c_{12}(\tau) + c_{12}^*(\tau) = 2|c_{12}(\tau)| \cos\omega\tau$$

Hence, provided that we observe the light intensity at Q with a detector which has a response time very much greater than the coherence times (self and mutual) of the sources (so that the time averages are valid), then we may write the intensity at Q as:

$$I_Q = I_1 + I_2 + 2|c_{12}(\tau)| \cos\omega\tau \tag{5.3}$$

As we move along x we shall effectively increase τ, so we shall see a variation in intensity whose amplitude will be $2|c_{12}(\tau)|$ (i.e., twice the modulus of the mutual coherence function) and which varies about a mean value equal to the sum of the two intensities (Fig. 5.9). Thus we have an experimental method by which the mutual coherence of the sources, $c_{12}(t)$, can be measured.

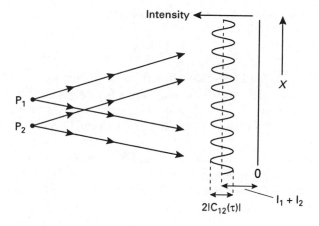

Fig. 5.9 Mutual coherence function ($|c_{12}(\tau)|$) from the two-source interference pattern.

If we now define a fringe *visibility* for this interference pattern by:

$$V = \frac{I_{\max} - I_{\min}}{I_{\max} + I_{\min}}$$

which quantifies the contrast in the pattern, i.e. the difference between maxima and minima as a fraction of the mean level, then, from (5.3):

$$V(\tau) = \frac{2\,|c_{12}(\tau)|}{(I_1 + I_2)}$$

and with, as previously defined:

$$\gamma(\tau) = \frac{|c_{12}(\tau)|}{|c_{12}(0)|}$$

we note that:

$$|c_{12}(0)| = \left| \int_0^\infty f_1(t) f_2^*(t)\,dt \right|$$

$$= K\langle E_1 \rangle \langle E_2 \rangle = (I_1 I_2)^{1/2}$$

and thus:

$$\gamma(\tau) = \frac{|c_{12}(\tau)|}{(I_1 I_2)^{1/2}}$$

Hence the visibility function $V(\tau)$ is related to the coherence function $\gamma(\tau)$ by:

$$V(\tau) = \frac{2(I_1 I_2)^{1/2}}{(I_1 + I_2)}\gamma(\tau)$$

and if the two intensities are equal, we have:

$$V(\tau) = \gamma(\tau)$$

i.e., the visibility and coherence functions are identical.

From this we may conclude that, for equal-intensity coherent sources, the visibility is 100% ($\gamma = 1$); for incoherent sources it is zero; and for partially coherent sources the visibility gives a direct measure of the actual coherence.

If we arrange that the points P_1 and P_2 are pinholes equidistant from and illuminated by a single source S, then the visibility function clearly measures the self-coherence of S. Moreover, the Fourier transform of the function will yield the source's power spectrum, via the Wiener-Khinchin theorem.

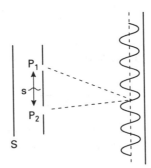

Fig. 5.10 Extended-source interference.

Suppose now that the two holes are placed in front of an extended source, S, as shown in Fig. 5.10, and that their separation is variable. The interference pattern produced by these sources of light now measures the correlation between the two corresponding points on the extended source. If the separation is initially zero and is increased until the visibility first falls to zero, the value of the separation at which this occurs defines a spatial coherence dimension for the extended source. And if the source is isotropic, a coherence area is correspondingly defined. In other words, in this case any given source point has no phase correlation with any point which lies outside the circular area of which it is the centre point.

5.5 PRACTICAL EXAMPLES

In order to appreciate more fully the practical importance of the concept of optical coherence we shall conclude with three examples of the concept in action.

5.5.1 Michelson's stellar interferometer

The concept of the spatial coherence of a light source is used in an instrument known as Michelson's stellar interferometer to measure the angular diameter of (nearer) stars (Fig. 5.11).

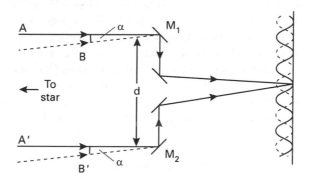

Fig. 5.11 Michelson's stellar interferometer.

If the star subtends an angle α at the two mirrors M_1, M_2, spaced at distance d, then the two monochromatic (with the aid of an optical filter) rays A, A′ (essentially parallel due to the very large distance of the star) from one point at the edge of the star will be coherent and will produce an interference pattern with visibility equal to 1. Similarly, so also will the two rays B, B′ from a diametrically opposite point at the other edge of the star. If the distance d between the mirrors is such that the ray B' is just one wavelength closer to M_2 than B is to M_1 then the second interference pattern, from BB', will coincide with the first from AA′. And all the intermediate points across the star produce interference patterns between these two to give a total resultant visibility of zero. Hence the value of d for which the fringe visibility first disappears provides the angular diameter of the star as λ/d (in fact, due to the circular rather than rectangular area, it is $1.22\lambda/d$). This method was first used by Michelson in 1920 to determine the angular diameter of the star Betelgeuse as 0.047 seconds of arc. Distances between mirrors (d) of up to 10 m have been used. (Betelgeuse (pronounced 'Beetle Juice') is a large star in the constellation of Orion and is quite close to the earth ($\sim$ 4 light years). The vast majority of stars are too distant even for this very sensitive method to be of any use.

5.5.2 The Mach-Zehnder interferometer

Consider now the two-arm optical-fibre Mach-Zehnder arrangement (see section 2.9) of Fig. 5.12. A measurand (quantity to be measured) M in a Mach-Zehnder interferometer causes a phase change in arm 1 which is detected by means of a change in the position of the interference pattern resulting from the recombination of the light at point R. Interference can only occur if the recombining beams have components of the same polarization, and if the difference in path length between the two arms is less than the source coherence length. This is not practicable with an LED, which has a coherence length $\sim 0.02\,\mathrm{mm}$, but even a modest semiconductor laser has a coherence length $\sim 1\,\mathrm{m}$ (coherence time $\sim 5\,\mathrm{ns}$) and can easily be used in this application. A single mode He-Ne laser has a coherence length of several kilometres. It is clear that, in order to make an accurate measurement of M, it is necessary to choose a source with a fairly large coherence length. However, if the coherence length is too large, every reflection in the system interferes with every other, and an unwanted interference 'noise' results. This is yet another problem for the opto-electronics designer: the coherence of the source must be optimized for the system in question.

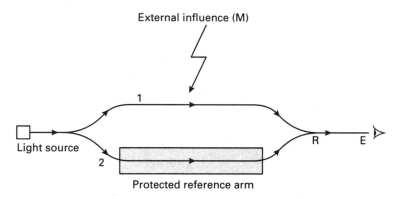

Fig. 5.12 An optical-fibre Mach-Zehnder interferometer.

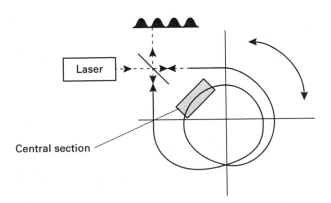

Fig. 5.13 The optical-fibre gyroscope.

This Mach-Zehnder arrangement is widely used in optical measurement technology [2].

5.5.3 The optical-fibre gyroscope

A rather more sophisticated example of the effect of coherence occurs in the optical-fibre gyroscope (Fig. 5.13). The principle of the gyroscope [3] is essentially the same as for the Mach-Zehnder interferometer we have just considered, the only important difference being that the two interfering arms now exist in the same fibre; the distinction between them lies in the fact that the light propagates in opposite directions in the two arms. The single source launches light in each of the two directions around a fibre loop and the light which emerges from the two ends of the loop interferes to produce a pattern on the screen. If the loop now rotates about an axis normal to its plane in, say, a clockwise direction, the light travelling in that direction will see its exit end receding from it, while the counterpropagating light will see its end approaching. Consequently the two components will traverse different optical paths and will emerge with a different relative phase relationship compared to that when the loop is stationary. The effect of the rotation is thus to cause a shift in the interference pattern, the magnitude of which is a measure of the rotation. Clearly, the greater is the rotational velocity the greater will be the path length shift, and the coherence length of the source must be large enough to embrace the range of rotation which is to be measured. However, if the coherence length is too great another problem arises. Some of the light will be backscattered, by the fibre material, as it propagates, and light which is backscattered from a region around the half-way point will itself interfere for the two directions (Fig. 5.13), and construct its own interference pattern. This will generate a noise level which will degrade the device performance. Clearly, the greater the coherence length of the source the greater will be the region around the midpoint from which this can occur and thus the greater will be the noise level. Hence a compromise or 'trade-off' has to be struck, as it always does in device and system design.

The optical-fibre gyroscope is described in detail in Chapter 10.

5.5.4 Birefringence depolarization of polarized light

Any birefringence in an optical fibre will delay one polarization direction with respect to the orthogonal direction. If this delay exceeds the coherence time of the source, then the light becomes effectively unpolarized. Indeed, this is a means by which depolarization of polarized light may be performed: linearly polarized low coherence light is launched at 45° to the birefringence axes, for example, (Fig. 5.14), and a length of fibre is used which is sufficient to ensure that the required condition obtains.

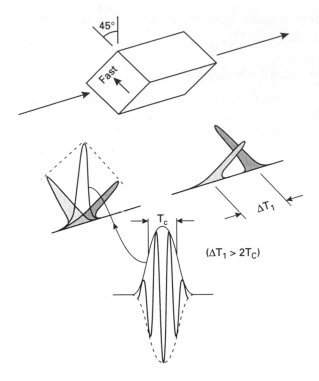

Fig. 5.14 Depolarization, using birefringence.

Depolarization of polarized light is often necessary in order to eliminate the effects of unwanted polarization selectivity in optical components, especially those used in detection systems.

5.5.5 Coherence properties of lasers

From the last two examples it has become clear that coherence in a light source can have disadvantages as well as advantages. A good further example of this is that a page of script is almost impossible to read by the light from a He-Ne laser owing to 'speckle-pattern' interference. This is essentially a complex three-dimensional interference pattern produced from the multitude of reflections from the page; the large coherence means that any reflection will interfere with any other. Consequently, any movement of the eye moves to another part of this complex pattern and the eye has to try to refocus each time. Laser systems produce excellent interference patterns, but these are not always wanted!

Coherence properties of lasers are complicated by their mode structure. A laser can have many transverse and many longitudinal natural modes of oscillation. Clearly the coherence function will peak whenever the modes come into phase in the 'delay-average' process, as one wave train slides

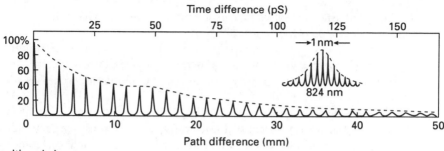

(a) A multimode laser

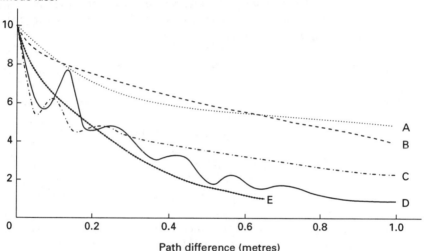

(b) Various single-mode lasers (A, B, C, D, E)

Fig. 5.15 Coherence functions for semi-conductor lasers.

across a replica of itself. The modes are regularly spaced in wavelength (just as are those for a stretched string), so the coherence functions show peaks at regular spatial and temporal intervals. Some typical coherence functions for multimode and single-mode semiconductor lasers are shown in Fig. 5.15. It follows from this that the existence and extent of any mode structure can be measured via a measurement of the coherence function (e.g., Fig. 5.15(b), curve D).

5.6 CONCLUSION

In this chapter we have looked at the conditions necessary for optical waves to interfere in a consistent and measurable way, with themselves and with other waves. We have seen that the conditions relate to the extent to which properties such as amplitude, phase, frequency and polarization remain constant in time and space or, to put it another way, the extent to which knowledge of the properties at one point in time or space tells us about these properties at other points.

Any interference pattern will only remain detectable as long as coherence persists and, by studying the rise and fall of interference patterns, much can be learned about the sources themselves and about the processes which act upon the light from them.

Coherence also relates critically to the information-carrying capacity of light and to our ability to control and manipulate it sensibly. The design and performance of any device or system which relies on interference or diffraction phenomena must take into account the coherence properties of the sources which are to be used; some of these work to the designer's disadvantage, but others do not.

PROBLEMS

5.1 A damped optical wave is described by the following function:

$$f(t) = A \exp[-(\alpha t - i\omega t)], \qquad t \geq 0$$

$$f(t) = 0, \qquad t < 0$$

What is its power spectrum? What is its coherence function?

5.2 In a Michelson stellar interferometer the mirror spacing is 8 m. What is the smallest angular diameter it could measure at a wavelength of 633 nm? What problems would you anticipate in trying to design a Michelson stellar interferometer to operate at γ-ray wavelengths ($\sim 10^{-12}$ m)?

5.3 A Young's double-slit arrangement is set up and illuminated by a sodium lamp placed behind a small circular hole of diameter 0.1 mm. The wavelength of the light is 589.3 nm and the distance between the source and the slits is 1 m. How far apart must the slits be for the interference pattern to (just) disappear (i.e. to have zero visibility)?

5.4 An FM radio station operates at a frequency of 91.5 MHz and its carrier frequency has a stability of one part in 10^7. What is the coherence length of its transmitter?

5.5 For a gas which is emitting radiation, the radiative processes are interrupted by molecular collisions. The result of this is that each individual emission line has the following (Lorentzian) intensity profile:

$$I(\omega) = \frac{A}{(\omega - \omega_0)^2 + b^2}$$

Derive the coherence function for this lineshape.

REFERENCES

[1] Panter, P. F. (1965) *Modulation, Noise and Spectral Analysis*, McGraw- Hill.

[2] Dakin, J. P. and Culshaw, B. (1990) *Optical-Fibre Sensors*, Vol. II, Artech House.

[3] Lefévre, H. (1993) *The Fiber-Optic Gyroscope*, Artech House, chap. 2.

FURTHER READING

Hecht, E. (1987) *Optics*, Addison-Wesley, 2nd edn, chap. 12

Born, M. and Wolf, E. (1975) *Principles of Optics*, Pergamon Press, 5th edn, chap. VII.

6

Some essential physics of radiation
and solids

6.1 INTRODUCTION

In the following three chapters we shall deal with the way in which light is produced, modulated, detected and guided. These processes are crucial to the methods by which we seek to manipulate and use light, for they control and limit the methods by which information is impressed upon, and subsequently extracted from, light signals.

Many of the basic ideas which are needed to understand these processes have already been introduced in earlier chapters; in this chapter, however, we shall need to extend and refine the coverage to the point where we can understand the structure and operation of specific optoelectronic devices. These devices depend, in most cases, on the physics of optical radiation and of the solid state in a fairly detailed way.

It is thus necessary to gain a more detailed familiarity with this physics, since it lies at the heart of our subject, and without this knowledge it would be impossible either to understand properly existing optoelectronics or to progress to new devices and new systems beyond our present-day thinking.

We begin by looking at the physics of radiation.

6.2 RADIATION

6.2.1 Black-body radiation

All matter, provided that it has a temperature other than absolute zero, emits radiation. This is a consequence of the fact that a temperature above absolute zero implies that the atoms or molecules are in motion, and are thus colliding with each other constantly. These collisions transfer kinetic energy of motion, but also sometimes excite the atomic system to a higher state, from which it may relax by emitting a photon. This is a consequence of a very basic principle in physics, the 'law of equipartition of energy', which states that the energy of a system, in equilibrium, will be distributed equally among all possible degrees of freedom: the kinetic energy of a material is one such degree, excited states represent another. By assigning a temperature to a body we are requiring it to be in thermal equilibrium with its surroundings (i.e., there is no net heat gain from, or loss to, the surroundings over time), so equipartition must apply.

The first question which naturally arises now is: how much radiation is emitted by a body at a given temperature? And the second (perhaps not quite as naturally!) is: what is the distribution of this emitted radiation over the wavelength spectrum?

In answering these questions we shall explore ideas which are valuable for a whole range of topics in optoelectronics, and more general physics, so it is worth while taking some time over them.

Classical thermodynamics assumed that atoms emitted light as a result of radiation by electrons which were oscillating at natural resonant frequencies within the atoms. It further assumed (it had no reason to assume otherwise) that these oscillations could occur with any amount of energy, depending on the strength of the stimulus.

The other piece of information which the classical thermodynamicists needed before they could proceed was the Boltzmann factor; we have already encountered this (section 4.4.1). It tells us the ratio of the number of atoms which have energy E_1 compared to those with energy E_2 in a system in equilibrium at absolute temperature T, and takes the form:

$$\frac{N_1}{N_2} = \exp\left(-\frac{E_1 - E_2}{kT}\right)$$

where k is Boltzmann's constant, with value 1.38×10^{-23} JK^{-1}. This factor had already been derived, using classical statistical thermodynamics, by means of an exquisite argument (which we do not have the space to develop, but see any text on statistical mechanics).

Let us now derive the classical result for the radiation emitted by a 'perfect' body, i.e., a body capable of emitting or absorbing radiation of any wavelength, and thus containing oscillators (atoms or molecules) capable of oscillating at any frequency. Such a body is called a 'black' body since it absorbs, rather than reflects, all light falling upon it, and therefore looks 'black' (until it emits!). Such a body is a valuable idealization, since we can categorize 'real' bodies according to how closely they approximate to it.

Let us suppose that within this black body the oscillators can have any energy and (for reasons which will become clear later) these energies will be described as a set distributed as:

$$0, dE, 2dE, ..., ndE, ...$$

where dE is infinitesimally small, so that the distribution is continuous. The ratios of numbers of oscillators with each of these energy bands will comprise the set (according to the Boltzmann factor):

$$1 : \exp\left(-\frac{dE}{kT}\right) : \exp\left(-\frac{2dE}{kT}\right) : \dots : \exp\left(-\frac{ndE}{kT}\right) : \dots$$

Thus if there are N oscillators with zero energy, the total number of oscillators will be:

$$N_T = N\left[1 + \exp\left(-\frac{dE}{kT}\right) + \exp\left(-\frac{2dE}{kT}\right) + \dots + \exp\left(-\frac{ndE}{kT}\right) + \dots\right]$$

This is a geometrical progression which is easily summed to give:

$$N_T = \frac{N}{1 - \exp\left(-\dfrac{dE}{kT}\right)} \tag{6.1}$$

Also, the total energy of all the oscillators is given by multiplying each term by its energy allocation:

$$E_T = N[dE \exp\left(-\frac{dE}{kT}\right) + 2dE \exp\left(-\frac{2dE}{kT}\right) + \dots$$
$$+ ndE \exp\left(-\frac{ndE}{kT}\right) + \dots]$$

giving, on summation,

$$E_T = \frac{NdE}{\exp\left(\dfrac{dE}{kT}\right)\left[1 - \exp\left(-\dfrac{dE}{kT}\right)\right]^2} \tag{6.2}$$

On dividing (6.2) by (6.1) we obtain the mean energy per oscillator as:

$$\bar{E} = \frac{dE}{\exp\left(\dfrac{dE}{kT}\right) - 1} \tag{6.3}$$

We may now let $dE \to 0$ to discover the physical value for this mean energy, whereupon we find (expanding the exponential in the denominator):

$$\bar{E} = \lim_{dE \to 0} \frac{dE}{1 + \frac{dE}{kT} + \frac{1}{2}\left(\frac{dE}{kT}\right)^2 + \dots - 1}$$

or

$$\bar{E} = kT \tag{6.4}$$

(This is entirely in accord with other 'equipartitional' approaches which allow $\frac{1}{2}kT$ of energy per degree of freedom. In our case there are two degrees of freedom per oscillator, one for kinetic energy, the other for potential energy, giving kT in all.)

The final piece of information we need is the number of independent oscillations which can occur within a given volume of material. Clearly the fact that the volume is finite means that there are boundaries and these impose boundary conditions on the oscillations, just as a string stretched between two fixed points is bounded by the fact that any oscillation of the string must have zero amplitude at the points of fixation.

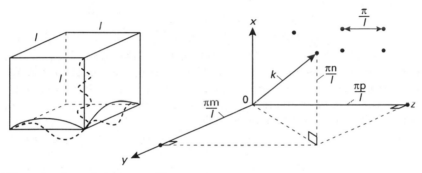

(a) Waves in a bounded cube, and 'k-space'

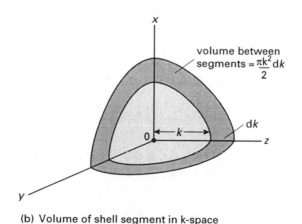

(b) Volume of shell segment in k-space

Fig. 6.1 k-space diagram.

Let us simplify things by taking the volume to be a cube of side l (Fig. 6.1(a)). Suppose now that oscillations occur within the cube and that the velocity with which these propagate is c. The walls of the cube impose a zero-amplitude boundary condition for these oscillations, so the resonant oscillations can only occur parallel with the sides of the cube with frequencies $nc/2l$, n being a positive integer and $c/2l$ the fundamental.

Now waves can, of course, travel in many directions and we can best represent any given wave by its wave vector $\mathbf{k}$, which has the direction of the wave and amplitude $2\pi/\lambda$, where λ is the wavelength. We may now write the frequencies of the waves travelling parallel to the sides of the cube as:

$$f_n = \frac{nc}{2l}$$

and the wavenumbers as:

$$k_n = \frac{2\pi}{c} f_n = \frac{\pi n}{l} \tag{6.5}$$

Let us now take axes Ox, Oy, Oz parallel with the sides of the cube and plot in three dimensions a lattice of points corresponding to all the k_n in (6.5). It is easily seen that *any* oscillation for the cube can be represented by its wave vector from the origin of axes to one of the points we have plotted. The plot is often is often called 'k-space', for obvious reasons.

We now ask the question: how many oscillations can the cube support with wavelengths between k and $k + \mathrm{d}k$? This, we can see, will correspond to the number of points on our plot which lie in the volume between spheres of radius k and $k + \mathrm{d}k$, respectively. This volume in k-space is, of course, $4\pi k^2 \mathrm{d}k$ (Fig. 6.1(b)). However, since only positive values of n are valid we only need that octant of the spherical shell where all the axes are positive, which is one-eighth of the total, i.e., $\frac{1}{2}\pi k^2 \mathrm{d}k$. To find the number of points in this volume we divide by one elementary volume in our lattice, defined by the interval between points, i.e., a cube of side π/l , volume π^3/l^3.

Hence the number of oscillations between k and $k + \mathrm{d}k$ is

$$N_0' = \frac{\frac{1}{2}\pi k^2 \mathrm{d}k}{\pi^3/l^3} = \frac{k^2 l^3 \mathrm{d}k}{2\pi^2}$$

and if we allow two orthogonal linear polarizations per oscillation (and we know from Chapter 3 that any electromagnetic wave can always be resolved into two such components) this becomes:

$$N_0 = \frac{k^2 l^3 \mathrm{d}k}{\pi^2}$$

Since the volume of our original cube is l^3 we can express this in the form of a number of oscillations per unit volume:

$$N_v = \frac{k^2 \mathrm{d}k}{\pi^2} \tag{6.6a}$$

It is now more convenient to write N_v in terms of frequency (since frequency is more directly related with energy). We have

$$k = \frac{2\pi}{\lambda} = \frac{2\pi f}{c}$$

Hence

$$N_v = \frac{8\pi f^2}{c^3} \mathrm{d}f \tag{6.6b}$$

This is an important result in itself and appears in many aspects of laser theory; it should be noted carefully.

We may use it immediately for our present purposes to derive the classical result for the energy spectrum of a black-body radiator. From (6.4) we saw that each oscillation has mean energy kT. Hence the energy density (i.e., energy per unit volume) lying between frequencies f and $f + \mathrm{d}f$ is given by:

$$\rho_f \mathrm{d}f = \frac{8\pi f^2}{c^3} kT \mathrm{d}f \tag{6.7}$$

This is the classical result, the so-called Rayleigh-Jeans equation. *It is wrong!* It has to be. We can see this immediately by calculating the total energy density emitted over all wavelengths of the spectrum:

$$\rho_T = \int\limits_0^\infty \frac{8\pi f^2}{c^3} kT \mathrm{d}f$$

$$= \frac{8\pi kT}{c^3} \left[\frac{f^3}{3} \right]_0^\infty \rightarrow \infty \; !$$

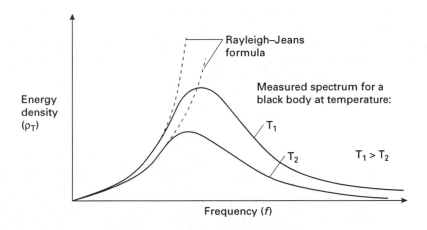

Fig. 6.2 The black body spectrum.

The answer is thus that an infinite amount of total energy is emitted by any black body; this is, of course, quite impossible. This result caused much head scratching amongst classical physicists around the turn of the century. When the spectrum of a black body (or as close to one as could be realized in practice) was measured, the shape was as shown in Fig. 6.2. The agreement with the Rayleigh-Jeans expression (equation 6.7) was good at low frequencies, but the two diverged wildly at the higher frequencies: the problem was thus dubbed the 'ultra-violet [i.e., high frequency] catastrophe'.

6.2.2 The quantum result

Max Planck, in 1900, saw a very simple way to avoid the above problem. He suggested that an oscillator could not possess *any* value of energy but only a value which was an integer times a certain minimum value. If this latter was ε, then the only possible values for the energy of the oscillation were ε, 2ε, 3ε ... $n\varepsilon$. This changed things completely, as we shall now show.

We can now, very conveniently, return to equation (6.3) for the mean energy of an oscillator:

$$\bar{E} = \frac{\mathrm{d}E}{\exp\left(\dfrac{\mathrm{d}E}{kT}\right) - 1}$$

$\mathrm{d}E$ can now be identified with ε, and now it does *not* tend to zero, but always remains non-zero.

Now ε is the minimum energy of the oscillator which emits radiation and thus can be identified, in turn, with the quantity $h\nu$, where ν is the lowest frequency of radiation it emits, remembering that the oscillator can have any of the energies $nh\nu$, where n is any positive integer. (We now use ν for frequency, rather than f, to remind ourselves that we are dealing with quantum phenomena rather than continuous events. h is the quantum constant (Planck's constant) with value 6.626×10^{-34} joule.seconds.)

Thus we have, in the quantum case:

$$\bar{E} = \frac{h\nu}{\exp\left(\dfrac{h\nu}{kT}\right) - 1}$$

rather than kT, and the energy density lying between ν and $\nu + \mathrm{d}\nu$ now will be, using (6.7):

$$\rho_\nu d\nu = \frac{8\pi\nu^2}{c^3} \frac{h\nu}{\exp\left(\frac{h\nu}{kT}\right) - 1} d\nu \tag{6.8}$$

This is the celebrated Planck radiation formula, and it solves all our problems, for it agrees with the experimental spectrum (Fig. 6.2).

If integrated over all frequencies it remains finite, and gives the result:

$$E_T = \frac{2\pi^5 k^4}{15c^2 h^3} T^4 \qquad (6.9)$$

Equation (6.2) represents the Stefan-Boltzmann law for the total energy emitted by a black body; classical thermodynamics was able to show that this quantity should be proportional to the fourth power of the absolute temperature, but was unable to predict the value of the constant of proportionality; quantum physics has provided the answer to this.

Similarly, classical thermodynamics was able to prove Wien's displacement law, which states that the value of the wavelength associated with the energy maximum in the spectrum (Fig. 6.2) is inversely proportional to the absolute temperature, i.e.,

$$\lambda_m = \frac{\Omega}{T}$$

but was unable to determine the value of the constant Ω. By differentiating equation (6.8) we easily find that:

$$\Omega = \frac{ch}{4.9651k}$$

The above results had a profound effect. Although Planck at first felt that his quantum hypothesis was no more than a mathematical trick to avoid the ultra-violet catastrophe, it soon became clear that it was fundamentally how the universe did, in fact, behave: quantum theory was born.

6.2.3 'Black-body' sources

The concept of a black body is that of a body which emits and absorbs all frequencies of radiation. We know now that the quantum theory requires us to limit the frequency to multiples of a certain fundamental frequency, but, in practice, owing to the particular molecular structure of any given body, the quantum (and classical) 'black body' remains an idealization, and real bodies, when hot, will not yield a spectrum in strict accordance with Planck's radiation law but only an approximation to it (sometimes a very close approximation, however).

Nevertheless, we can very conveniently measure the temperature of a radiating body by measuring the wavelength at which the spectrum peaks, using Wien's law, or, if the peak is not at a convenient (for our detector) position in the spectrum, by measuring the total energy emitted (using a bolometer) and applying the Stefan-Boltzmann law. Very often we require a source which emits over a broad range of frequencies, and a convenient way to obtain this is to create a discharge in a gas. An electrical discharge creates a large number of free, energetic electrons which cause a large range of atomic excitations, thus giving rise to radiation over a broad frequency range. Intensities can be quite high, so that the experimenter or designer can then pick out those frequencies that are needed, with frequency-selective optical components such as prisms or diffraction gratings.

However, the importance of the idealization known as a black body lies primarily in the fact that it allows an insight into the fundamental nature of electromagnetic radiation and the quantum laws which it obeys. This is crucial to our understanding of its role in optoelectronics, and especially to our understanding of laser radiation, which is the next topic for consideration.

6.2.4 The theory of laser action

(i) *The rate equations and the gain mechanism*

The elements of laser action were introduced in section 4.4.2. Lasers are extremely important in optoelectronics, as has been stressed, and it is necessary now to deal with laser action in more quantitative detail.

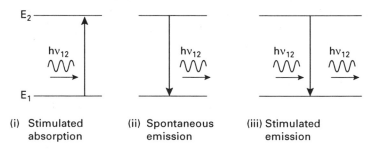

(i) Stimulated
absorption

(ii) Spontaneous
emission

(iii) Stimulated
emission

Fig. 6.3 Two-level photon transitions.

Let us consider two energy levels of an atomic system E_1 and E_2, with $E_2 > E_1$ (Fig. 6.3).

We know that the system can be raised from E_1 to E_2 by absorption of a photon with frequency ν_{12}, where:

$$h\nu_{12} = E_2 - E_1$$

and we also know that the system, after having been excited, will eventually, spontaneously, revert to its ground state E_1 by emitting a photon of energy $h\nu_{12}$.

However, we saw in section 4.4.2 that the excited state E_2 can also be stimulated to decay to the state E_1 by the action of another photon of energy $h\nu_{12}$. This process is called stimulated emission. Thus we now are considering three distinct processes:

(a) stimulated absorption $(E_1 \rightarrow E_2)$

(b) spontaneous emission $(E_2 \rightarrow E_1)$

(c) stimulated emission $(E_2 \rightarrow E_1)$.

(There can be no spontaneous 'absorption' since this would violate the law of conservation of energy.)

In order to calculate the relationships between atoms and radiation in equilibrium (i.e., black-body radiation) Einstein used a very simple argument: consider the atoms to be in equilibrium with each other and with the radiation in a closed system. The rate (per unit volume) at which atoms are raised to the upper state is proportional to the density of photons, ρ_ν, at energy $h\nu_{12}$ and to the density of atoms N_1 (number per unit volume) in state E_1, i.e.,

$$R_{12} = N_1 \rho_\nu B_{12} \quad \text{(stimulated absorption)}$$

where B_{12} is a constant.

Similarly, the rate at which atoms in state E_2 are stimulated to return to state E_1 is given by:

$$R_{21} = N_2 \rho_\nu B_{21} \quad \text{(stimulated emission)}$$

where N_2 is the density of atoms in state E_2. Now spontaneous emission from state E_2 to E_1 occurs after a characteristic delay determined by the detailed atomic characteristics, and is governed by quantum rules. Its rate therefore is proportional simply to N_2, the constant of proportionality, comprising, essentially, the reciprocal of the decay time. Thus we have

$$S_{21} = N_2 A_{21} \quad \text{(spontaneous emission)}$$

The constants A_{21}, B_{12}, B_{21} are called the Einstein coefficients.

Clearly, in equilibrium, we must have:

$$N_1 \rho_\nu B_{12} = N_2 \rho_\nu B_{21} + N_2 A_{21} \tag{6.10}$$

since the total upward and downward transition rates must be equal.

Hence from (6.10):

$$\rho_\nu = \frac{(A_{21}/B_{21})}{(B_{12}N_1/B_{21}N_2) - 1}$$

But we know from the Boltzmann relation (4.13) that:

$$\frac{N_1}{N_2} = \exp\left(-\frac{E_1 - E_2}{kT}\right)$$

and also that $E_2 - E_1 = h\nu_{12}$.

Hence, generalizing from ν_{12} to ν:

$$\rho_\nu = \frac{(A_{21}/B_{21})}{(B_{12}/B_{21})\exp\left(\frac{h\nu}{kT}\right) - 1} \qquad (6.11)$$

Now it was shown in section 6.2 that for equilibrium (black-body) radiation (equation 6.8):

$$\rho_\nu = \frac{8\pi h\nu^3}{c^3}\frac{1}{\exp\left(\frac{h\nu}{kT}\right) - 1}$$

Hence it follows, by comparing this with (6.11):

$$B_{12} = B_{21} \qquad (6.12a)$$

$$A_{21} = B_{21}\frac{8\pi h\nu^3}{c^3} \qquad (6.12b)$$

Relations (6.12) are known as the Einstein relations, and are very important determinants in the relationships between atoms and radiation. For example, it is clear that, under these conditions, the ratio of stimulated to spontaneous emission from E_2 to E_1 is given by:

$$S = \frac{R_{21}}{S_{21}} = \frac{\rho_\nu N_2 B_{21}}{N_2 A_{21}} = \frac{\rho_\nu c^3}{8\pi h\nu^3}$$

and using the expression for ρ_ν from equation (6.8):

$$S = \frac{1}{\exp\left(\dfrac{h\nu}{kT}\right) - 1}$$

If, for example, we consider the specific case of the He-Ne discharge at a temperature of 370 K with $\lambda = 632.8\,\text{nm}$ ($\nu = 4.74 \times 10^{14}\,\text{Hz}$) then we find

$$S \approx 2 \times 10^{-27}$$

Stimulated emission is thus very unlikely for equilibrium systems.

Another point worthy of note is that, for given values of N_2 (density of atoms in upper state E_2) and ρ_ν (density of photons) the rate of stimulated emission (B_{21}) is proportional to $1/\nu^3$. This follows from equation (6.12b) since

$$B_{21} = \frac{A_{21}c^3}{8\pi h\nu^3}$$

and A_{21} is an atomic constant, representing the reciprocal of the spontaneous decay time.

This means that the higher the frequency the more difficult is laser action, for this depends upon stimulated emission (section 4.4.2). Ultraviolet, X-ray and γ-ray lasers present very special problems which, hopefully, will preclude the possibility of 'death-ray' weapons (X-rays and γ-rays are very damaging to living tissues).

However, we do wish to use lasers at lower frequencies, visible and infrared for example, for purposes of communication, display and measurement, and the equation for R_{21} tells us that the way to increase the stimulated emission is to increase the values of N_2 and ρ_ν.

We know that, in equilibrium, $N_2 < N_1$, from the form of the Boltzmann factor, and ρ_ν is given by equation (6.8). Hence we shall have to disturb the equilibrium to achieve significant levels of stimulated emission.

One way in which this can be done is to inject radiation at frequency ν, so that ρ_ν is increased above its equilibrium value. Suppose that this is done until the stimulated emission greatly exceeds the spontaneous emission (which does not, of course, depend upon ρ_ν), i.e., until:

$$N_2\rho_\nu B_{21} \gg N_2 A_{21}$$

The condition for this, clearly, is that

$$\rho_\nu \gg \frac{A_{21}}{B_{21}}$$

which, from (6.12), means that:

$$\rho_\nu \gg \frac{8\pi h\nu^3}{c^3}$$

However, increasing ρ_ν does also increase the stimulated absorption. In fact, equation (6.10) becomes, when ρ_ν is large:

$$N_1\rho_\nu B_{12} = N_2\rho_\nu B_{21}$$

But we also know from equations (6.12) that $B_{12} = B_{21}$; hence $N_1 = N_2$ under these conditions. In other words, an incoming photon at frequency ν is just as likely to cause a downward transition (stimulated emission) as it is to cause an upward one (stimulated absorption). Hence we cannot increase the population N_2 above that of N_1 simply by pumping more radiation,

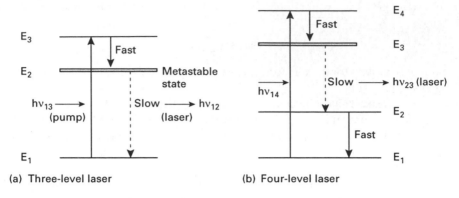

(a) Three-level laser   (b) Four-level laser

Fig. 6.4 Energy level diagrams for laser action.

at frequency ν, into the system. Clearly, we must change tack if we are to enhance the stimulated emission and produce a laser.

Consider a three-level rather than a two-level system (Fig. 6.4(a)). Suppose that light at frequency ν_{13} is injected into this system, so that there is a large amount of stimulated absorption from E_1 to E_3. Spontaneous decays will occur from E_3 to E_2 and then $E_2 \rightarrow E_1$ with also $E_3 \rightarrow E_1$; but if the levels are chosen appropriately according to the quantum rules, the $E_3 \rightarrow E_2$ decay can be fast and the $E_2 \rightarrow E_1$ relatively much slower. Clearly the result of this will be that atoms will accumulate in level E_2. Now the really important point is that, unlike the previous two-level case, atoms in level E_2 are immune from stimulated emission by photons at frequency ν_{13}. Hence we can now increase the numbers of atoms in level E_2, at the expense of those in E_1, by increasing the intensity of the radiation at frequency ν_{13}. We can thus soon ensure that:

$$\boxed{N_2 > N_1}$$

and we have an 'inverted population' (i.e., more atoms in a higher energy state than a lower one) as a result of the 'pump' at frequency ν_{13}. This inverted population can now be exploited to give optical amplification at frequency ν_{12}.

Let us quantify this amplification via the rate equations we have developed. Suppose that photons at frequency ν_{12} are injected into the medium in a certain direction. These will meet the inverted population in energy state E_2 and will stimulate the downward transition $E_2 \rightarrow E_1$, producing more photons at frequency ν_{12} in so doing (this is, of course, the origin of the amplification). We assume, quite confidently, that the medium is being sufficiently strongly pumped for the stimulated photons to be well in excess of any spontaneous emission from E_2 to E_1. Now suppose that, under these conditions, the number of photons per unit volume when the injected radiation enters the system is p_{12}. Then the rate at which p_{12} increases will depend upon the difference between upward and downward transition rates between levels 1 and 2, and hence we write:

$$\frac{dp_{12}}{dt} = N_2 \rho_{\nu_{12}} B_{21} - N_1 \rho_{\nu_{12}} B_{12}$$

Now $\rho_{\nu_{12}}$ is the energy density of photons, hence

$$\rho_{\nu_{12}} = p_{12} h \nu_{12}$$

Also we know that $B_{12} = B_{21}$ (from (6.12)) and thus:

$$\frac{1}{h\nu_{12}} / \frac{d\rho_{\nu_{12}}}{dt} = B_{12} \rho_{\nu_{12}} (N_2 - N_1) \tag{6.13}$$

We shall now write $\rho_{\nu_{12}}$ as ρ_ν to avoid cluttered equations and, integrating equation (6.13):

$$\rho_\nu = \rho_{\nu,0} \exp\left[h\nu B_{12}\left(N_2 - N_1\right)t\right]$$

where $\rho_{\nu,0} = \rho_{\nu_1}$ at $t = 0$.

If the injected wave is travelling at velocity c in the medium we can transfer to a distance parameter via $s = ct$ and obtain:

$$\rho_\nu = \rho_{\nu,0} \exp\left[\frac{h\nu}{c} B_{12}(N_2 - N_1)s\right]$$

This is to be compared with the standard loss/gain relation for propagation in an interactive medium, i.e.,

$$I = I_0 \exp(gx)$$

and it is clear that the gain coefficient g can be identified as:

$$g = \frac{h\nu}{c} B_{12}(N_2 - N_1) \tag{6.14a}$$

which is the gain coefficient for the medium (fractional increase in intensity level per unit length) and will be positive (i.e., gain rather than loss) provided that $N_2 > N_1$, as will be the case for an inverted population. Hence this medium is an optical amplifier. The injected radiation at frequency ν_{12} receives gain from the optical pump of amount:

$$G = \frac{I}{I_0} = \exp(gs)$$

so that it increases exponentially with distance into the medium. Clearly g in equation (6.14) is proportional to $(N_2 - N_1)$. In a three-level system such as we are considering the lower level of the amplifying transition is the ground state, which is initially heavily populated. It follows that more than half the atoms must be excited by the pump before population inversion can be achieved ($N_2 > N_1$). It is quite hard work for the pump to excite all these atoms. Consider, however, the *four*-level system shown in Fig. (6.4(b)). Here the pump is at ν_{14}, there is a quick decay to level 3 and a slow one to levels 2 and 1. The decay from 2 to 1 is again fast. Clearly the consequence of this is that it is relatively easy to provide level 3 with an inverted population over level 2, since level 2 was not well populated in the first place (being above the ground state), and atoms do not accumulate there since it decays quickly to ground. Hence we can ensure, fairly easily, that:

$$N_3 \gg N_2$$

with much less pump power than for $N_2 > N_1$ in the three-level case. The amplification at ν_{32} is thus much more efficient, and the four-level system makes for a more efficient amplifier.

(ii) *The laser structure*

Having arranged for efficient amplification to take place in a medium it is a relatively straightforward matter to turn it into an oscillator, i.e., a laser source. To do this for any amplifier it is necessary to provide positive feedback, i.e., to feed some of the amplified output back into the amplifier in reinforcing phase.

As has been described in section 4.4.2 this is done by placing parallel mirrors at each end of a box containing the medium, to form a Fabry-Perot cavity (section 2.9). (We should also remember the 'stable' resonator configuration (section 2.11) which is valuable for many types of laser design.) The essential physics of this process is that any given photon at ν_{12} will be bounced back and forth between the mirrors, stimulating the emission of other such photons as it does so, whereas without the mirrors it would make only one such pass.

An important condition for any system to oscillate under these circumstances is that the gain should be in excess of the loss for each cycle of oscillation. The total loss for a photon executing a double passage of the cavity (Fig. 6.5) will depend not only on the loss per unit length in the medium (due to scattering, excitations to other states, wall losses, etc.) but also on the losses at the mirrors, and, it must be remembered, one of the mirrors has to be a partial mirror in order to let some of the light out, otherwise we couldn't use the laser oscillator as a source! Hence the condition for oscillation:

$$\frac{I_f}{I_i} = R_1 R_2 \exp[(g - \alpha)2l] > 1 \tag{6.14b}$$

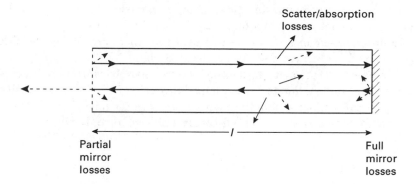

Fig. 6.5 Loss mechanisms in a laser cavity.

where I_f and I_i refer to the final and initial intensities for the double passage of the cavity, R_1 and R_2 are, respectively, the reflectivities for the two mirrors, α is the loss per unit length in the medium and l is the cavity length. (The factor 2 in the exponential refers of course to the double passage of the photon.)

One further word of warning: the value of g must correspond to the population inversion whilst oscillation is taking place, not the value before the feedback is applied. Clearly the value of N_2/N_1 will be very different, once stimulated emission starts to occur, from its value when the system is simply being pumped into its inverted state. This has implications for pumping rates and the balancing of rate equations which we shall not pursue: the principles, hopefully, are quite clear, however.

The simple arrangement of a pumped medium lying between two parallel mirrors (one partial) will, under the correct pump conditions, therefore lead to radiation emerging with the following properties:

(a) narrow linewidth, since only one energy of transition is involved in the laser action; and the mirrors, if wavelength selective, will block any spontaneous light which is emitted in addition.

(b) the output direction of the light will be exactly normal to the (accurately parallel) planes of the mirrors and thus will be highly collimated in one direction.

(c) When a photon is emitted via stimulated emission by another photon, it is emitted with the same phase as the original photon (remember the driving force/resonating system analogy), thus all the laser photons are locked in phase: we have coherent light (within the limitations only of the linewidth of the transition).

(d) The light can be very intense since all the 'light amplification by stimulated emission of radiation' from a long length of medium with small cross-sectional area can be collimated into the one direction.

The above important features summarize the basic properties of laser light:

it is pure (in wavelength and phase), intense, well-collimated light. It is thus easy to control and modulate; it is a powerful tool.

In order to enhance its usefulness as a tool there are two quite simple additions which can be made to the basic design:

The Fabry-Perot cavity formed by the two parallel mirrors will possess defined longitudinal 'modes' as explained in section 2.9. Waves propagating in opposite directions within the cavity, normal to the mirrors, will interfere and reinforce to give rise to an allowable stable mode only when

$$2L = m\lambda$$

where L is the length of the cavity and m is an integer.

From this we can also write

$$\lambda = \frac{2L}{m}; \quad f = \frac{cm}{2L}$$

At all other wavelengths there is destructive interference. Now, of course, the stimulated emission occurs over a small range of wavelengths. This range is determined by the spectral width of the downward transition. The width depends upon a number of factors but primarily (unless cooled to very low temperatures) on the Doppler shift caused by the thermal motion of the molecules. Clearly, at any given time, some molecules will be moving towards the stimulating photon and others away, leading to a spread of Doppler shifts around the central line for the stationary molecule (at absolute zero of temperature!).

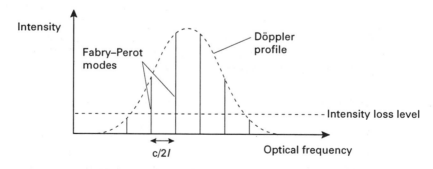

Fig. 6.6 Laser-cavity spectrum.

The output spectrum of the laser light thus is the result of combining these two features, as shown in Fig. 6.6. Here we can see the Fabry-Perot mode structure enveloped by the natural linewidth of the transition. In order to fix ideas somewhat, let us insert some real numbers into this.

Suppose we have an He-Ne gas laser with length 0.5 m. Since, in a gas at less than atmospheric pressure, we have $c \sim 3 \times 10^8 \, \mathrm{ms^{-1}}$, we see that:

$$\frac{c}{2L} = 300 \, \mathrm{MHz}$$

which is the separation of the modes along the frequency axis. Now the Doppler line width of the 632.8 nm transition at 300 K is $\sim 1.5 \, \mathrm{GHz}$; hence the number of modes within this width is

$$\frac{1.5 \times 10^9}{3 \times 10^8} \sim 5$$

so that we have just five modes in the output spectrum.

So far we have dealt only with longitudinal modes; but off-axis rays also may interfere (Fig. 6.7). The reinforcement condition now depends also on the angle which the ray makes with the long axis, and the result is a variation in intensity over the cross-section of the cavity, and thus over the cross-section of the output laser beam (Fig. 6.7). (The notation used to classify these variations will be described in more detail when we deal with wave guiding (Chapter 8) but TEM stands for 'transverse electromagnetic' and the two suffixes refer to the number of minima in the pattern in the horizontal and vertical directions, respectively.)

(iii) *Mode-locking*

Let us return now to the longitudinal mode structure of the laser cavity. Normally, these longitudinal modes are entirely independent, since they result from wholly independently-acting interference conditions. Suppose, however, that they were to be locked into a constant phase relationship. In that case we would have a definite relationship, in the frequency domain, between the phases and the amplitudes of the various components of the frequency spectrum. If we were to translate those relationships into the time domain, by means of a Fourier transform, the result would be a series of pulses spaced by the reciprocal of the mode frequency interval, with each pulse shape the Fourier transform of the mode envelope (Fig. 6.8). All we are really saying here, in physical terms, is that if each frequency component bears a constant phase relationship to all the others then, when all frequency components are superimposed, there will be certain points in time where maxima occur (the peaks of the pulses) and others where minima occur (the troughs between pulses). If there is no fixed phase relationship between components both maxima and minima are 'washed out' into a uniform-level, randomized continuum.

Now a series of evenly spaced pulses is often a very useful form of laser output, so how can it be achieved?

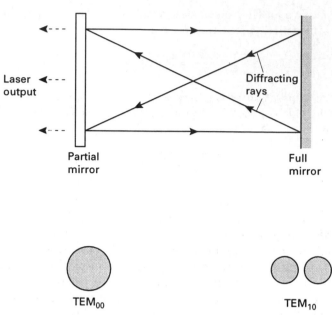

Fig. 6.7 Transverse cavity modes.

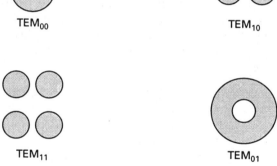

Fig. 6.8 Mode-locking Fourier transform: spectrum v. pulse train.

We must lock the phases of the longitudinal modes. One way of doing this is to include, within the cavity, an amplitude modulator, and then modulate (not necessarily sinusoidally) the amplitudes of the modes at just the mode frequency interval, $c/2L$. Then, each mode generates a series of sidebands at frequencies $mc/2L$, which corresponds to the frequencies of the other modes. The result of this is that all the modes are 'pulled' mutually into phase by the driving forces at the other frequencies, and complete phase locking occurs. The inserted modulator thus has the effect of producing, from the laser output, a pulse stream with pulse repetition rate $2L/c$. For example, with the He-Ne laser quoted in the preceding section, the repetition rate is 300 MHz, and each pulse has a width

$$\sim \frac{1}{1.5 \times 10^9} \sim 0.67\,\text{ns} \quad (\text{Fig. 6.8})$$

The laser is now said to be 'mode-locked' and the pulse stream is a set of 'mode-locked pulses'. Sometimes when a laser is being pumped quite hard and the output levels are high, the laser will 'self-mode lock'. This is due to the fact that the medium has been driven into the non-linear regime (see Chapter 9) and the modes generate their own harmonics as a result of the induced optical non-linearities. Clearly, this will depend upon the medium as well as the driving level, since it will depend on which particular non-linear threshold is exceeded by the pumping action.

(iv) Q-switching

The 'Q' or 'quality factor' of an oscillator refers to its purity, or 'sharpness of resonance'. The lower the loss in an oscillator the narrower is its resonance peak and the longer it will oscillate on its own after a single driving impulse. The equivalent quantity in a Fabry-Perot cavity (an optical oscillator) is the 'finesse' (see section 2.9) and the two quantities are directly related. From these ideas we can readily understand that if the loss in a resonator is varied then so is its 'Q'.

Suppose we have a laser medium sitting in its usual Fabry-Perot cavity but with a high loss; this means that a large fraction of the light power oscillating between the mirrors is lost per pass: we might, for example, have one of the mirrors with very low reflectivity.

Now the oscillator can only oscillate if the gain which the light receives per double pass between the mirrors exceeds the loss per double pass (section 6.4(ii)), and we shall suppose that the loss is very high, so that as we pump more and more molecules of the medium up into the excited state of the inverted population, the loss still exceeds the gain for as hard as our pump source can work. The result is that the inversion of the population becomes very large indeed, for there are very few photons to cause stimulated emission down to the lower state - they are all being lost by other means (e.g., a poor mirror at one end). Having achieved this very highly inverted population suppose that the loss is now suddenly reduced by means of an intercavity switch ('Q' switch) by, for example, speedily

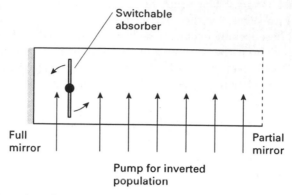

(a) Inversion of the atomic population before lasing

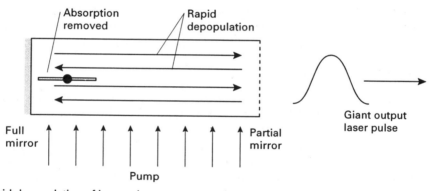

(b) Rapid depopulation of inverted states on removing absorption

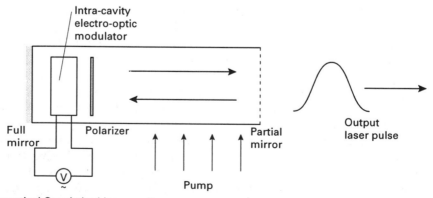

(c) A practical Q-switched laser cavity

Fig. 6.9 Q-switching.

rotating to a high-reflectivity mirror (Fig. 6.9). The result is that there is suddenly an enormous number of photons to depopulate the inverted population, which then rapidly de-excites to emit all its accumulated energy in one giant laser pulse - the Q-switched pulse. In this way is obtained the means by which very large energy, very high intensity pulses can be obtained, albeit relatively infrequently ($\sim$ 25 pps).

Three further points should be noted concerning Q-switching:

(a) At the end of the pulse the lasing action ceases completely, since the large number of photons suddenly available completely depopulates the upper laser state.

(b)The switching to the low loss condition must take place in a time which is small compared with the stimulated depopulation time of the upper state, so as to allow the pulse to build up very quickly.

(c)The pumping rate must be large compared with the spontaneous decay rate of the upper state so as to allow a large population inversion to occur.

Q-switching can produce pulses with several millijoules of energy with only a few nanoseconds duration. Thus, peak powers of several megawatts can result. Such powers take most media into their non-linear regimes (many will be evaporated!), so Q-switching is very useful for studying the non-linear optical effects which will be considered in Chapter 9.

Both mode-locking and Q-switching require intracavity modulation devices. These can take a variety of forms, and in Chapter 7 we shall be considering these.

Having taken a good look at the physics of photons, and, in particular, laser action, attention now will be turned to the other half of optoelectronics, to electrons. Just as our primary concern with photons is with the way they interact with matter, so it is with electrons. An understanding of the behaviour of electrons in solid materials is crucial to our subject.

6.3 ELECTRONS IN SOLIDS

In order to understand the mechanisms involved in the operation of important solid state devices such as semi-conductor lasers, light-emitting diodes, various types of photodetectors, light modulators, etc., it is necessary to look into some of the rather special features of the behaviour of electrons in solid materials, and this is the subject of the present section.

A solid is a state of matter where the constituent atoms or molecules are held in a rigid structure as a result of the fact that the inter-molecular forces are large compared with the forces of thermal motion of the molecules. This can only be true if the molecules are close together, for the molecules are electronically neutral overall, and forces can only exist between them if there is significant overlap among the wavefunctions of the outer electrons.

This overlap leads to another important consequence: the energy levels in which the electrons lie are shared levels; they are a property of the material as a whole rather than of the individual molecules, as is the case for a gas, for example.

In order to gain a physical 'feel' for the effect of the strong interaction on the energy level structure in a solid, consider what happens when two simple oscillators, such as two pendulums, interact. If two pendulums each of the same length, and thus with the same independent frequency of oscillation, f, are strung from the same support bar, they will interact with each other via the stresses transmitted through the bar, as they swing. For the combined system there are two 'eigenmodes', that is to say two states of oscillation which are stable in time. These are (Fig. 1.7) the state where the two pendulums swing together, in phase, and that where they swing in opposition, in antiphase. For all other states the amplitudes and relative phases of the two pendulums vary with time. The two eigenstates have difference frequencies f_p (in phase) and f_a (in anti-phase) and we find that

$$f_p > f > f_a$$

The original frequency f is not now a characterizing parameter of the system, having been replaced by two other frequencies, one higher and one lower. If just one of the two pendulums is set swinging it will do so at a frequency in the range f_p to f_a and will set the other pendulum swinging. The second pendulum will acquire maximum amplitude when the first has come to a stop and then the process will reverse. The energy will continuously transfer between the pendulums at a frequency $(f_p - f_a)$. Consider now *three* identical pendulums strung from the same bar. Now there are three eigenstates: (i) all in phase; (ii) outer two in phase, central one in antiphase; (iii) left- or right-hand two in phase, right- or left-hand one in antiphase. Each of these states has its own frequency of oscillation, so we now have three frequencies. It is an easy conceptual extrapolation to n pendulums, where there will be n frequencies centred on the original f, i.e., the original single frequency has become a band of n frequencies. If n is very large, as it is with the number of molecules in a solid, the frequencies are so close together as to comprise essentially a continuous *band* of frequencies, and thus also of electron energy levels. Thus we can expect each discrete energy level of the isolated atom or molecule to form a separate band of allowable energies, and the bands will be separated by gaps which represent energies forbidden to electrons (Fig. 6.10). This feature is crucial to the behaviour of electrons in solids and accounts for most of the properties which are important in optoelectronics. It is, therefore, necessary to study it in more detail before looking at why, exactly, it is so important to us.

6.3.1 Elements of the band theory of solids

Having understood why energy bands form in solids, it is necessary

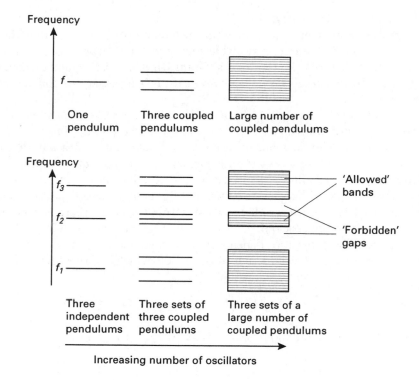

Fig. 6.10 Band structure resulting from the coupling of oscillators.

now to understand the ways in which electrons occupy them. This is as crucial to the understanding of the optoelectronic properties of solids as the formation of the bands themselves.

First, it is necessary to remember that electrons obey quantum rules. Associated with each electron is a wave (section 1.8) whose wavenumber ($\sim$ reciprocal of wavelength) is related to the momentum (p) of the electron. If the electron is propagating freely the relationship is:

$$p = \frac{h}{\lambda} = \frac{hk}{2\pi}$$

and since the kinetic energy of a particle of mass m is related to its momentum p by:

$$E_k = \frac{p^2}{2m}$$

We have, in the case of the free electron,

$$E_k = \frac{h^2 k^2}{8\pi^2 m} \tag{6.15}$$

We shall need this shortly.

Another consequence of the quantum behaviour of electrons is that they distribute themselves among available energy levels in a rather special way. We say that their distribution obeys 'Fermi-Dirac' statistics, and although it is not necessary to go very deeply into this, it is necessary to understand the basic ideas.

All fundamental particles, such as electrons, protons, neutrons, mesons, quarks, etc., are indistinguishable particles, i.e., there is no way in which an electron, say, can be 'labelled' at one time or place, in such a way that it is possible to recognize it as the same particle at another time or place; and this is not just a 'labelling' problem: it is quite fundamental - a consequence of quantum physics. Hence if two identical (indistinguishable) particles are interchanged in any energy distribution within a system there can be no change in any of the observable macroscopic properties of the system. Now these observable properties depend only on the square of the modulus of the system's overall wavefunction (section 1.8), which is formed from all of the individual electron wavefunctions, i.e.,

$$\psi = \psi(1)\psi(2)...\psi(n)$$

If electrons 1 and 2 are interchanged then $|\psi|^2$ must remain the same, i.e.,

$$|\psi_{12}|^2 = |\psi_{21}|^2$$

hence

$$\psi_{12} = \pm\psi_{21}$$

This presents two possibilities: either the interchange leaves the sign of the wavefunction the same, or it reverses it. Particles which leave the sign the same are called symmetrical particles; particles which reverse it are called anti-symmetrical particles.

Now comes the vital point: two anti-symmetrical particles cannot occupy the same quantum state, since the interchange of two identical particles occupying the same quantum state cannot alter the wavefunction in any way at all, not even its sign. Hence no two anti-symmetrical particles can have the same set of 'quantum numbers', numbers which define the quantum state uniquely. This is known as the Pauli exclusion principle. Electrons are anti-symmetrical particles and thus obey the Pauli exclusion principle. In fact all particles with 'half-integral spin', $(n + \frac{1}{2})h/2\pi$, obey the principle, e.g., electrons, protons, neutrons, μ-mesons; these are called fermions (note the small f now!). Particles with integral spin, $nh/2\pi$, are symmetrical particles and obey 'Bose-Einstein' statistics: e.g., photons, α-particles, π-mesons; these are called bosons.

The fact that no two electrons can occupy the same quantum state is profound, and is the single most important feature of the behaviour of electrons, in regard to the optoelectronic properties of solids. It means that the available electrons will fill the available quantum states progressively and systematically from bottom to top, like balls in a vertical tube whose diameter is just sufficient to take one ball at a time.

Let us examine this 'filling' process in more detail.

Each allowed energy level in any system contains (in general) more than one quantum state. The number of states which it contains is called the 'degeneracy' of the energy level. (Remember also that each of the bands in the solid state energy structure results from a large number of closely spaced energy levels, so there is also a kind of multiple degeneracy within a band).

Now suppose, firstly, that the electrons within a given energy band are completely free to move around as if they were an electron 'gas' in the solid. This is approximately true for electrons in a metal, and the only restriction really is that the electrons are not free to leave the solid. How should we calculate the energy states available to the electrons in this case?

Well, fortunately, most of the work necessary for calculating the number of electron states which lie between energies E and $E + dE$ for this case has already been done, in section 6.2, for atomic oscillators which give rise to electromagnetic waves; the analogy between electromagnetic waves in a box and electrons in a box is very close. The electron waves are restricted to the same set of discrete values by the box boundaries as were the electromagnetic waves. The only difference is that whereas we had to allow for two polarization states in the electromagnetic case, we now have to allow for two spin directions (e.g., up and down) in the electron case. In both cases we must multiply by a factor of 2, so that equation (6.6(a)) remains valid; i.e., the number of electron states with k values between k and $k + dk$ is $g(k)$ where

$$g(k)dk = \frac{k^2 dk}{\pi^2}$$

$g(k)$ is known as the degeneracy function. All that is necessary now is to express this in terms of the energy by substituting for k and dk from equation (6.15), i.e.,:

$$
\begin{aligned}
g(E_k)dE_k &= \frac{1}{\pi^2} \frac{8\pi^2 m}{h^2} E_k \left(\frac{8\pi^2 m}{h^2}\right)^{1/2} \tfrac{1}{2} E_k^{-1/2} dE_k \\
&= \frac{4\pi}{h^3} (2m)^{3/2} E_k^{1/2} dE_k
\end{aligned}
\tag{6.16a}
$$

This function is shown in Fig. 6.11, and, in solid state parlance, is usually called the 'density of states' function. It represents the number of states between energies E_k and $E_k + dE_k$.

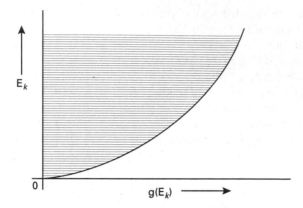

Fig. 6.11 'Density of states' function for a metal.

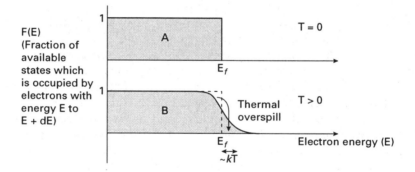

Fig. 6.12 Fermi-Dirac 'occupancy' distributions.

Hence these are the states which are going to be filled from the bottom up. Each range of E_k to $E_k + dE_k$ will be filled sequentially like the balls in the tube. What, then, is the occupancy of these states? How do the electrons actually distribute themselves among them?

If each energy range is filled in turn, the best way to express this is to plot the fraction of the $g(E_k)$ levels which is filled by a total of N_T electrons. At the absolute zero of temperature this occupancy function will look like variation A in Fig. 6.12. All the states will be filled up to the level at which the electrons are exhausted. Hence up to that level the fractional occupancy is 1; above that level, it is zero. This level is known as the Fermi level, E_F, and is easily calculated if the total number of electrons, N_T, is known, for it is necessary only to integrate (6.16a) between 0 and E_F:

$$N_T = \frac{4\pi}{h^3}(2m)^{3/2} \int_0^{E_F} E_k^{1/2} dE_k$$

Hence

$$E_F = \left(\frac{3N_T}{8\pi}\right)^{2/3} \frac{h^2}{2m} \qquad (6.17)$$

Using (6.17) the density of states function (6.16a) may now conveniently be expressed in the form:

$$g(E)\mathrm{d}E = \frac{3}{2}N_T \frac{E^{1/2}}{E_F^{3/2}}\mathrm{d}E \qquad (6.16b)$$

Suppose now that the temperature rises above absolute zero to a small value $T\,(>0)$. This makes available to each electron an extra energy $\sim kT$ ($\ll E_F$). However, most of the electrons cannot take advantage of this because the next available empty level for them is more than kT away. The only electrons which can gain energy are those at the top of the distribution, for they have empty states above them. The distribution thus changes to the one shown as B in Fig. 6.12 for temperature T. Hence the electrons behave very differently from a gas, say, where the average energy of *all* the molecules would increase by kT. (It is for this reason that the specific heat of metals is much smaller ($\sim 1\%$) than was predicted on a free electron theory of metallic conduction; this discrepancy was a great puzzle to physicists in the early years of this century.)

The function which describes the occupancy of the levels at a given temperature is known as the Fermi-Dirac function and is derived fully in Appendix VII.

It is given by:

$$F(E) = \frac{1}{\exp\left(\frac{E-E_F}{kT}\right)+1} \qquad (6.18a)$$

Note that it has the behaviour which has already been described:
For $T=0$ and

$$E < E_F: \quad \exp\left(\frac{E-E_F}{kT}\right) \to 0, \quad F(E) \to 1$$

For $T=0$ and

$$E > E_F: \quad \exp\left(\frac{E-E_F}{kT}\right) \to \infty, \quad F(E) \to 0$$

This clearly corresponds to variation A in Fig. 6.12.

As the temperature rises, the topmost electrons move to higher states and the function develops a 'tail', whose width is $\sim kT$ (curve B in Fig. 6.12). The energy E_F in this case corresponds to the energy for which $F(E) = 0.5$.

Now we are in a position to make the final step: the density of electrons within a given small energy range will be the product of the density of quantum states and the actual occupancy of these states. It will be the product of the density of states function (6.16b) and the Fermi-Dirac function (6.18a), i.e., $n(E)\mathrm{d}E = g(E)F(E)\mathrm{d}E$ or

$$n(E)\mathrm{d}E = \frac{3}{2}N_T \frac{E^{1/2}}{E_F^{3/2}} \frac{\mathrm{d}E}{\exp\left(\frac{E - E_F}{kT}\right) + 1} \tag{6.19a}$$

where $n(E)\mathrm{d}E$ is the number per unit volume of electrons with energies between E and $E + \mathrm{d}E$. This function is shown in Fig. 6.13 for $T = 0$ and for $T \neq 0$.

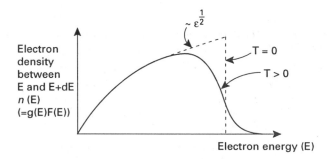

Fig. 6.13 Electron density distributions for a metal.

It is interesting to note, before leaving this, that the Fermi-Dirac distribution is a prevalent feature primarily because, in a solid, the number of electrons is comparable with the number of quantum states, and therefore the electrons must be carefully packed according to the quantum rules. If the number of quantum states far exceeds the number of identical particles, as it does in a gas for example, the quantum rules are scarcely noticeable. To see this suppose that, in equation (6.16b), $g(E) \gg N_T$, then $E^{1/2} \gg E_F^{3/2}$ and hence $E \gg E_F$. Equation (6.19a) becomes:

$$n(E)\mathrm{d}E = \frac{3}{2}N_T \frac{E^{1/2}}{E_F^{3/2}} \exp\left(-\frac{E}{kT}\right) \mathrm{d}E \tag{6.19b}$$

Expressed in terms of molecular velocity, v, and remembering that the molecular energy in this case is purely kinetic energy of motion, i.e.,

$$E = \tfrac{1}{2}mv^2$$

we have:

$$n(v)\mathrm{d}v = Av^2 \exp\left(-\frac{mv^2}{2kT}\right) \mathrm{d}v$$

which is the Maxwell-Boltzmann gas velocity distribution as deduced from classical (i.e., non-quantum) statistical thermodynamics (see, e.g., [1]).

It turns out that equation (6.19b) often also represents a useful approximation in solid state physics. In all cases where the electron distribution is being considered well above the Fermi level (i.e., $E \gg E_F$) the Fermi-Dirac distribution function of equation (6.18a) approximates to:

$$F(E) = \exp\left(-\frac{E}{kT}\right) \qquad (6.18b)$$

which is, of course, the Boltzmann factor. We shall have several occasions to use this later.

6.3.2 Metals, insulation and semiconductors

We are now in a position to understand, qualitatively at first, what it is that distinguishes metals, insulators and semiconductors. It all depends upon the position of the Fermi level.

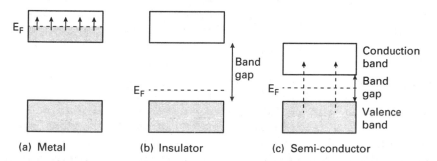

(a) Metal   (b) Insulator   (c) Semi-conductor

Fig. 6.14 The Fermi level and the classification of solids.

Consider the solid state band structure in Fig. 6.14. Suppose, first, that a solid consists of atoms or molecules with just one electron in the outermost energy shell. This shell forms a band of energy levels in the solid, as we have seen, and the total number of available states will be $2N$ per unit volume, where N is the number of atoms per unit volume (i.e., two electron spin states per quantum state). But there will be only N electrons per unit volume since there is only one electron per atom. Hence the band is half-filled, and the Fermi level lies halfway up the second band, as in Fig. 6.14(a). The electrons at the top of the Fermi-Dirac distribution have easy access to the quantum states above them and can thus move freely in response to, for example, an applied electric field, by gaining small amounts of kinetic energy from it; they can also move to conduct heat energy quickly and easily: we have a metal.

Suppose, secondly, that there are two electrons in the outer shell of the atoms or molecules comprising the solid. The band formed from the shell is now just full and the Fermi level is above the top of the band, as in Fig. 6.14(b). The electrons at the top of the band now can only increase their energies by jumping up to the next band. If the energy gap is quite large, neither moderate temperatures nor moderate electric fields can provide sufficient energy for this to happen. Hence, the material does not conduct electricity at all easily: we have an insulator.

Finally, consider the case shown in Fig. 6.14(c), again a case where the uppermost level is just full (which will, clearly, be the case for any even number of electrons in the outer shell). Here the Fermi level lies about halfway up the gap between the valence and conduction bands and the gap is now relatively small, say less than $100kT$ for room temperature. (For example the element silicon has a band gap of $1.1\,eV$, compared with a value for kT, at room temperature, of $\sim 2.5 \times 10^{-2}\,eV$. An electron would gain an energy of $1\,eV$ in falling through a potential difference of $1\,V$.) In this case, although at low temperature the Fermi-Dirac 'tail' does not extend into the upper, conduction band, at higher temperatures it does, giving a small number of electrons in the conduction band. These electrons can then move easily into the abundance of empty states now available to them in this band. Thus the room temperature electrical conductivity is low but measurable; furthermore, it is clear that it will increase quite rapidly with temperature, as the 'tail' extends. We have here a 'semiconductor', more precisely an *intrinsic* semiconductor (it will become clear later why this adjective is necessary). For obvious reasons the upper band is called the conduction band and the lower one the valence band (since it is the stability of electrons in the lower band which provides the atomic forces holding the solid together.) There is another important point to be made for the intrinsic semiconductor. When thermal agitation raises an electron from the valence band to the conduction band, it leaves behind an empty state. This state can be filled by another electron, in the valence band, which can then gain energy and contribute to the electrical conduction. These empty states, comprising, as they do, the absence of negative electric charge, are equivalent to positive 'holes' in the valence band, and they effectively move like positive charges as the electrons in the valence band move in the opposite direction to fill them. Positive holes in the valence band comprise an important feature of semiconductor behaviour, and we shall be returning to them shortly.

Before moving on it should be emphasized that the description above is a greatly simplified one, in order to establish the ideas. Solids are complicated states of matter and are of course three dimensional, so in general we must deal not just with a single Fermi level but with a 3-D Fermi surface, which will have a shape dependent upon the variation of the material's properties with direction. Many important properties of solids depend upon the particular shape which this surface assumes. Especially important is the fact that two energy bands can sometimes overlap, so that it is possible for some elements to behave as metals even though each of their atoms possesses an even number of electrons (the lower band feeds electrons into the middle of the upper band); examples are beryllium, magnesium and zinc. However, this is the stuff of pure solid state physics and, for more, interested readers must refer to one of the many specialist texts on solid state physics (e.g., [2]).

It has become clear then that the position of the Fermi level in relation to the band structure for a particular solid material is vitally important. It is important not only for distinguishing between metals, insulators and semiconductors but also for understanding the detailed behaviour of any particular material.

We have seen how to calculate the Fermi level for the case of electrons moving freely within a solid. It is necessary now to discover how to calculate it when we are faced with the restrictions of a band structure. To do this we need to know about Brillouin zones.

6.3.3 Brillouin zones

The first problem to be addressed now is that of determining the density of states for a solid with band structure. Equation (6.16a,b) represents the density of states function for electrons which are completely free of restrictions in an 'electron gas', but in a solid material the lattice structure imposes quite severe restrictions, in the form of forbidden energy bands, so that we must expect the calculation in this case to be somewhat more complex.

Having determined the density of states function, the Fermi-Dirac distribution can again be used to derive the actual electron energy distribution, upon which all physical behaviour depends.

For the free electron, the energy and wavenumber are related by:

$$E = \frac{h^2 k^2}{8\pi^2 m}$$

but for the electrons confined within the band structure this is no longer true. It is clear, indeed, that it cannot be true, because E is a continuous function of k in the above expression and we know that the band structure forbids certain energies.

The forbidden bands can be regarded as being due to Bragg reflection, i.e., to reflection of those electrons for which an integral number of half wavelengths is equal to the lattice spacing, since, in this case, the waves reflected back from the atoms are all in phase (Fig. 6.15) and hence strongly reinforce. This means that the forward propagating electron wave

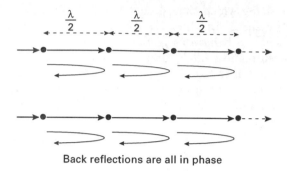

Back reflections are all in phase

Fig. 6.15 'Bragg reflection' explanation of forbidden bands.

is strongly reflected and hence cannot progress easily. The condition which must be obeyed for this is:

$$\tfrac{1}{2}n\lambda = a$$

i.e.,

$$k = \frac{n\pi}{a} \tag{6.20}$$

where a is the lattice spacing and n is a positive or negative integer.

In practice what has happened is that the electrical potential wells formed by the atoms in their lattice have maximum effect on the electrons which satisfy (6.20), and their kinetic energy is changed by the potential energy of the interaction. The forward propagating waves are reduced in energy, the backward ones increased; there is a gap in energy at the wavenumber which exactly corresponds to the condition (6.20). It is also true that electrons whose wavenumbers are close to condition (6.20) will be affected to some extent, but less strongly the greater is the deviation of their wavenumber from one of the $n\pi/a$. The result, then, in one particular energy band is that the energy of the electron is periodic with wavenumber (Fig. 6.16(a)), the period being $2\pi/a$. Another way of saying this is that, whenever the wavenumber increases by $2\pi/a$ the relationship between the electron wave and the lattice is, physically, essentially the same, since the phase relationships between the backscattered wavelets have recurred, and hence the energy is the same. Thus the relationship between E and k is completely characterized by the variation over one such period, i.e., between k values $-\pi/a$ and $+\pi/a$ (Fig. 6.16(b)). Note that the variation approximates to the free space variation (E/k^2) near the midpoint of the band, but close to the edges it deviates markedly from it.

For the next higher electron band the same arguments apply. The electrons now have larger potential energy, since they are farther from the nucleus, but their kinetic energy is governed by the same considerations as for the first band, and hence the total energy will again be periodic in

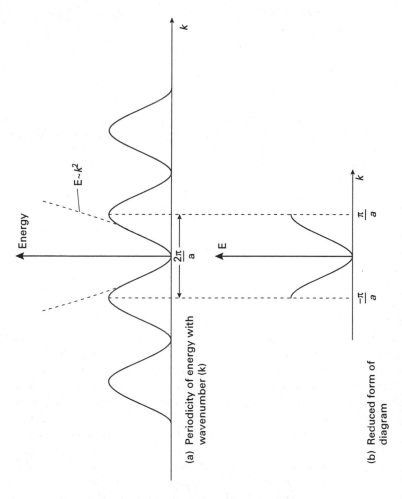

(a) Periodicity of energy with wavenumber (k)

(b) Reduced form of diagram

Fig. 6.16 Energy versus wavenumber in the lowest band for a lattice.

k, and again with period $2\pi/a$. It follows that the E/k variations for all the energy bands can be restricted to the range of k from $-\pi/a$ to $+\pi/a$, and these are shown in Fig. 6.17(b). This diagram is known as the 'reduced Brillouin zone' diagram, and each of the energy bands is now called a Brillouin zone. The Brillouin zone diagram has some important physical consequences. One of these relates to the effective masses of the electrons close to the edge of a zone. In order to understand the variation of effective electron mass consider, again, the relationship between energy and wavenumber for a free electron:

$$E = \frac{h^2 k^2}{8\pi m} \tag{6.21}$$

Near to the edges of the zone there is a large deviation from this relationship, as shown in Fig. 6.17(a). What conclusions can be drawn from this? Well, h is a fundamental constant and k is the independent variable, so it must be the mass (m) which is causing the deviation. The independent variable can be removed by differentiating (6.21) twice, i.e.,

$$\frac{d^2 E}{dk^2} = \frac{h^2}{4\pi^2 m}$$

or

$$m^* = \frac{h^2}{4\pi^2} \Big/ \frac{d^2 E}{dk^2} \tag{6.22}$$

m^* is now an 'effective mass' which differs from that of a free electron whenever $d^2 E/dk^2$ differs from that for a free electron, as it does near the edge of a Brillouin zone (Fig. 6.17(a)). In fact, near the upper edge of a zone, $d^2 E/dk^2$ is very small and *negative*, so m^* becomes very large and negative. Hence it is very difficult to move this very large mass and the contribution to the conductivity by the electrons near to the top of the band is close to zero, as we have already seen from the qualitative argument in section 6.5, i.e., they cannot move, since there are no energy levels available for them to move into.

Further, if, in the case of a relatively small band gap, such an electron manages to jump to the conduction band, it leaves behind a hole which is the absence of negative charge, and the absence of negative mass, i.e., the hole has positive charge and positive mass! It is for this reason that it is convenient and practical to think of these holes as, effectively, positively charged particles in all kinds of semiconductor considerations.

Conversely, near to the bottom edge of a Brillouin zone the effective mass is small and positive, so the electrons respond readily to electric fields. It follows that any electrons which do manage to jump from the top of the valence band to the bottom of the conduction band have large mobility when they arrive there and thus are able to contribute significantly to the

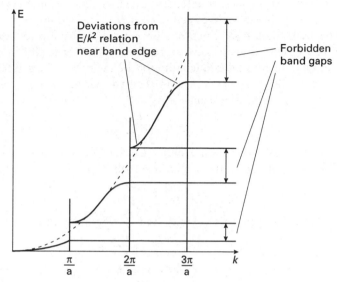

(a) Brillouin zone diagram : E/k for a lattice

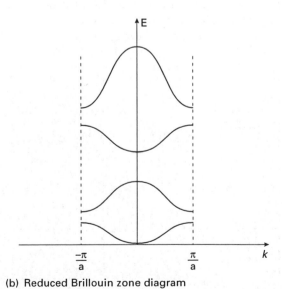

(b) Reduced Brillouin zone diagram

Fig. 6.17 Brillouin zone diagrams for a solid lattice.

material conductivity. The holes they leave behind also contribute to the conductivity, but to a lesser extent, since their effective mass is large. These ideas comprise the essence of semiconductor behaviour.

(The term 'mobility' which has been used rather loosely above, in fact has a strict definition in solid state theory, and is useful in a variety of calculations concerning semiconductor behaviour. If an electric field E acts upon a charge carrier in a solid then the carrier will experience a force Ee, and hence an acceleration:

$$\alpha = \frac{Ee}{m^*}$$

If the mean time between atomic collisions for the carrier is τ_c then the velocity it acquires in time τ_c as a result of α will be:

$$v_d = \frac{Ee}{m^*}\tau_c$$

This is known as the drift velocity and its value for unit field:

$$\mu = \frac{v_d}{E} = \frac{e\tau_c}{m^*}$$

is known as the mobility. Clearly this is a property of the semiconductor and is inversely proportional to the effective mass m^*, which is the sense in which it was used in the preceding paragraph.)

6.3.4 Electron energy distribution in semiconductors

What conclusions can be drawn from the Brillouin zone diagram in regard to the electron energy distribution and the position of the Fermi level in semiconductors? The full quantitative treatment of this topic is complex and tends, in any case, to mask the physical ideas. The treatment given here, therefore, will concentrate on the latter.

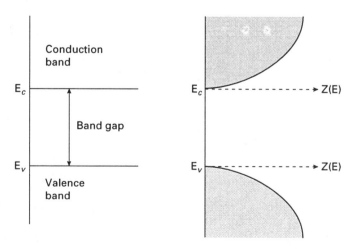

Fig. 6.18 Density of states function ($Z(E)$) for a band structure.

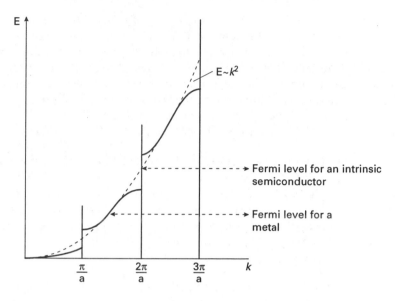

Fig. 6.19 Fermi levels on the Brillouin zone diagram.

It is clear that the calculation of the density of states function will be more complex than for the free electron case (see equation (6.16(b)) since the Brillouin zone diagram tells us that the simple E/k relationship for electrons, on which the free-electron calculation was based, is no longer valid. In fact, each band of allowable energies effectively acts as a containing 'box' for the electrons, so that the density of states function rises parabolically from the edge of a band in much the same way as the free-electron function did, from zero energy, in equation (6.16a). The result is that the density of states function takes the form, near the top edge of the valence band and the bottom edge of the conduction band, shown in Fig. 6.18. Now in order to determine the electron energy distribution among these states, the density of states function, $Z(E)$ say, must be multiplied by the Fermi-Dirac occupancy function as in equation (6.19(a)). Hence the electron energy distribution will be given by:

$$n(E)\mathrm{d}E = Z(E)F(E)\mathrm{d}E$$

and if there is a total number of electrons per unit volume, N_T, then

$$N_T = \int_0^\infty Z(E)F(E)\mathrm{d}E \qquad (6.23)$$

Remember that:

$$F(E) = \cfrac{1}{\exp\left(\cfrac{E - E_F}{kT}\right) + 1}$$

so that equation (6.23) is an expression for E_F, the Fermi level, in terms of N_T. Now N_T varies with the number of atoms/molecules per unit volume and the number of electrons per atom/molecule, so that it varies from material to material. Hence E_F can be calculated for each material, but remember the caution, emphasized in section 6.5.2, that solids are complex; lattice spacings vary with direction, hence the Brillouin zones will vary with direction. Hence $Z(E)$ varies with direction and then so does E_F. We have to deal, in general, with complex, three-dimensional, Fermi surfaces.

Having calculated the Fermi level for a particular set of Brillouin zones (for a given direction in a crystal, say), this can be drawn on the Brillouin zone diagram (Fig. 6.19). If it lies in the middle of a zone, the topmost electrons are following an $E \sim k^2$ variation, and are thus behaving like free electrons (although, of course, *only* the topmost electrons can do so) and we are dealing with a metal. This is the justification for identifying the 'free electron gas' with the metallic state.

If the Fermi surface lies at the top of a zone or somewhere within a band gap then the topmost electrons cannot move from the filled zones unless they are given sufficient energy (thermally or otherwise) to jump the gap into the next band. And if they do this they leave 'holes', which are positively charged and, as we noted in the preceding section, also have positive mass, since they represent the absence of electrons with negative effective mass. This is the intrinsic semiconductor.

Finally, we consider the actual position of the Fermi level in an intrinsic semiconductor. From equation (6.23) it was noted that the position of the Fermi level was dependent on the density of states function, $Z(E)$. This varies from material to material and, in a solid, will be zero in the energy band gaps. Remember that this function does not describe the actual distribution of electrons, only the states available to them. Hence the position of the Fermi level is determined by the available states in conjunction with the Fermi-Dirac distribution, which conditions how they will be distributed amongst those states. In a semiconductor or an insulator, at absolute zero, the valence band is full and the conduction band is empty. Clearly, the Fermi level lies somewhere in the gap between the bands, but where? Its position will be dependent on the empty, available states in the conduction band as well as the filled levels in the valence band. A simple calculation will give substance to these ideas and also fix, roughly, the position of the Fermi level.

Suppose that the widths of the valence and conduction bands are assumed equal and small compared with the band gap: effectively they are assumed to be at energies E_v and E_c respectively (Fig. 6.20). The density of states also is assumed equal for the two bands, and equal to Z. In this

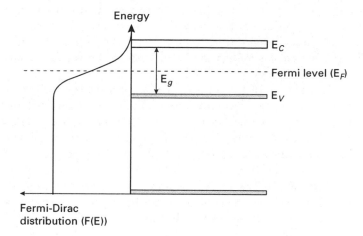

Fig. 6.20 Simplified energy level diagram for a semiconductor.

case the two electron densities are given by:

$$n_v = \frac{Z}{\exp\left(\frac{E_v - E_F}{kT}\right) + 1} \tag{6.24a}$$

for the valence band, and

$$n_c = \frac{Z}{\exp\left(\frac{E_c - E_F}{kT}\right) + 1} \tag{6.24b}$$

for the conduction band.

Now, of course, as the temperature rises, some electrons are able to jump from the valence band to the conduction band, via the Fermi-Dirac 'tail'. But if, at absolute zero, the valence band is full and the conduction band is empty, the total number of electrons (per unit volume) in the system is just equal to the number of available states in the valence band, Z. At any other temperature whatever electrons there are in the conduction band must have come from the valence band so that:

$$n_v + n_c = Z$$

and, using (6.24(a)) and (6.24(b)), this gives:

$$E_F = \tfrac{1}{2}(E_v + E_c) \tag{6.25}$$

In other words the Fermi level lies exactly halfway between the valence band and the conduction band. This approximation is quite a good one but, owing to the fact that the effective mass of the electrons elevated to the conduction band is less than it was when they were at the top of the valence band (and they are thus more mobile) the Fermi level in practice lies a little closer to the conduction band than to the valence band, i.e., it rises somewhat.

Essentially what all of the above is saying is that since the tail of the electron distribution must be symmetrical about the Fermi level, and since the electrons in the conduction band must have come from the valence band and thus leave an equal number of holes there, the Fermi level must be symmetrically placed with respect to the two bands, i.e., it must lie about halfway between them.

The position of the Fermi level in semiconductors has several consequences for their practical behaviour, as we have come to expect, and the consequences for the electron energy distribution are detailed in Appendix X.

The rigorous result, also derived in Appendix X, gives for the value of the Fermi energy in a semiconductor:

$$E_F = \tfrac{1}{2}(E_v + E_c) - \tfrac{1}{2}kT \ln \frac{p}{n} - \tfrac{3}{4}kT \ln \frac{m_e^*}{m_h^*} \tag{6.26}$$

where p and n are the concentrations of holes in the valence band and electrons in the conduction band, respectively.

Now for the intrinsic type of semiconductor material we have been dealing with, holes in the valence band result from the thermal excitation of electrons to the conduction band, and hence:

$$n = p = n_i \text{ (say)} \tag{6.27a}$$

and thus

$$np = n_i^2 \tag{6.27b}$$

n_i is known as the intrinsic carrier density and is a constant for a given material at a given temperature.

Equation (6.27(b)) will always be true since it represents an equilibrium between holes and electrons in the two bands. It will be true no matter what the origin of the electrons and holes, so that it is as true for an extrinsic semiconductor (see next section) as for an intrinsic one. For the former case we shall have:

$$np = n_i^2, \text{ but } n \neq p$$

These points are treated rigorously in Appendix X.

It now follows that, since $p = n$ for an intrinsic semiconductor, the second term in equation (6.26) is zero for this type of material. Consequently, the only difference between equations (6.25) and (6.26) lies in the difference between m_e^* and m_h^*, for, if these were equal, the third term in (6.26) also would be zero and the equations would be identical. However, we have seen in section 6.3.3 that $m_h^* > m_e^*$, since the two types of carrier have different mobilities near their respective Brillouin zone edges. Hence $\ln(m_e^*/m_h^*)$ is negative and the Fermi level moves slightly upwards, from the midpoint of the band gap, towards the conduction band, giving proper weight to the greater contribution of the more mobile electrons to the material's electrical conductivity.

We shall now move on to extend all of these ideas to 'doped' or 'extrinsic' semiconductors.

6.3.5 Extrinsic semiconductors

Finally we must consider another very important type of semiconductor material. This is the doped semiconductor, otherwise known as the 'extrinsic' semiconductor. In these materials the semiconducting properties can be both enhanced and controlled by adding specific impurities in carefully judged quantities. The effect of this is to alter the electron energy distribution in a controlled way.

We begin by considering a particular intrinsic semiconductor, silicon, since, with germanium, it is one of the two most commonly used materials for doping in this way. Both materials have a diamond-like structure, with each atom surrounded symmetrically by four others. Silicon is tetravalent, having an even number of electrons in its valence shell. There will thus be $4N$ available electrons per unit volume. The first valence energy band will be filled with $2N$ electrons and the second band also with $2N$ electrons; thus both lower bands are full and the next higher one is empty, at absolute zero (Fig. 6.21). The gap between the upper valence band and the conduction band is quite small, only 1.1 eV, compared with a room temperature value of kT of $\sim 2.5 \times 10^{-2}$ eV so, although silicon is an intrinsic semiconductor, its semiconductivity is moderate, and it increases exponentially with temperature.

Suppose now that the silicon is doped with a small fraction (between 1 atom in 10^6 and 1 in 10^9) of a pentavalent (valency of five) impurity atom such as phosphorus. This atom sits quite comfortably in the silicon lattice but has one extra electron compared with the silicon atoms by which it is surrounded, and this electron is easily detached to roam through the lattice and add to the conductivity. In energy level parlance we say that it needs little energy to raise it into the conduction band; in fact for this particular case it needs only 4.5×10^{-2} eV, equivalent to only about $2kT$ at room temperature. So the energy level structure looks like Fig. 6.21 with the 'donor' level, E_D, just below the conduction band. Note that the level remains sharp, since the dopant atoms are scarce and spaced well

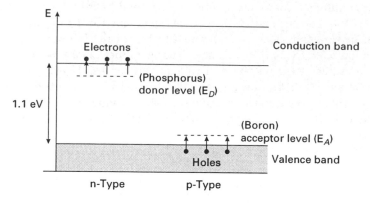

Fig. 6.21 Energy level diagram for doped silicon.

apart in the lattice, and thus their wavefunctions cannot overlap to form
a band structure, as do the atoms of the host lattice. Since, in this case,
the effect of the phosphorus is to donate electrons (negative charges) to the
conductivity, this is called an n-type semiconductor. The really important
point is that the conductivity now is entirely under control, via control of
the doping level. The greater the concentration of the dopant atoms, the
greater will be the concentration of electrons in the conduction band.

Consider, on the other hand, what happens if we dope the silicon with
a tervalent (valency of three) impurity such as boron. In this case the
impurity atom has one electron fewer than the surrounding silicon atoms,
and electrons from the valence band in the silicon can easily move into the
space so created. These 'absent' (from the valence band) electrons create
positive holes, as we have seen, and these also are effective in increasing the
conductivity. For obvious reasons this is now called a p-type semiconductor
(Fig. 6.21) and the corresponding energy level is an 'acceptor' level. The
'majority carriers' are holes in this case; in an n-type material the majority
carriers are, of course, electrons.

Usually, the donor or acceptor dopings dominate the semiconductor behaviour. In other words, it is normally the case that the dopant concentrations exceed the intrinsic carrier concentration n_i (equation 6.27). If the
dopant concentrations are N_d for donor and N_a for acceptor, it must be
that for charge neutrality of the material:

$$n + N_a = p + N_d \tag{6.28}$$

However, it is also true that, for all circumstances, equation (6.27(b)) holds:

$$pn = n_i^2$$

Hence, for an acceptor doping (p-type material) we have: $N_a \gg N_d, n_i$ and
thus from (6.28) and (6.27):

$$p = N_a$$

$$n = \frac{n_i^2}{N_a}$$

For a donor material (n-type): $N_d \gg N_a$, n_i and thus:

$$n = N_d$$

$$p = \frac{n_i^2}{N_d}$$

It is clear, then, that a knowledge of n_i and the dopant level fixes the carrier concentrations and thus allows the main features of behaviour to be determined. Where are the Fermi levels in these extrinsic semiconductors? We know that in the case of intrinsic semiconductors, the Fermi level lies about halfway between the valence and conduction bands. In an n-type semiconductor the valence band is almost full and most of the conduction is due to electrons donated from the donor levels. Hence it follows that the '50% electron occupancy' level, i.e., the Fermi level, will now lie about half way between the donor level and the bottom of the conduction band, since the top of the 'valence' band can now be identified with the donor level (Fig. 6.22).

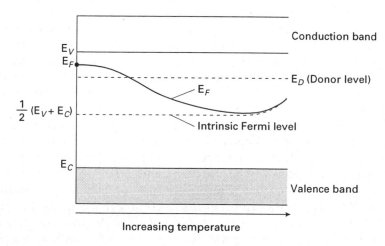

Fig. 6.22 Fermi level in an n-type extrinsic semiconductor and its variation with temperature.

Similarly, for p-type semiconductors, it will lie midway between the top of the valence band and the acceptor energy level. However, this can only be the case as long as the donor or acceptor mechanisms dominate. At higher temperatures most of the donor and acceptor sites will have been exhausted and the true valence band then starts to dominate the conduction mechanism. Hence the Fermi level will vary with temperature as shown in Fig. 6.22, until at high enough temperatures, it reverts to the intrinsic value in both p and n cases.

6.3.6 Binary and ternary semiconductors

We cannot leave semiconductors without a mention of some important, relatively new, materials. These are alloys, made from two or more elements in roughly equal proportions and from different groups in the periodic table, and thus with differing numbers of electrons in the outermost shell.

The best known of these is gallium arsenide (GaAs) which, since Ga has a valency of III and As of V, is an example of what is called a III-V compound.

An important aspect of these compounds is that we can 'tailor' the band gap by varying the mix. The eight electrons in the two outer shells are shared to some extent and create some ionic bonding, through absence from the parent atom (i.e., it creates a positive ion). The band gap of GaAs when the two elements are present in equal proportions (i.e., same number of molecules per unit volume) is 1.4 eV, but this can be varied by replacing the As by P (GaP, 2.25 eV) or Sb (GaSb, 0.7 eV) for example. Furthermore, the materials can be made p-type or n-type by increasing the V(As) over the III(Ga) component, or vice versa.

Another very important aspect of GaAs is that it is a direct-band-gap material: the minimum energy in the conduction-band Brillouin zone occurs at the same k value as the maximum energy of the valence band (Fig. 6.23(a)). This means that electrons can make the transition between the two bands without having to lose or gain momentum in the process. Any necessary loss or gain of momentum must always involve a third entity, a 'phonon' (quantum of vibration) for example, and this renders the transition much less probable. Hence a direct-band-gap material is much more efficient than an indirect-band-gap material (Fig. 6.23(b)) and the processes are much faster, leading to higher device bandwidth.

Quite frequently even finer control is required over the value of the band gap and, for this, 'band-gap engineers' turn to ternary alloys, i.e., those involving three elements, where the ratio of III to V composition is still approximately 1:1. An example is the range of alloys which is described by the formula $Al_x Ga_{1-x} As$. By varying x one can move along the line between GaAs and AlAs on Fig. 6.24 and thus vary the band gap appropriately. Fig. 6.24 illustrates other materials which can be tailored in this way.

One final difficulty is that, in order to grow the required material one needs a substrate on which to grow it, from either the gas phase (gasphase epitaxy) or the liquid phase (liquid-phase epitaxy), and this requires that the desired material has approximately the same lattice spacing as the substrate. For the example of $Al_x Ga_{1-x} As$ there is little difficulty since the

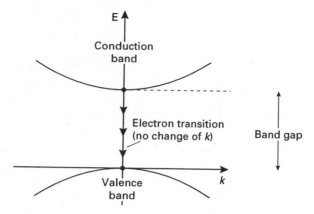

(a) Direct band-gap semiconductor (eg. GaAs)

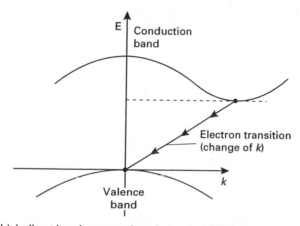

(b) Indirect band-gap semiconductor (eg. Si)

Fig. 6.23 Brillouin zone diagrams for direct and indirect band-gap semiconductors.

GaAs/AlAs line is almost vertical (Fig. 6.24), and thus the lattice spacing always is close to that of GaAs, which can thus be used as the substrate. For other materials, for example, $InAs_y Sb_{1-y}$, this is clearly not the case since the InAs/InSb line is almost horizontal. This problem can be solved by going one stage yet further, to quaternary alloys which lie in the regions bounded by the lines in Fig. 6.24. An example of a quaternary alloy is $In_x Ga_{1-x} As_y P_{1-y}$ and one of these alloys is shown marked x on Fig. 6.24. This has a lattice spacing similar to that of InP, which can thus be used as a satisfactory substrate.

Thus a band-gap engineer generally will choose a substrate, then a suit-

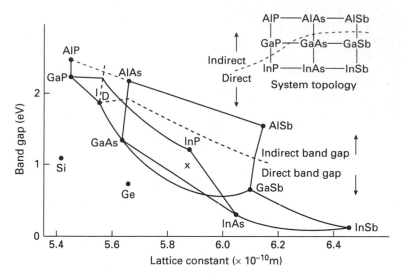

Fig. 6.24 Band-gap diagram.

able quaternary alloy which has the required band gap, probably making sure it is a direct-band-gap material, and then grow the semiconductor.

Band-gap engineering is now a sophisticated, and extremely valuable, technology for the provision of materials for optoelectronic devices; their behaviour and performance depend critically, of course, on the material from which they are made.

6.4 CONCLUSION

Solid state physics, even the basis thereof, is not an easy subject: solids are complex states of matter, with their range of overlapping atomic wave-functions. Many of the ideas are broadly unfamiliar to a non-specialist physicist and, in some aspects, even mildly uncomfortable. However, many optoelectronic devices and systems rely on the behaviour of electrons in solids and we needed to have some understanding of this. This chapter is the kind that becomes a lot easier on the second reading! We certainly shall need to draw repeatedly upon the essentials of what it contains.

PROBLEMS

6.1 Derive the Rayleigh-Jeans formula for the energy density distribution with frequency for a black-body radiator:

$$\rho_f \mathrm{d}f = \frac{8\pi f^2}{c^3} kT \mathrm{d}f$$

Why must it fail at high frequencies? What is the reason for its failure?

6.2 Describe the physical principles for the action of three-level and four-level lasers. Why is the four-level system more efficient? For a laser operating at a wavelength of $10.6\,\mu$m, at what temperature are the rates of spontaneous and stimulated emission equal?

6.3 The lifetime of the upper level of a laser is $5 \times 10^{-8}\,$sec. What is the population inversion that is required to give a gain of $5 \times 10^{-2}\,\mathrm{m}^{-1}$, if its output wavelength is 488 nm? (Ignore line broadening.)

6.4 Sketch a practical design for a four-level gas laser. Indicate, on your sketch, how your laser could be Q-switched.

A four-level laser system is being pumped by a source of power, P, and optical frequency ν_p. The efficiency with which the upper lasing level is being populated is η, and the relaxation from the upper to the lower lasing levels is exponential, with time constant τ. The energy difference between the lasing levels is $h\nu_L$, where ν_L is the lasing frequency, and h is the quantum constant.

Lasing is inhibited by a closed Q switch until population equilibrium is attained for the system. The Q switch is then opened in a time which is very small compared with τ. How much energy is released in the Q-switched laser pulse?

6.5 Derive, from first principles, the Fermi-Dirac energy distribution function:

$$F(E)\mathrm{d}E = \frac{\mathrm{d}E}{\exp\left(\dfrac{E - E_F}{kT}\right) + 1}$$

What are the probabilities of finding occupied electron levels at energies $E_F + 0.01\,$eV and $E_F - 0.05\,$eV at temperatures of 0 K and 300 K?

6.6 Explain the differences between metals, insulators and semiconductors in terms of the band theory of solids. Explain also the differences between intrinsic and extrinsic semiconductors.

6.7 Explain the significance of the Fermi level for a semiconductor, and derive an expression for its value in terms of the densities, and effective masses, of the electrons and holes.

If the electron density is 5% more than that of the holes and the electron effective mass is twice that of the holes, where does the Fermi level lie in relation to the band gap at room temperature (295 K)? Under what circumstances does the Fermi level lie within the conduction band?

6.8 Why are III-V semiconductors important and what is meant by 'band-gap engineering'? By referring to the periodic table of elements, devise as many III-V semiconductors as possible, and speculate on their semiconductor properties.

REFERENCES

[1] Zemansky, M. W. (1968) *Heat and Thermodynamics*, 5th edn., Mc-Graw Hill, chap. 6.
[2] Kittel, C. (1968) *Introduction to Solid State Physics*, 3rd edn., John Wiley and Sons.

FURTHER READING

References [1] and [2] above, plus:
Siegman, A. E. (1986) *Lasers*, University Science Books.
Solymar, L. and Walsh, D. (1993) *Lectures on the Electrical Properties of Materials*, 5th edn., Oxford Science Publications.
Sze, S. M. (1981) *Physics of Semiconductor Devices*, 2nd edn., John Wiley and Sons.

7

Optical sources, modulators
and detectors

7.1 INTRODUCTION

Chapter 6 has provided the background necessary to understand a variety
of optoelectronic devices, the most important of these being optical sources,
modulators and detectors. The source provides the light in the appropriate
form, the modulator controls it or impresses information upon it, and the
detector receives the light signal and allows that information to be extracted
in the form of an electronic signal.

Most optoelectronic devices nowadays are of the solid state variety, for
solid devices can be compact, robust, reliable and readily manufacturable
(and, therefore, cheap!). However, there are still some important gas lasers,
and we must not ignore these.

In this chapter optical sources, modulators and detectors will be dealt
with in turn. The emphasis will be on practical aspects of the devices, so
that the way in which the basic principles are put into practice will begin
to become clear. We shall begin with optical sources.

7.2 OPTICAL SOURCES

The requirements demanded of optical sources by optoelectronic re-
searchers, designers and users are many and varied. The source may be
required to be broadband or narrowband; tunable or fixed in frequency; coherent, partially coherent or incoherent; polarized or unpolarized; continuous wave (CW) or pulsed; divergent or collimated. It is hardly surprising
that as no single type of source can provide all of these features, a range of
sources has been developed for optoelectronic use. One way of providing
a quite versatile source is to use a broadband black body, or 'grey' body,
source and then to use a variety of external components and devices to manipulate the light in order to provide what is needed. For example, filters
can select a limited wavelength range, and these can be made tunable if
necessary via frequency-selective components, such as prisms, diffraction
gratings or Fabry-Perot interferometers. The polarization can be selected

with the aid of polarizing prisms and retardation plates, the collimation controlled with lenses, and pulses provided by fast electro-optic switches.

There are two objections to this approach. Firstly, each one of the selective operations is lossy and significantly reduces the available power; secondly, the source system becomes complex, cumbersome, difficult to align and stabilize, heavy, bulky and expensive.

The efforts, therefore, have been in the direction of developing sources which are both compact and intense but which also have inherent, desirable characteristics, sometimes controllable sometimes not necessarily so, for particular applications. We have already learned that laser sources have many of the required characteristics, as natural features. Hence our study of optical sources should, naturally, begin with lasers.

7.2.1 Laser sources

(i) *Introduction*

Basic laser design was introduced in Chapter 4 but it is necessary to expand somewhat on what was considered there.

Earlier (equation (6.14)) we derived the expression for the gain which was obtained for unit length in a medium pumped to a state of inverted population as:

$$g = \frac{h\nu}{c} B_{12}(N_2 - N_1)$$

This medium can be used as an optical amplifier, since light entering the medium at the frequency corresponding to the inverted transition will cause stimulated emission at the same frequency, and thus will emerge from the medium with increased power. By providing feedback, the amplifier may be made to oscillate and become a laser. All of these ideas were considered in some detail in Chapter 4. In particular, in section 4.4.2, the basic features of He-Ne laser design were considered. This was a three-level system, and it was pointed out that, although the principles of laser action are relatively straightforward, the transition pathways for any particular laser are often complex. We shall now try to cement these ideas by looking at some aspects of practical design for specific lasers.

(ii) *The argon laser: a four-level system*

The first system for consideration is a four-level system. Remember that the advantage of this is that the lower level of the laser transition is not the ground state, hence the level is relatively sparsely populated, inversion

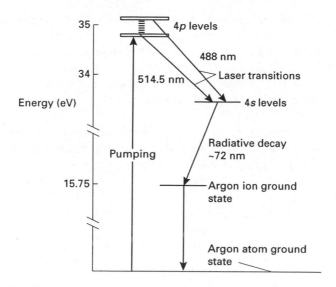

Fig. 7.1 Simplified energy-level diagram for the argon laser.

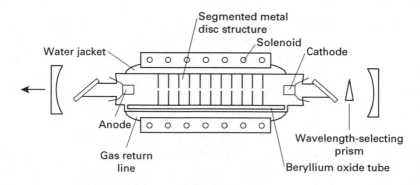

Fig. 7.2 Schematic for the design of an argon laser.

of the population is therefore easier for the pump to achieve, and the laser can thus be more efficient.

The energy level diagram is shown in Fig. 7.1. Argon is an inert gas but can be ionized by passing a large current (~ 50 A) through it (by striking a gas discharge). In its ionized state the atom is then excited up to the 4p levels by successive electron collisions, and these levels become inverted with respect to the 4s state, which is sparsely populated, being above the Ar^+ ground state, which is itself above the neutral Ar ground state. Thus efficient lasing occurs between 4p and 4s, most strongly at 514.5 nm (green) and 488 nm (blue) wavelengths. The design of the laser is shown in Fig. 7.2. Two main features should be noted: first, since the large discharge current generates a lot of heat, this must be dissipated with the aid of metal-disc heat exchangers and water cooling; secondly, in order to reduce ionic de-excitation via wall collisions (rather than the required laser transitions) the tube is enclosed within a solenoid, which provides a longitudinal magnetic field whose action is to force the ions away from the walls.

Finally, a prism is usually provided inside the cavity, to select the required wavelength.

Argon-ion lasers can provide several watts of CW power and can also be both mode-locked and Q-switched, conditions which were discussed in Chapter 6. Mode-locking provides microsecond pulses at several kilowatts of peak power, whilst Q-switching provides megawatts of peak power (for this particular laser).

Each type of laser presents its own design problems as a result of its own special set of energy-loss pathways.

The argon laser is a valuable high-power source of CW light in the middle of the visible spectrum.

(iii) *The dye laser*

So far we have given examples only of gas lasers, but there are very important lasers which use the liquid and solid phases. Looking first at the liquid phase we shall consider the dye laser. The advantage of liquid and solid lasers over gas lasers is the greater density of the medium, which means more photons per unit volume. The disadvantage is that the more dense media are less homogeneous, so that the laser output is less pure and, especially, of greater linewidth.

However, liquids are more homogeneous than solids (under normal circumstances), so they provide a convenient compromise from some points of view.

One very important type of liquid laser is the dye laser, and it is important because its output is wavelength tunable over quite a large range (~ 100 nm).

Dyes are complex organic molecules within which the atoms can quite easily vibrate and rotate. As a result, the main molecular quantum states are split into a large number of associated rotational and vibrational sub-states whose energy degeneracy is removed, when the dye is dissolved in a liquid, by the electric fields of the solvent molecules. The result is a virtual continuum of states around the main quantum levels (Fig. 7.3). If now the solution is irradiated with light, absorption takes place over a broad band, as does fluorescence, down to the large number of lower levels. It is thus quite easy to arrange for efficient laser action between, for example, the many states around S_1 down to the upper states of S_0 (Fig. 7.3), the latter of which will be sparsely populated compared with the lower states

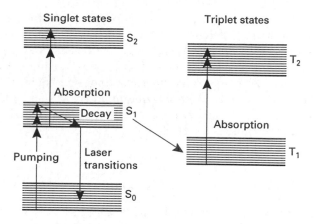

Fig. 7.3 Schematic energy-level diagram for a dye laser.

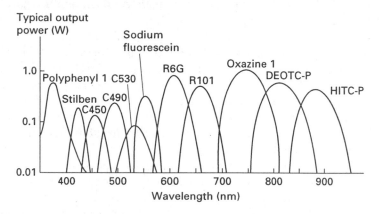

Fig. 7.4 Tuning ranges for commonly-used laser dyes.

of S_0. The only difficulty lies in the so-called 'triplet' states (states with two electrons whose spins are parallel). Transitions from singlet (paired, i.e., anti-parallel, spins) to triplet states are 'forbidden', which means that quantum rules render them improbable, but not impossible. In fact, transitions from S_1 to T_1 can occur (especially as a result of molecular collision) and, since $T_1 \rightarrow S_0$ is forbidden, there is an accumulation of molecules in T_1. Now the $T_1 \rightarrow T_2$ transition is allowed by the quantum rules, and this corresponds to wavelengths close to the lasing wavelengths, which are thus absorbed, reducing the laser gain. The special solution for this problem is to flow the dye solution through the cavity, thus removing the T_1 states from the lasing region. By the time the dye returns to the cavity the T_1 states will have decayed (if the flow rate is chosen correctly) and all will be well. Clearly, in order to tune this system, a wavelength-selective element, such as a prism or a diffraction grating, is required within the cavity, and is varied as required. Different dyes have different tuning ranges and, by varying the dyes, a very large tuning range can be achieved (see Fig. 7.4).

The disadvantage of this type of system is that it requires a continuous flow of messy, corrosive and, sometimes, carcinogenic liquid in the laboratory (or wherever). Liquids are easily contaminated, so there is the problem of purification as well. However, if the large tuning range is needed in the application, all the effort is worthwhile.

(iv) *The Nd-YAG laser: a solid state system*

There are many types of laser which use solid media. We shall give one example now and another when we have, very shortly, become familiar with some more of the relevant solid state physics.

The present example is the neodymium/yttrium/aluminium garnet, or Nd-YAG, laser.

The big advantage, as has been stated, of a solid state laser is that the large density of atoms means that large outputs per unit volume can be achieved. The laser medium is in the form of a Nd-YAG rod. Nd is a rare earth element. The important characteristic of the rare earth elements is that their atoms all have the same outer electronic structure. As the atomic number increases, moving up the rare earth series, electrons are added to inner, rather than outer, levels. The result is that all the rare earths have very similar chemical properties. From our present point of view, however, the most important thing is that transition between levels can occur deep in the atom, and thus these transitions are shielded, by the outer electrons, from the atom's environment. Hence, the Nd^{3+} ion can sit in a YAG lattice (substituting for the yttrium atoms to the extent of $\sim 1.5\%$) and the transitions to be used for lasing action will be scarcely broadened at all by the fierce fields of the solid lattice structure.

The levels used are shown in Fig. 7.5. The system again is a four-level one and thus is quite efficient. The levels above $4F_{3/2}$ all, conveniently, decay to $4F_{3/2}$, so this level can be readily populated (level 2) by a broad-spectrum pump source, which is usually a krypton or xenon flashtube. This is positioned axially, parallel with the Nd-YAG rod, as shown in Fig. 7.6. The laser transition is from $4F_{3/2}$ (level 2) to $4I_{11/2}$ (level 1) which, being above the ground state (level 0), is very sparsely populated. (It is interesting to note that $F \rightarrow I$ transitions are normally highly forbidden, but the action of the YAG lattice field is to increase their probability in this case). The $1 \rightarrow 0$ transition ($4I_{11/2} \rightarrow 4I_{9/2}$) is allowed and is thus very rapid, as required for efficient population inversion (section 6.2.4). Since the depopulation of the inverted-population state is very rapid (a consequence of the dense material) and, provided that a suitably large

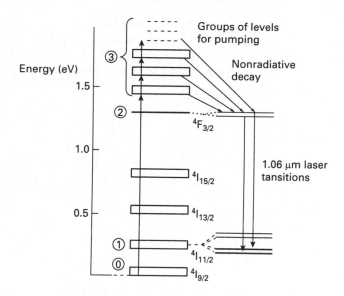

Fig. 7.5 Simplified energy-level diagram for the Nd-Yag laser.

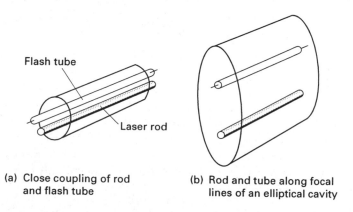

(a) Close coupling of rod and flash tube

(b) Rod and tube along focal lines of an elliptical cavity

Fig. 7.6 Flash-tube pumping geometries for a Nd-Yag laser rod.

amount of pump power is provided, the output can be up to a kilowatt of CW power at 1064 nm ($4F_{3/2} \rightarrow 4I_{11/2}$). The Nd-YAG rod becomes very hot while generating all this power, and must be cooled either with water (for the higher output powers) or with air.

This laser can also readily be Q-switched (section 6.2.4(iv)), in which state it can provide pulses with energies of several millijoules in a time $\sim 10\,\mathrm{ns}$, implying peak pulse powers ~megawatts. The Nd-YAG laser is thus a convenient source of high-energy pulses in the near infrared. The pulses are relatively short and thus the peak power of the pulses is very high. We shall see later that this is the prescription for ready entry into the non-linear optical regime where, for example, harmonics of the fundamental optical frequency can be conveniently generated. The Nd-YAG laser also can be used for manufacturing processes such as laser cutting, welding and laser forming of metals.

(v) *Other types of laser*

There are many other types of laser: for example, the CO_2 laser which provides large amounts of power in the infrared; the excimer (*excited dimer*) laser which provides large amounts of power in the ultraviolet; the Er-doped fibre laser which provides a very convenient, lower-power source for optical-fibre communications (see Chapter 10); the Ti-sapphire laser which can provide very short, intense pulses of light ($\sim 100\,\mathrm{ps}$), etc. Each has its own special advantages and disadvantages. Hence each has its own application area.

We have neither the time nor the space to deal with the full range of lasers which are currently in use. However, there is one more which is of paramount importance to our subject: the semiconductor laser. This will be dealt with very soon now (section 7.2.2), just after dealing with its close companion the light-emitting diode (LED).

7.2.2 Semiconductor sources

In section 6.5.4 we considered p-type and n-type extrinsic semiconductors. These concepts are needed now in order to understand the operation of semiconductor lasers (SLs) and light emitting diodes (LEDs). As might be expected, these two devices are closely related. We shall establish the most important ideas by dealing first with the LED. Before dealing with either of them, however, it is first necessary to understand what is meant by a $p - n$ junction.

(i) *The $p - n$ junction*

Let us take a piece of p-type and a piece of n-type semiconductor in the same intrinsic material. In other words, one piece of, say, silicon has been doped with trivalent atoms (e.g., boron or gallium) and the other with pentavalent atoms (e.g., phosphorus or arsenic). Suppose now that flat faces of these two pieces of semiconductor are placed in close contact, to form what is called a '$p - n$ junction'. Of course we remember that the Fermi levels for the two types of material will be different, since in the p-type material it will lie about halfway between the top of the valence band and the acceptor level and, in the n-type semiconductor about halfway between the donor level and the bottom of the conduction band. Hence there is a difference in Fermi levels given by:

$$E_{\mathrm{J}} = \tfrac{1}{2}(E_{\mathrm{C}} + E_{\mathrm{D}}) - \tfrac{1}{2}(E_{\mathrm{A}} + E_{\mathrm{v}})$$

where E_{C} is the energy at the bottom of the conduction band, E_{D} is the donor level, E_{A} is the acceptor level and E_{v} is the energy at the top of the valence band. Hence

$$E_{\mathrm{J}} = \tfrac{1}{2}E_{\mathrm{g}} + \tfrac{1}{2}(E_{\mathrm{D}} - E_{\mathrm{A}})$$

where E_{g} is the material's 'energy gap' $(E_{\mathrm{C}} - E_{\mathrm{v}})$, and E_{J} will thus be of order half the energy gap (Fig. 7.7(a)). It must be remembered, however, following the discussions in section 6.5.4, that, as the temperature rises and the acceptor/donor levels become exhausted, the Fermi levels for both p and n materials tend to revert to the value for the intrinsic material, i.e., to become the same value for each type of material, and thus $E_{\mathrm{J}} \rightarrow 0$. Clearly, the electrons are the majority carriers in the n-type material and the minority carriers in the p-type material, whereas the reverse is true for the holes. It is also clear that both types of carrier must come to equilibrium across the junction.

Now the Fermi level, as was explained in section 6.5.1, lies at the centre of the Fermi-Dirac 'tail', i.e., at the position where the occupancy of the electron states is 0.5. (At $T = 0$ it represents the highest occupied energy level.) When the two types of semiconductor are placed in contact, some electrons in the conduction band of the n-type material have available to them some new states (holes) in the valence band of the p-type material. Hence they can fall into them. Clearly, they must diffuse into the p-type material in order to do this and the reverse is true for the holes, so, in a region around the area of contact, holes and electrons recombine, leading to a dearth of charge carriers. The electrical resistance in the region around the junction, which it is known as the 'depletion' region, becomes high as a result of this.

In this depletion region a new equilibrium electron distribution is established. Clearly, the system can only come to equilibrium when the two Fermi levels have equalized (Fig. 7.7(b)), for there can only be one Fermi level for a given equilibrium electron distribution. However, under these conditions, negative electrons will have left the n-type material in order to neutralize positive holes in the p-type material. Since the materials must have been electrically neutral to start with, each must become charged, n-type positively, p-type negatively. Hence a voltage must exist across the depletion region, equal to E_{J}, the original difference between the Fermi levels, and thus of the order of half the band gap (~ 0.1 to $0.3\,\mathrm{V}$ depending on the materials); this is called the 'contact' potential. So we see that the

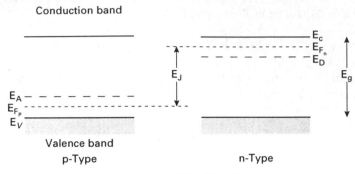

(a) Difference between p- and n-type Fermi levels

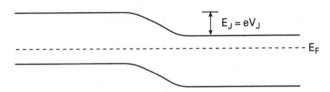

(b) Equilibrium voltage (V_J) across a p-n junction

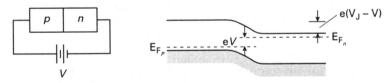

(c) A forward-biased p-n junction

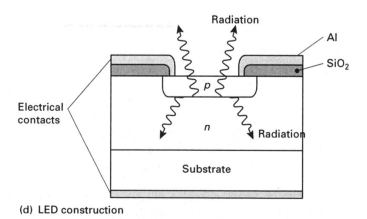

(d) LED construction

Fig. 7.7 The $p-n$ junction and the light-emitting diode (LED).

result is a dynamic equilibrium, with electrons diffusing from n-type to p-type material, but this 'diffusion current' being opposed by the voltage which develops as a result of the diffusion. The voltage is said to lead to an opposing 'drift current', since it creates an electric field, across the depletion region, which acts on the electrons and holes. These two currents will be equal at equilibrium, and quite small since, as has been noted, the electrical resistance is high in the depletion region.

The contact potential also can be expressed in terms of either electron or hole concentrations either side of the junction. Let us first consider the electrons, with concentration n_n in the n-type material (majority carriers) and n_p in the p-type material (minority carriers). With the Fermi level close to the centre of the band gap, the bottom of the conduction band and the top of the valence band are sufficiently far from the Fermi level to allow use of the Boltzmann approximation for the tail of the Fermi-Dirac distribution (equation (6.18b)) and it thus follows that n_n and n_p are related, in thermal equilibrium at temperature T, by:

$$\frac{n_p}{n_n} = \exp\left(-\frac{eV_C}{kT}\right)$$

where V_C is the contact potential. Hence

$$V_C = \frac{kT}{e} \ln\left(\frac{n_n}{n_p}\right)$$

Of course, the same argument can be applied to the hole concentrations p_p and p_n, so that also:

$$V_C = \frac{kT}{e} \ln\left(\frac{p_p}{p_n}\right)$$

(notice the $p - n$ inversion between the two equations).

It must follow that:

$$p_p n_p = p_n n_n$$

which, from equation 6.27(b), we know to be true.

This contact potential can easily be expressed in terms of the dopant concentrations. Suppose that the donor (n-type) and acceptor (p-type) concentrations are N_d and N_a, respectively. To a very good approximation in most cases $n_n = N_d$, i.e., all the electrons in the n-type extrinsic semiconductor are effectively provided by the donor atoms. Also, from equation 6.27(b):

$$pn = n_i^2$$

so that, in the p-type material

$$n_p = \frac{n_i^2}{p} = \frac{n_i^2}{N_a}$$

since, in this case, all the holes are created by the acceptor atoms. Inserting these values for n_n and n_p into the above expression for V_C we find:

$$V_C = \frac{kT}{e} \ln \frac{N_d N_a}{n_i^2}$$

as the voltage across the depletion region.

How wide is the depletion region? Clearly, this must depend upon the ease with which the electrons can diffuse, on the positions of the Fermi levels, and on the carrier concentrations; but the width will be, typically, a few micrometres. Furthermore, since the electrical resistance of the depletion region is high compared with that of the bulk material on either side of it, almost all of the junction voltage appears across the junction, leaving the rest of the material unaffected. The Fermi levels quite quickly (but smoothly) revert to their original levels either side of the junction.

(ii) *The light emitting diode (LED)*

We are now in a position to understand how the $p - n$ junction can be used as a light source. Suppose that an external voltage is applied across a $p - n$ junction, in opposition to the contact potential. This means that the n-type material is the negative terminal and the p-type the positive one (Fig. 7.7(c)). This arrangement is thus referred to as one of 'forward bias'.

The result, as shown in Fig. 7.7(c), is that the equilibrium is again disturbed and the Fermi levels are again relatively displaced. Under these conditions more electrons from the n-type material, and holes from the p-type material, can enter the depletion region. The result, inevitably, when they co-exist in the same volume of material, is that they recombine: electrons fall down through the band gap to cancel out the holes, releasing energy in the process; this process is known as forward-carrier injection. If this flow of electrons from n to p material is maintained by the external voltage source, the electrons will be descending continuously from the conduction band to holes in the valence band, and this means that energy is being continuously released. This energy can be dissipated in a variety of ways, but if the correct choices are made, it will be released in the form of photons, i.e., light energy. These choices will be discussed shortly.

It is easy to calculate the optical frequency of such photons. The photons, clearly, will have an energy of the order of the band gap, E_g, and thus a frequency given by ν where:

$$h\nu = E_g; \; \nu = \frac{E_g}{h}$$

There will be a spread of frequencies around this value since electrons will make transitions from within the conduction band to within the valence band (i.e. not just lower edge to upper edge) and also between bands and donor/acceptor levels, but we can expect the spectrum to be fairly narrow and centred on E_g/h.

If the semiconductor material is a 'direct band-gap' material (with the minimum of the conduction-band Brillouin zone at the same k number as the maximum of the valence-band zone (section 6.3.5)), all of the energy difference is taken up by the release of a photon. If we are dealing with an 'indirect band-gap' material then there will be a discrepancy in the initial and final k values for the electron, and a phonon (lattice vibration) must also be involved. In this latter case the downward transition is much less likely, since three particles are now involved (electron, hole, phonon) and both energy and momentum must be conserved overall for the three-particle process.

Forward bias on a direct band gap $p - n$ junction therefore is an efficient generator of photons, with energy equal to the difference between the conduction band and the valence band.

We can now use this emission process to design an optical source. A simple design is shown in Fig. 7.7(d). The Al acts as a direct contact across the junction and the SiO_2 as an insulator. The photons emitted in the junction region can escape as shown in the diagram, giving radiation in the direction of escape. We have the geometry for a light-emitting diode (LED).

What materials should we use? We know that for efficient generation of photons a direct band-gap semiconductor material is required. The best of these probably is GaAs.

GaAs, as mentioned in section 6.3.5, is an example of what is known as a III-V material. Gallium (Ga) has a valency of III, arsenic (As) has a valency of V. There are several important points to make about such materials. First, since the three outer electrons of Ga (valency 3) join with the five of the As to form a stable octet, some of the compound's bonding is ionic, which is stronger than the 'shared' electron valence band of, for example, Ge and Si. The result of this is that the energy gap between the conduction and valence bands is larger, leading to higher energy radiation as a result of the transition, across the gap. The gap in this case is 1.44 eV, giving optical radiation at 860 nm and thus quite close to the upper limit of the visible range $(400 - 700 \, nm)$. Secondly, it is a direct band-gap semiconductor, as we require. Thirdly it can readily be made p-type or n-type by doping with excess Ga or As (for example) as needed. Finally, and very importantly, it can be alloyed with other elements in groups III or V of the periodic table, in order to vary the strength of the bonding and thus the band gap. Hence, we have $GaAs_{0.6}P_{0.4}$ giving radiation at 650 nm (red) or GaP at 549 nm (green). This alloying, together with impurity doping, was discussed in section 6.3.5, where it was seen to be an example of band-gap engineering, a topic which has led to many recent advances in the availability of light sources suitable for given applications. II-VI and I-VII compounds also are important in this context and have their own special advantages, but there is not time, unfortunately, to deal with these (although, again, they were touched on in section 6.3.5).

What kind of performance can we expect from a typical GaAs LED? The device can be very compact, with an emitting area of diameter only $\sim 0.25\,\text{mm}$. The LED can be driven with a forward-bias voltage $\sim 2\,\text{V}$ and will take $\sim 50\,\text{mA}$. Such a device will emit $\sim 0.1\,\text{mW}$ of optical power fairly uniformly over a solid angle of 2π. Thus it is not very efficient: it converts, on the above numbers, $0.1\,\text{W}$ electrical to $0.1\,\text{mW}$ optical power, a conversion efficiency of 0.1%. Larger LEDs do, of course, exist. The example given represents only what is very easily achievable, and readily obtainable commercially.

(iii) *The semiconductor laser diode (SLD)*

The basic principles on which the semiconductor laser diode is based are very similar to those just discussed for the LED. Forward-carrier injection is again used. However, there are also important differences.

When a $p-n$ junction is created from the two types of extrinsic semiconductor material, the initial condition, with one Fermi level (n-type) lying above the other (p-type), corresponds quite closely to the inverted population which we know is required for optical amplification and, with suitable feedback, lasing action. Obviously, this state is a non-equilibrium one and does not last long (perhaps a few nanoseconds) but we have also learned that it is possible to maintain one Fermi level above the other by applying a forward-bias voltage across the junction. In this case we are, effectively, 'pumping' with the external source of current to maintain an inverted population, of the n-states of the conduction band over the p-states in the valence band, in the volume of material around the junction. The electron energy transitions from the conduction and to the valence band either can occur spontaneously, in which case we have an LED, or can be stimulated to emit by incoming photons of the same frequency E_g/h in which case we have a semiconductor optical amplifier. Applying positive feedback to this latter arrangement can be expected, under the correct circumstances, to produce a semiconductor laser. What are these correct circumstances?

First, it will be necessary to produce a large population inversion in order to provide a high gain (equation 6.14), since the losses in a solid material are expected to be quite high, owing to scattering by the dense medium, and other factors. In order to achieve this we should ensure that the dopant levels are high, so that there are large concentrations of electrons and holes in the depletion region. In fact, for SLD devices, dopant levels (e.g., excesses of Ga or As atoms above 'stoichiometric' values) up

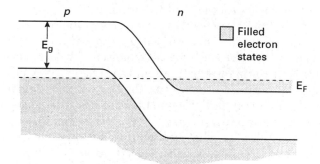

(a) Energy level diagram for an unbiased 'degenerate' p-n junction

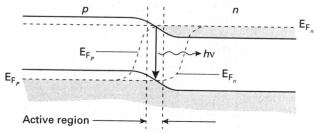

(b) Forward-biased degenerate p-n junction leading to stimulated emission

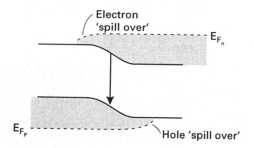

(c) Excess of forward bias, leading to lossy 'spill over'

Fig. 7.8 The degenerate $p - n$ junction.

to ~ 0.01% (equivalent to ~ 10^{24} atoms m^{-3}) are used to give rise to what are known as 'degenerate' semiconductor donor and acceptor levels. So great, in fact, are the donor and acceptor densities that very many of the available levels at the bottom of the conduction band are filled with electrons, and very many of those at the top of valence band are occupied by holes. The consequence of this is that the two Fermi levels can move into the conduction band (*n*-type) and the valence band (*p*-type) respectively, since the Fermi level refers to that energy level whose states are just 50% occupied. So great, in other words, are the extra quantities of charge carriers that we are approaching the condition which applies to metallic conductors (i.e., Fermi level halfway up the conduction band). In thermal equilibrium the energy level diagram for the junction will take the form shown in Fig. 7.8(a). If, now, the junction is forward-biased with a voltage of the order of the band-gap energy itself, the diagram changes to that of Fig. 7.8(b). In this case what has happened is that the region in which there is an excess of conduction-band electrons in the n-type material, and the region in which there is an excess of valence-band holes in the *p*-type material have moved closer to the point of overlap, so that the electrons can fall into the holes and create photons of frequency E_g/h; and they will do this especially readily if a direct-band-gap semiconductor, such as GaAs, is being used.

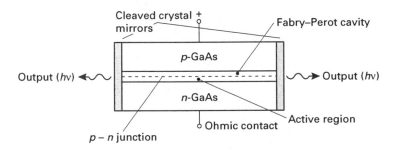

Fig. 7.9 Homojunction GaAs injection laser design.

As the large numbers of electrons and the holes combine they create correspondingly large numbers of photons, and the electrons and holes are replaced constantly by drawing current from the biasing-voltage source (just as for the LED), so light is generated continuously. We have created an effective population inversion in the semiconductor, the inversion being 'pumped' by the injected current.

In order to get the forward-biased $p-n$ junction to lase it is necessary, as we know from section 6.2.4, to provide positive feedback. This is done by arranging the geometry of the junction as shown in Fig. 7.9 and polishing two opposite crystal faces (facets) of the material so that they act as laser mirrors, via the Fresnel reflection coefficient. Provided that the optical gain exceeds the loss, the device will lase by emitting radiation, at the frequency corresponding to the band gap, in a direction normal to the polished facets.

Clearly, the population inversion, and thus the optical gain, can be increased by increasing the bias voltage. This increases the drift current with respect to the diffusion current and thus more electrons combine with holes, leading to more photons. Eq. (6.14b) tells us when the system will lase. If the reflectivities of the two facet mirrors are each equal to R, then lasing occurs when

$$R\exp(\gamma - \alpha)l \geq 1 \qquad (7.1)$$

where γ is the gain, α the loss and l the distance between the mirrors. So we have to increase the voltage until lasing occurs. The current density which flows at the onset of lasing is known as the 'threshold' current density and, for the type of geometry just described, is quite high, at $\sim 400\text{A mm}^{-2}$. This is due to the fact that the losses are quite high. What are these losses? First, there are of course the losses at the reflecting surfaces, represented by R in equation (7.1). These are moderately high since we are relying on the Fresnel reflection (equation (2.13a)) which results from the difference in refractive index between GaAs and air (at the lasing frequency). This is of the order of 3.35, leading to a value for $[(n_1 - n_2)/(n_1 + n_2)]^2$ of 0.292. This can be increased by using coatings of various forms or by using Bragg reflections (of which more later (Chapter 10)) but equation (7.1) tells us that we shall, in any case, achieve no more than a linear reduction in the threshold current with R, whereas, for a major improvement, we would be better advised to concentrate on the exponential term. It turns out that γ and α are not independent, for, if the bias voltage (and thus γ) is increased, the Fermi levels on each side rise and fall, respectively (Fig. 7.8(c)) to the point where majority carriers on each side can 'spill over' into empty levels in the bulk material, and thus diffuse away from the 'junction' transition region to be lost from the photon-generating process.

Another important source of loss lies in the fact that the difference in refractive index between the junction region and the rest of the semiconductor is quite small. Consequently, the light bouncing back and forth between the facet mirrors within the junction region (Fig. 7.9) is not well confined, laterally, to this region, and when propagating outside the region will simply be absorbed by exciting some valence band electrons up to conduction band levels, a process known as 'free-carrier absorption'.

All of these losses mean that large injection currents are required to overcome them, so the threshold current for lasing is, as has been noted, quite high. The high value of threshold current means, of course, that the device consumes a lot of power, and thus will suffer a potentially ruinous rise in temperature unless operated only in short pulses. This is a severe limitation: we need continuous sources of optical power in addition to pulsed sources. How can the threshold current be lowered?

We have understood the sources of loss in the junction device: this usually means that we should be able to see a way to reduce them. Proper understanding is our most powerful weapon in physics and in optoelectronic design.

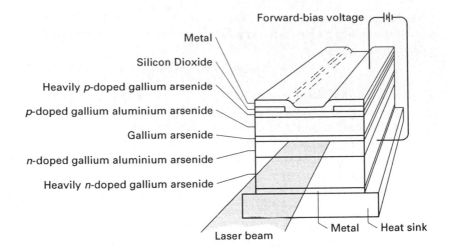

Fig. 7.10 Design for a GaAs heterojunction laser.

Consider, then, the structure shown in Fig. 7.10. Here we have a multi-layered structure known as a heterojunction laser, to distinguish it from the type of device we have been considering up to now, which is called a homojunction laser.

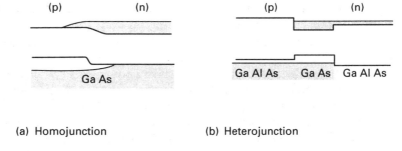

(a) Homojunction   (b) Heterojunction

Fig. 7.11 Carrier distributions in types of $p - n$ junction.

In the heterojunction laser each layer has a well-defined function (Fig. 7.10). The silicon dioxide simply isolates the semiconductor from the electrical contacts; the heavily p/n-doped layers on each side act as 'ohmic' semiconductors which interface crystallographically with the 'junction' p and n degenerate GaAlAs layers which themselves sandwich an all-important thin layer of GaAs. This layer is only ~ 200 nm thick and, since it has a higher refractive index than the surrounding GaAlAs layer, acts to confine the optical radiation within itself. One of the loss problems is thus alleviated. But the thin GaAs layer performs another important function. A study of Fig. 7.11 illustrates this. Since GaAs has a smaller energy gap than GaAlAs, the majority carriers can no longer spill over into the material on the other side of the gap. Thus loss by carrier diffusion is all but eliminated. Such a design provides laser threshold currents $\sim 10\,\mathrm{A\,mm^{-2}}$, an improvement more than an order of magnitude over the homojunction, and it allows CW operation to be achieved. The heterojunction semiconductor laser was an important breakthrough in optoelectronics for it provided a source of light, pulsed or continuous, which was rugged, compact, coherent, intense, monochromatic, operated with a low voltage, and was easily mass produced (and therefore cheap). Consumer optoelectronics really came of age with this device's availability (which is why it has been considered in some detail).

The GaAs version of the device provided laser light at 840 nm wavelength, but band-gap engineering (see section 6.3.5) allows a range of wavelengths including those suitable for optical-fibre communications. This is a very convenient, versatile and commercially important source of light. It is used, for example, in CD players (see Chapter 10), in supermarket bar-code readers and in many display functions. This most important light source is where we shall finish the discussion of sources of light and move on to consider how it is possible to impress information on the light which they provide.

7.3 OPTICAL MODULATORS

Thus far in this chapter we have been studying optical sources. It has been noted several times in the preceding text that the usefulness of optical sources in practical optoelectronics relies upon our ability to impress information upon, and subsequently to extract that information from, the light which they produce. The device which performs this function is the optical modulator.

Our next task then is to deal with the most important types of optical modulator. Fortunately, we have already covered almost all of the physical ideas which are needed for an understanding of these.

A purely sinusoidal optical wave carries no information at all for it extends over all space and all time. If it suddenly ceases to exist (either spatially or temporally), then we have some information conveyed (the time or position of the cessation) but it then ceases also to be a pure sine wave, since a truncation must introduce other frequencies (Fourier!). Clearly, then, there must be a relationship between coherence and modulation (which we shall not explore in detail).

In order to convey information by means of an optical signal we must modify a characterizing parameter of the wave in a way which allows a definite, deterministic relationship to exist between the modification and the information. Thus we might modify at the transmitter end the amplitude, intensity, frequency, phase, polarization state, direction or coherence of the wave. Which one of these we choose to modify depends very much on the nature of the information, the nature of the source, the nature of the environment and how much money we are allowed to spend! The criteria are complex and very applications dependent.

Clearly, the first consideration must be the extent to which the conveyed information can be corrupted by the environment: it may be, for example, that we wish to communicate through cloud, or dust, or a particular gas/liquid. In such circumstances environmental amplitude perturbations will, perhaps, be stronger than phase perturbations, so we would not chose amplitude modulation: the received signal probably would be too noisy. Again, it is necessary to extract the information at the receiving end, and this must be done as efficiently and as cheaply as possible. Hence we must choose a modulation scheme which allows this. Amplitude (power) detection is (as we shall see) the easiest and cheapest, but if this proves too noisy, we turn perhaps to phase or polarization detection with its increased complexity and cost. System design is the art of optimization and compromise. Clearly a range of modulators is needed to give us the flexibility to effect these optimizations. In the next few sections the most important and prevalent of these are described.

7.3.1 The electro-optic modulator

The electro-optic effect was introduced briefly in section 3.9(a). It is necessary now to investigate it in more detail in order to understand its special usefulness in electro-optic modulators.

When an electric field is applied to a medium the effect is to induce a linear birefringence, so that there will be two linear, orthogonal optical polarization states which propagate without change of form, but at different velocities. If the medium is naturally isotropic (i.e., non-crystalline or symmetrically crystalline), then these two eigenstates lie respectively parallel and orthogonal to the electric field direction. In a crystalline material the eigenstate polarization directions will depend on the particular symmetry possessed by the crystal structure.

Two different velocities in a material are, of course, characterized by a difference in refractive index, Δn. The two most important electro-optic effects are distinguished by the dependence of Δn on the electric field in each of their cases: in one case the dependence is linear, in the other, quadratic, i.e.,

$$\Delta n = PE \quad \text{Pockels effect} \tag{7.2a}$$

$$\Delta n = KE^2 \quad \text{Kerr effect} \tag{7.2b}$$

where E is the applied electric field and P and K are the Pockels and Kerr coefficients, respectively. (K is often replaced by λB in the literature, where λ is the wavelength of the light and B is then called the Kerr coefficient;

the justification for this seems to be that a phase change over a distance l can then be written $(2\pi/\lambda)\lambda B E^2 l$ and thus become independent of λ, to first order (B still depends on λ, however, to some extent, depending on the dispersion of the medium at the wavelength used)).

How can we use this modification of Δn to design a modulator?

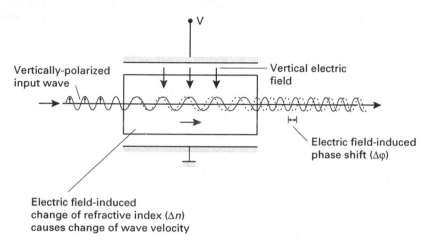

Fig. 7.12 Phase modulation via the electro-optic effect.

In the first place it is clear that a phase modulation is very easy. Fig. 7.12 shows how it can be done. Linearly polarized light is launched into a material to which a transverse electric field is applied. The polarization direction lies parallel to one of the birefringence eigenaxes produced by the applied electric field. The variation in optical phase caused by the field is given by:

$$\Delta\varphi = \frac{2\pi}{\lambda}\Delta n\, l$$

where l is the length of the optical path in the material and λ is the wavelength of the light.

Hence, from equation (7.2a) we see that the Pockels effect will then provide

$$\Delta\varphi = \frac{2\pi}{\lambda}lPE$$

and if E is due to a voltage V being applied across a width d of material

$$\Delta\varphi = \frac{2\pi}{\lambda}\frac{l}{d}PV$$

and the phase change is proportional to the applied voltage. It is, of course, quite common for the information we wish to transmit to be in the form of a voltage waveform.

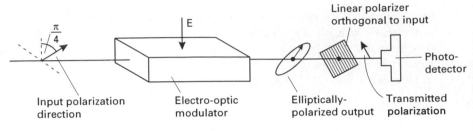

Linear polarizer
orthogonal to input

E

Photo-
detector

Input polarization
direction

Electro-optic
modulator

Elliptically-
polarized output

Transmitted
polarization

(a) Electro-optic amplitude modulation

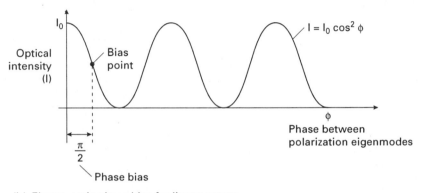

I_0

Optical
intensity
(I)

Bias
point

$I = I_0 \cos^2 \phi$

ϕ

Phase between
polarization eigenmodes

$\frac{\pi}{2}$

Phase bias

(b) Electro-optic phase bias for linear output

Fig. 7.13 Linear electro-optic amplitude modulation.

It is also possible, however, to use the electro-optic effect to modulate the amplitude of the wave. To see this, consider Fig. 7.13(a). Here, the linearly polarized wave is launched at 45° to the eigenaxes produced by the applied field. The result now is that a phase difference is inserted between the two polarization components corresponding to the directions of the induced birefringence eigenaxes, so that the linearly polarized wave now emerges elliptically (in general) polarized. The light has been polarization modulated!

Suppose that a polarization analyser now is placed so that the emerging light has to pass through it before falling on a photodetector. Suppose also that the acceptance direction of this analyser is placed at 90° to the original, input, polarization direction so that, in the absence of an applied field, no light passes through the analyser to the detector. On application of the electric field the polarization becomes elliptical, and some light must now pass through the analyser, since there will be a component in the acceptance direction (a true ellipse has a non-zero amplitude in any given direction). Let us give proper quantification to this important case. We suppose, for simplicity, that the electro-optic crystal material which is used has no natural birefringence. The electric field induces birefringence axes in the Ox, Oy directions, while the light propagates in the Oz direction. The polarization direction of the input linearly polarized light lies at $45°$ to these axes, so the components in each of the two eigenaxis directions may be written as:

$$e_x = \frac{e_0}{\sqrt{2}} \cos \omega t$$

$$e_y = \frac{e_0}{\sqrt{2}} \cos \omega t$$

where $e_0 \cos \omega t$ describes the input wave. On emergence from the crystal to which an electric field E is applied the waves may be written as:

$$e'_x = \frac{e_0}{\sqrt{2}} \cos(\omega t + \delta + \varphi) \qquad (7.3a)$$

$$e'_y = \frac{e_0}{\sqrt{2}} \cos(\omega t + \delta) \qquad (7.3b)$$

where δ is the phase delay introduced by the crystal in the absence of a field (and is thus the same for both axes) and φ is given by:

$$\varphi = \frac{2\pi}{\lambda} \Delta n \, l$$

with

$$\Delta n = K E^2$$

or

$$\Delta n = P E$$

depending on whether the Kerr or Pockels effect is in action.

The polarization analyser at $90°$ to the original, input, polarization direction will pass (see Fig. 7.13(a)):

$$e_p = \frac{e'_x}{\sqrt{2}} - \frac{e'_y}{\sqrt{2}} = \frac{e_0}{2} \cos(\omega t + \delta + \varphi) - \frac{e_0}{2} \cos(\omega t + \delta)$$

from equations (7.3). Hence, using elementary trigonometry:

$$e_p = e_0 \sin \tfrac{1}{2}\varphi \sin(\omega t + \delta + \tfrac{1}{2}\varphi)$$

and thus the *intensity* of the light striking the detector will be given by:

$$I \sim |e_p|^2 = e_0^2 \sin^2 \tfrac{1}{2}\varphi$$

Now with φ proportional to E (Pockels) or E^2 (Kerr), this is not a very satisfactory relationship between I and E, for the term renders it very non-linear and I would not faithfully represent a voltage signal applied to the modulator. What is needed is a direct proportionality between I and E. This can be arranged! Suppose that a bias phase delay of $\pi/2$ is inserted between the polarization components before they reach the analyser. In this case the light, in the absence of an applied electric field, reaches the analyser circularly polarized. Then, when an electric field is applied:

$$I \sim e_0^2 \sin^2 \left(\tfrac{1}{2}\pi + \tfrac{1}{2}\varphi \right) = e_0^2 \cos^2 \tfrac{1}{2}\varphi$$

Hence

$$I \sim \tfrac{1}{2}e_0^2 (1 - \sin \varphi)$$

and if φ is small,

$$I \sim \tfrac{1}{2}e_0^2 (1 - \varphi)$$

Now the photodetector 'sees' a steady light intensity proportional to $\tfrac{1}{2}e_0^2$, but any *changes* in I are due to φ and thus to E or E^2, i.e.,

$$\Delta I = \tfrac{1}{2}e_0^2\varphi$$

What we have done is to choose a bias point which lies on a region of the $\cos^2$ variation which is linear, to first order (Fig. 7.13(b)).

For the Pockels effect, ΔI is now proportional to E and thus to voltage, as before. Clearly the changes in optical power received by the detector are just ΔI multiplied by the effective area of the detector.

For the Kerr effect these changes are proportional to the square of the field (and thus the voltage). However, we can use the same kind of biasing ploy to correct this (if necessary). This time the bias is applied to the external electric field. If a steady bias field E_0 is applied we have

$$\varphi = \frac{2\pi}{\lambda}lK(E_0 + E)^2 = \chi(E_0 + E)^2$$

say, where E is the signal (modulating) field. Then

$$\varphi = \chi(E_0^2 + 2E_0E + E^2)$$

and, if $E_0 \gg E$,

$$\varphi \approx \chi E_0^2 + 2\chi E_0 E$$

Hence any changes in φ are now proportional to E, as required. This biasing technique is a useful general principle for regularizing inconvenient square-law relationships for practical purposes, and should be kept in mind.

Thus, by biasing the phase delay for the Pockels effect material and both the phase delay and the electric field for the Kerr effect material, it is possible to devise an electro-optic modulator which provides an optical intensity proportional to the applied voltage. Which type of material should be chosen?

By and large, it is most convenient to use materials which have a strong
linear (Pockels) electro-optic effect. Examples of such materials are am-
monium di-hydrogen phosphate (ADP), potassium dihydrogen phosphate
(KDP), lithium niobate (LiNbO$_3$), lithium tantalate (LiTaO$_3$), zinc sul-
phide (ZnS) and gallium arsenide (GaAs). Satisfactory responses can be
achieved for only a few (1-10) applied volts. However, as was pointed out
in section 3.9, only non-centro-symmetric crystals exhibit a Pockels effect
while *all* materials exhibit a Kerr effect, so, in practice, the latter is often
more accessible. Furthermore, the Kerr effect in amorphous materials, such
as fused silica, is usually very fast, the response of the material to the external field taking place in only a few femtoseconds ($\sim 10^{-15}$ s). It follows
that very fast switches can be devised with the aid of the Kerr effect. In
the case of a switch, of course, there is no necessity for the intensity to be
linear with applied field. The requirement is simply that the field causes a
phase delay of π, so that the linear polarization direction is rotated through
$\frac{1}{2}\pi$ and is then rejected by the analyser (Fig. 7.14), giving 'on' and 'off'
states. (The voltage required to do this for a given modulator is known as
the half-wave voltage, $V_{\lambda/2}$.)

Finally, let us look at the units for the electro-optic coefficients P and
K. Equations (7.2) define the quantities and, since Δn is dimensionless,
it is clear that P has dimensions of m V^{-1} and K of m^{2}V^{-2}. Lithium
niobate has a Pockels coefficient of 3.26×10^{-11} m V^{-1}; silica glasses have
Kerr coefficients in the range 2×10^{-22} to 9×10^{-22} m^2 V^{-2}.

As an example, suppose that the requirement is for a very fast two-state
optical switch for light of wavelength 633 nm using the Kerr effect in silica.
We have available a cubic block of silica with side 10 mm. What voltage is
required?

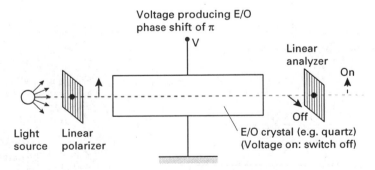

Fig. 7.14 A schematic two-state electro-optic switch.

We have

$$\Delta\varphi = \frac{2\pi}{\lambda}\Delta n\, l$$

and

$$\Delta n = K E^2 = K \frac{V^2}{l^2}$$

where V is the voltage and l the side of the cube. For $\Delta\varphi = \pi$ and $K = 5 \times 10^{-22} \, \text{m}^2 \text{V}^{-2}$ we find $V = 2.52 \, \text{MV}!$. Speed is not without its problems!

7.3.2 The magneto-optic modulator

In section 3.9 the Faraday magneto-optic effect was described. From what was learned in section 7.3.1 it should be clear how this can be used for light modulation. Fig. 7.15 shows how it can actually be done. Linearly polarized light enters a magneto-optic medium and emerges to pass first through a linear polarization analyser and thence on to a photodetector. Suppose that the acceptance direction of the analyser lies at angle ϑ to the polarization direction of the emerging light. Then the optical electric field amplitude passed by the analyser will be

$$e_p = e_0 \cos \vartheta \cos \omega t$$

and hence the light intensity striking the photodetector will be

$$I \sim |e_p|^2 = e_0^2 \cos^2 \vartheta \tag{7.4}$$

Suppose now that a uniform longitudinal magnetic field, H, is applied to the magneto-optic medium. This will induce a rotation, ρ, of the polarization direction, where:

$$\rho = V H l \tag{7.5}$$

where V is the Verdet (magneto-optic) constant and l is the length of the optical path in the medium.

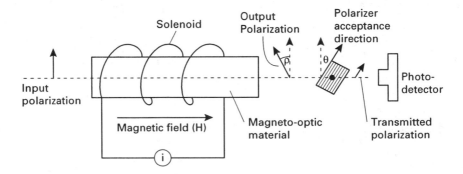

Fig. 7.15 Magneto-optic amplitude modulation.

Equation (7.4) now becomes:

$$I \sim e_0^2 \cos^2(\vartheta + \rho)$$

Taking our cue from the electro-optic treatment we set $\vartheta = \frac{1}{4}\pi$ (acceptance direction of analyser at 45° to the emerging light's polarization direction in the absence of a magnetic field) so that

$$I \sim e_0^2 \cos^2\left(\frac{\pi}{4} + \rho\right) = \frac{1}{2}e_0^2\left[1 + \cos\left(\frac{1}{2}\pi + 2\rho\right)\right]$$

hence:

$$I \sim \frac{1}{2}e_0^2 - \frac{1}{2}e_0^2 \sin 2\rho$$

and, again, if 2ρ is small (compared with $\frac{1}{2}\pi$),

$$\Delta I \sim 2\rho \tag{7.6}$$

Hence changes in light power at the detector are now proportional to ρ and thus (via equation (7.5)) to H. Now H is produced by electric current, so, for this modulator we require the signal information to be in the form of a current rather than a voltage. This is, of course, quite easy to arrange. However, the value of the Verdet constant is usually quite small and large values of current are needed to provide the magnetic fields required. The alternative, of course, is to use solenoids with a large number of turns, but this implies large electrical inductance, which, in turn, means sluggish response, and thus very limited frequency response. For these reasons the magneto-optic effect is not generally as convenient for use in optical modulation as the electro-optic effect. It is more useful for measurement of magnetic fields or of electric currents (see Chapter 10).

7.3.3 The acousto-optic modulator

The final modulator type which will be considered relies upon an effect which has not yet been mentioned: the acousto-optic effect.

When a longitudinal acoustic (i.e., sound) wave passes through a medium it does so by creating a series of compressions and rarefactions in the direction of propagation. Since these will vary the density of the medium they also will vary its refractive index because, from the discussions in section 4.2, it is clear that the refractive index is proportional to the number of atoms per unit volume. Hence the refractive index of the medium varies in sympathy with the amplitude of the acoustic wave.

Now when a light wave passes through a material medium it will of course suffer a phase change which depends upon the path length and the refractive index, so if a plane optical wave passes through a medium in a direction at right angles to the direction of propagation of an acoustic wave

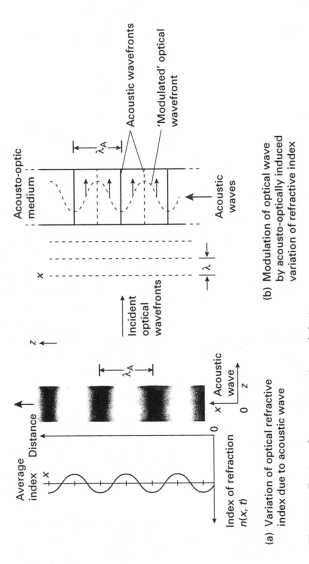

Fig. 7.16 Basics of acousto-optic modulation.

(a) Variation of optical refractive index due to acoustic wave

(b) Modulation of optical wave by acousto-optically induced variation of refractive index

the effect is to produce a spatial modulation of the optical phase across the wavefront aperture, again in sympathy with the acoustic wave. The effect of this is to give rise to maxima in the perturbed wave in directions other than forward and hence to a pattern of light which depends upon the amplitude and frequency of the acoustic and optical waves. The acoustic wave has produced a sinusoidal (it normally will be a sinusoid) 'phase plate' which then acts effectively as a diffraction grating. Since light can be deflected from the forward direction by the acoustic wave and since the angle of deflection and amplitude of deflected light can be controlled by controlling the characteristics of the acoustic wave, we have a means by which the light can be effectively modulated. Clearly, the exact nature of the deflection process will depend on the structure of the refractive index perturbation, and we are about to study this.

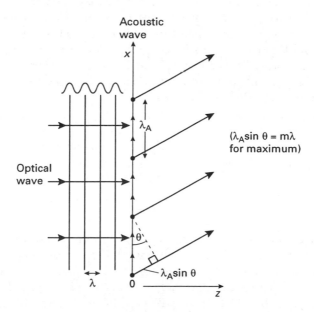

Fig. 7.17 Condition for constructive interference in Raman-Nath acoustic scattering.

There are two cases, or 'regimes', to consider. The first is where the acousto-optic medium is 'thin', so that the distance over which the light wave interacts with the acoustic wave is small. In this case the medium looks, to the optical wave, like a two-dimensional diffraction phase grating, with the phase of the optical wave emerging from the plate depending only on its point of entry. This is known as the 'Raman-Nath' regime. In the second case the medium is 'thick' and the light suffers continuous redistribution as it propagates through it. It is now essentially a three-dimensional diffraction phase grating. This is known as the 'Bragg' regime. We shall deal with each of these in turn.

(i) *The Raman-Nath regime*

Consider the arrangement shown in Fig. 7.16. The optical wave enters a medium through which a sinusoidal acoustic wave is propagating in a direction normal to the optical wave. Now if the medium has a thickness less than, say, a few hundred acoustic wavelengths (it is then called a 'thin' sheet in this context), the slab of the medium appears to the optical wave as a thin diffracting aperture, the aperture diffraction function being, of course, a sinusoidal phase function, caused by the acoustic wave. A diffraction grating which consists of a sinusoidal phase variation is surprisingly difficult to analyse exactly: it involves Bessel functions, since it must be expressed in the form of a sine (or cosine) function of a sine (or cosine) function and these will be considered a little later on. However, it is quite easy to derive the primary behavioural features by using our physical instincts. Fig. 7.17 allows us to exercise these. It is clear that if we take a set of points along the aperture spaced at a distance λ_A from each other (where λ_A is the acoustic wavelength), then the optical wave has the same phase at each of them.

The condition that rays from these points at angle ϑ to the normal to the grating (Fig. 7.17) will interfere constructively is then given by:

$$\frac{2\pi}{\lambda}\lambda_A \sin\vartheta = 2\pi m \qquad (7.7)$$

where λ is the optical wavelength and m is an integer, positive or negative.

For any other set of points, the phase of the refracted wave will vary from point to point along the aperture direction and thus, when averaged over the aperture, will interfere to give close-to-zero amplitude. Hence the directions, ϑ_m, for which there will be diffraction maxima will be given from (7.7) by:

$$\sin\vartheta_m = \frac{m\lambda}{\lambda_A} \qquad (7.8)$$

Clearly, this includes the case where $m = 0$, i.e., where some of the light is undeflected and thus comprises the 'straight through' component.

These arguments do not, of course, allow us to determine how much light is deflected into the various orders; for that information, a proper analysis needs to be done, and this follows shortly. Before doing this, there is another important aspect of this diffraction process which must be noted:

The acoustic wave is propagating through the medium, so that the acoustic grating is actually moving. Now when a wave is deflected by a moving surface a Doppler frequency shift takes place of an amount given by:

$$\Delta\omega = \frac{\omega v_A}{c}$$

where v_A is now the velocity with which the set of points is moving in the direction of propagation of the acoustic wave (and is thus equal to the acoustic velocity), and c is the velocity of light in the medium. Now, for light diffracted at angle ϑ_m, the component of the acoustic velocity in the direction of propagation of the light is $v_A \sin\vartheta_m$ (Fig. 7.18).

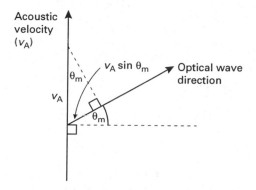

Fig. 7.18 Velocity resolution for Doppler shift.

Hence, with ϑ_m small, and using (7.8) we have:

$$\Delta\omega = \frac{\omega}{c} v_A \sin\vartheta_m = \frac{\omega}{c} v_A \frac{m\lambda}{\lambda_A}$$

i.e.,

$$\Delta\omega = m\Omega$$

where Ω is the acoustic angular frequency. Thus the shift in frequency is the order of the diffraction times the acoustic frequency. Note also that for negative deflections (i.e., ϑ_m is negative, m is negative) the frequency is reduced, and for positive deflections ('upwards') it is increased. Hence we have frequency modulation in addition to deflection modulation. The greater the angle of deflection, the greater is the frequency shift.

All of this can be properly quantified by considering the mathematics.

Suppose that the acoustic wave propagates in the Ox direction through the medium (Fig. 7.16) and as a result it perturbs the refractive index which is given by

$$n(x,t) = n_0 + \Delta n \sin(\Omega t - Kx) \tag{7.9}$$

where n_0 is the refractive index of the undisturbed medium, Δn is the refractive index perturbation which is proportional to the amplitude of the acoustic wave of angular frequency Ω and wavenumber K.

Now an optical wave, of wavelength l, travelling through the medium at right angles to the acoustic wave will suffer a phase perturbation according to:

$$\Delta\varphi = \frac{2\pi}{\lambda}n(x,t)l = \varphi_0 + \varphi_0 \sin(\Omega t - Kx)$$

say, where l is the optical path length in the medium and where, from equation (7.9):

$$\varphi_0 = \frac{2\pi}{\lambda}n_0 l$$

and

$$\varphi_0 = \frac{2\pi}{\lambda}\Delta n \, l \qquad (7.10)$$

It is φ_1 which, of course, is now proportional to the amplitude of the acoustic wave.

Suppose that the optical wave is written, in exponential form, on entering the medium as:

$$E = E_0 \exp\left[i(\omega t - kz)\right]$$

travelling, of course, in the direction Oz. On emerging from the medium with its phase perturbed in the Ox direction it becomes:

$$E = E_0 \exp\left\{i\left[\omega t - kz - \varphi_0 - \varphi_1 \sin(\Omega t - Kx)\right]\right\}$$
$$= E_0 \exp\left[i(\omega t - kz - \varphi_0)\right]\exp[-i\varphi_1 \sin(\Omega t - Kx)]$$

It is the second factor here which gives the sine (or cosine) of a sine (or cosine) and hence involves Bessel functions. In fact, we may use a well-known mathematical identity to expand E now into the following factors (see, for example, [1]):

$$E = E_0 J_0(\varphi_1)\exp[i(\omega t - kz - \varphi_0)]$$
$$- E_0 J_1(\varphi_1)[\exp i(\omega + \Omega)t - kz - Kx - \varphi_0]$$
$$- \exp[i(\omega - \Omega)t - kz + Kx - \varphi_0] \qquad (7.11)$$
$$+ E_0 J_2(\varphi_1)[\exp i(\omega + 2\Omega)t - kz - 2Kx - \varphi_0]$$
$$+ \exp[i(\omega - 2\Omega)t - kz + 2Kx - \varphi_0]$$

Here the $J_m(\varphi_1)$ are Bessel functions with argument φ_1, the phase perturbation, and they take the form shown in Fig. 7.19. Thus it is the $J_m(\varphi_1)$ factors which determine how much light is deflected into the mth order and, clearly, from the form of the $J_m(\varphi_1)$, the light amplitude does not always increase with increasing φ_1. Now φ_1 usually is small for the Raman-Nath

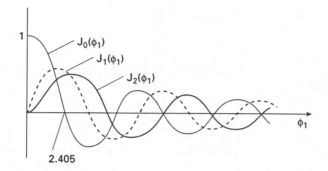

Fig. 7.19 Bessel functions (of the first kind): the first three orders.

regime provided that the acoustic powers are not too high, since the 'thin' medium leads to relatively small values of l, the optical path length, in equation (7.10). Hence the deflected light amplitudes are usually quite small, most of the light remaining undeflected. In fact, for $\varphi_1 \ll \pi/2$:

$$J_m(\varphi_1) \approx \frac{1}{m!} \left(\tfrac{1}{2}\varphi_1\right)^m$$

and then

$$E = E_0 \exp[i(\omega t - kz - \varphi_0)]$$
$$- \tfrac{1}{2}E_0\varphi_1[\exp i(\omega + \Omega)t - kz - Kx - \varphi_1]$$
$$+ \tfrac{1}{2}E_0\varphi_1[\exp i(\omega - \Omega)t - kz - Kx + \varphi_1]$$

So that the fraction of the power deflected into either one of these first-order modes, compared with that which is undeflected, is given by

$$\eta = \left(\frac{\tfrac{1}{2}E_0\varphi_1}{E_0}\right)^2 = \tfrac{1}{4}\varphi_1^2$$

which, from (7.10), is given by

$$\eta = \frac{\pi^2(\Delta n)^2 l^2}{\lambda^2}$$

Hence the power in these first deflected modes is, to first order, proportional to the square of the amplitude of the perturbation caused by the acoustic wave, and thus to the power in the acoustic wave.

   The deflection angles also are small. Let us calculate the value for a practical example. Suppose that $\lambda = 633\,\text{nm}$ and $\lambda_A = 0.5\,\text{mm}$ (corresponding to an acoustic frequency in silica of $\sim 10\,\text{MHz}$), then from equation (7.8) we find $\vartheta_1 = \pm 1.7\,\text{mrad} = 0.07°$.

Finally, look again at equation (7.11). Note that the optical frequency of each order m is given by

$$\omega \pm m\Omega$$

This confirms the conclusion arrived at earlier via the physical arguments involving the Doppler effect.

Finally, in relation to equation (7.11), and the form of the Bessel functions shown in Fig. 7.19, note that for certain values of φ_1, $J_0(\varphi_1) = 0$ and for these values there will be no undeflected light. The first value of φ_1 for which this happens is 2.405 and it requires a large acoustic power.

Hence, in the Raman-Nath regime the acousto-optic effect allows simultaneous modulation of the deflection, the amplitude and the frequency of the optical wave, via control of the amplitude and frequency of the acoustic wave. It is a valuable and versatile form of optical modulation, and is widely used.

However, as we have seen, the thickness of the diffracting material must be quite small for this type of diffraction and thus the acoustic-optic interaction length is correspondingly small, and the modulation is weak. This is especially true at the higher acoustic frequencies, with smaller acoustic wavelengths, since the thickness must be very small compared with the factor $\lambda_A^2/2\pi\lambda$ [2]. For the example given earlier, with $\lambda_A = 0.5\,\mathrm{mm}$ at an acoustic frequency of $10\,\mathrm{MHz}$, and an optical wavelength of $633\,\mathrm{nm}$, we find that the thickness $t \ll 63\,\mathrm{mm}$, so that a value $\sim 5\,\mathrm{mm}$ is acceptable and practicable. However at $50\,\mathrm{MHz}$ we have $t \ll 2.5\,\mathrm{mm}$, say $\sim 0.2\,\mathrm{mm}$. This is a very fragile piece of silica and could easily be shattered by the acoustic wave. Hence for stronger modulations, especially at higher frequencies, it is convenient to turn to the case where the material thickness can be much larger.

(ii) *The Bragg regime*

The Bragg acousto-optic regime uses thicker gratings wherein the light is effectively 'lensed' or multiply scattered by the acoustic wave in the medium. The result is a much stronger scattering of the optical power and thus a much more efficient optical modulator. This time the analysis is done by considering reflections from thin layers which have different refractive indices from their adjacent layers (Fig. 7.20).

Suppose that we have a sinusoidal variation of refractive index (again caused by a propagating acoustic wave) along the Oz direction and that the refractive index change across an infinitesimally thin slab of thickness dz in the direction Oz is given by:

$$dn = \Delta n_0 \cos k_A z\, dz$$

where k_A is the acoustic wavenumber.

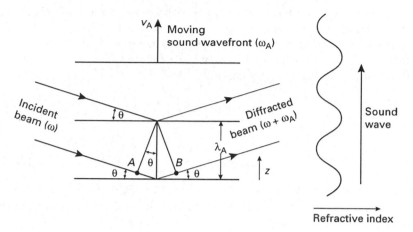

Fig. 7.20 Bragg reflections from successive layers.

Suppose also that an optical wave is propagating in the same direction (Oz) and that it interacts with this refractive index perturbation. The amplitude of the optical wave which is reflected back (along Oz) from the thin slab, at any given value of z, will be proportional to dn (Fresnel's equations!). The phase of the optical wave will be proportional to z and thus when the reflected wave gets back to $z = 0$, its phase will be proportional to $2z$ (Fig. 7.21). Hence the amplitude at $z = 0$ of the wave which has been reflected from the slab which lies between z and $z + dz$ is given by:

$$a_z = \alpha \Delta n_0 \cos k_A z \cos(\omega t - 2kz)\mathrm{d}z$$

where α is a constant and k is the optical wavenumber.

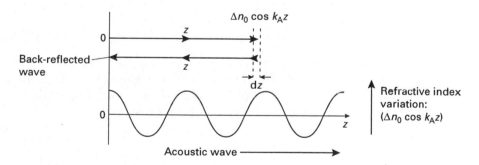

Fig. 7.21 Back-reflection from a parallel acoustic wave.

Thus, using the well-known trigonometrical identity for the product of two cosines:

$$a_z = \alpha \frac{\Delta n_0}{2}[\cos(\omega t - 2kz + k_A z) + \cos(\omega t - 2kz - k_A z)]\mathrm{d}z \qquad (7.12)$$

Now we can easily see how this behaves without performing a rather cumbersome integration. Averaged over all z, from 0 to ∞, a_z will be zero (average of a sinusoid in z is zero) unless one or more of the terms is independent of z. Since both k and k_A must be positive (a negative wavelength is physically meaningless) this can only be so if:

$$k_A = 2k \qquad (7.13a)$$

or

$$\lambda_A = \frac{\lambda}{2n} \qquad (7.13b)$$

where n is the undisturbed refractive index of the medium. In other words, the acoustic wavelength must be equal to half the optical wavelength in the medium (λ is the optical wavelength in free space).

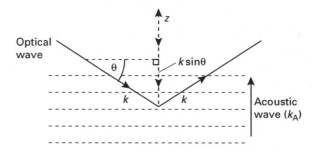

Fig. 7.22 Bragg reflection at an angle with respect to acoustic wave direction.

If equation (7.13a) holds, then the first term in (7.12) becomes independent of z and the average (over z) value of the reflected amplitude can be written

$$\langle a_z \rangle = \tfrac{1}{2}\alpha \Delta n_0 \cos \omega t \qquad (7.14)$$

This is the Bragg reflected amplitude. Suppose now, however, that the optical wave is incident upon the Bragg grating at angle ϑ (Fig. 7.22). In this case there will only be a reflection if equation (7.14) is satisfied by that component of k which lies in the Oz direction, so we have:

$$2k \sin \vartheta = k_A \qquad (7.15)$$

Using, again, $k = 2\pi n/\lambda$, where n is the (undisturbed) refractive index of the medium, and $k_A = 2\pi/\lambda_A$ where λ_A is the acoustic wavelength in the medium, equation (7.15) can be written in its more common form:

$$\sin \vartheta = \frac{\lambda}{2n\lambda_A} \qquad (7.16)$$

giving the angle at which the input light must be incident, with respect to the propagation of the sound wave in the medium, in order for deflection to occur, at the same angle. Note that there is only one order in this case, unlike Raman-Nath diffraction. In fact there will be a small range of angles around which diffraction will occur: this is due to the fact that the cross-section of the optical beam is finite, rather than infinite as was implicity assumed by the full averaging process which was considered in order to obtain equation (7.14). (The full integration of the equation between finite limits provides the sinc (ϑ) function which we have come to expect in these cases, and is left as an exercise for the reader.) Clearly, from equation (7.14), the amplitude of the deflected wave depends upon Δn_0 and thus upon the amplitude of the acoustic wave; hence the deflected beam is amplitude-modulated by the acoustic wave at the fixed angle ϑ (if the acoustic frequency is held constant). Again, analogously with the Raman-Nath case, the optical wave is being 'deflected' by a moving medium, as the acoustic wave propagates, and thus there is a Doppler frequency shift.

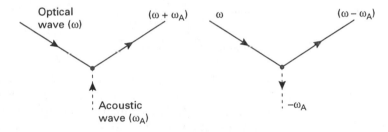

Fig. 7.23 Up-shifted and down-shifted optical waves.

Since there is only one order we have, using arguments very similar to before, that the angular frequency shift of the deflected beam is just ω_A, giving an angular optical frequency of $\omega + \omega_A$. In order to obtain $\omega - \omega_A$, it is necessary to reverse the direction of the acoustic wave with respect to the optical propagation direction (Fig. 7.23). From what has been said so far about acousto-optic deflection in the Bragg regime it might appear that the input angle must be arranged to satisfy quite accurately the Bragg condition (equation (7.16)) in relation to the acoustic wavelength before any modulation can occur. Fortunately, this is not strictly true. Provided that the angle is close to the Bragg angle, a variation in the acoustic frequency will vary the angle of deflection and, of course, the Doppler shift. It will also vary the amplitude of the deflected beam. Of course, the Bragg condition does not strictly apply in this case, but the wave vectors from each of the reflections, while not now exactly in phase (as for the Bragg condition) are close enough still to provide a sensible amplitude. (It is our old friend the sinc function, with steadily increasing phase differences between large numbers of amplitudes, yet again). Hence, provided that the violation of the Bragg condition is not large, a deflection/amplitude/frequency modulation can be achieved by varying just the frequency of the acoustic wave.

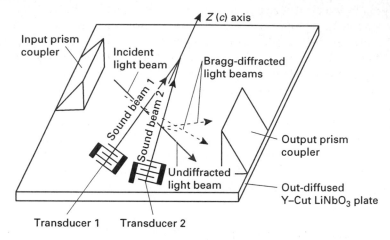

Fig. 7.24 Layout for an acousto-optic Bragg modulator.

A very important advantage of the Bragg modulator over the Raman-Nath modulator is that high acoustic frequencies can be used, since there is now no restriction on the thickness of the medium. For example, for silica at 500 MHz the acoustic wavelength is ~ 0.01 mm and with an optical wavelength of 633 nm and a refractive index of 1.5 we find from (7.16):

$$\sin \vartheta = 0.02, \quad \vartheta \sim 1.2°$$

Light at this angle is readily separated from the undeflected light, and it will be both frequency and amplitude modulated by the frequency and amplitude, respectively, of the acoustic wave. Furthermore, by switching the acoustic wave on and off, the optical wave may be switched between the two positions. The switching time will depend only on the transit time of the acoustic wave across the beam, and will be ~ 100 ns.

Thus for the Bragg case again it is possible to modulate simultaneously the amplitude, frequency and deflection of the optical wave but much more strongly than in the Raman-Nath regime. For this reason most acousto-optic modulators are of the Bragg type. A schematic arrangement for an acousto-optic Bragg frequency modulator is shown in Fig. 7.24.

The table below provides a list of materials commonly used in acousto-optic work, together with their relevant properties.

Table of properties of commonly-used acousto-optic materials (Ref. Yariv, A. (1976) *Optical electronics*, Holt Rinehart Winston)

Material	ρ (mg/m³)	v_s (km/s)	n	p	M_0
Water	1.0	1.5	1.33	0.31	1.0
Extra-dense flint glass	6.3	3.1	1.92	0.25	0.12
Fused quartz (SiO_2)	2.2	5.97	1.46	0.20	0.006
Polystyrene	1.06	2.35	1.59	0.31	0.8
KRS-5	7.4	2.11	2.60	0.21	1.6
Lithium niobate ($LiNbO_3$)	4.7	7.40	2.25	0.15	0.012
Lithium fluoride (LiF)	2.6	6.00	1.39	0.13	0.001
Rutile (TiO_2)	4.26	10.30	2.60	0.05	0.001
Sapphire (Al_2O_3)	4.0	11.00	1.76	0.17	0.001
Lead molybdate ($PbMO_4$)	6.95	3.75	2.30	0.28	0.22
Alpha iodic acid (HIO_3)	4.63	2.44	1.90	0.41	0.5
Tellurium dioxide (TeO_2) (Slow shear wave)	5.99	0.617	2.35	0.09	5.0

ρ = density; v_s = sound velocity; n = refractive index; p = photo-elastic constant; M_0 = efficiency of diffraction compared with H_2O

Having considered how to provide the light and to impress information upon it with modulators, attention must now be turned to the extraction of that information and its conversion into an electrical signal once more. We turn now to the subject of photodetection.

7.4 PHOTODETECTORS

The processes which enable light signals to be sensed and measured accurately depend directly, as we should expect, upon the interaction between photons and atoms. In most cases of quantitative measurement the processes rely on the photon to raise an atomic electron to a state where it can be measured directly as an electric current. We should also expect that there will be a variety of devices and geometries depending on the required speed of response, sensitivity, wavelength of the light, etc., and this is indeed so.

In this section we shall deal with a few of the most widely used photodetectors and also, more importantly, with those which illustrate well the principles on which all photodetectors are based.

7.4.1 Photoconductive detectors

We know that the conductivity of a semiconductor depends upon the number of charge carriers in the conduction band. A photon, when absorbed by a semiconductor atom, is capable of raising an electron from the valence band to the conduction band (provided that its energy, $h\nu$, is greater than the band-gap energy) and thus of increasing the conductivity. In doing so, an electron hole pair is created; this will eventually be annihilated, since it represents a pair of charges in excess of the thermal equilibrium value. However, while the extra charge exists, it can be used to contribute to an electric current, whose value will depend upon the light power incident upon the semiconductor. Thus, a small voltage across the photoconductive material layer will give rise to the required, measurable current in an external circuit (Fig. 7.25).

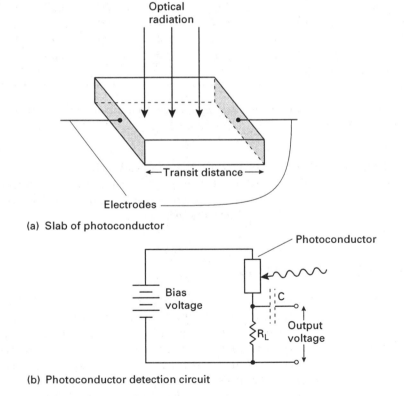

(a) Slab of photoconductor

(b) Photoconductor detection circuit

Fig. 7.25 Photoconductive detection.

From this simple arrangement it is straightforward to derive some simple, but important, relationships.

First, for the photon to yield an electron-hole pair its energy must satisfy $h\nu > E_g$, where E_g is the band-gap energy of the material. If, however, ν is too high, then all the photons will be absorbed in a thin surface layer, and the charge pairs will not be collected efficiently. Thus there is a frequency 'responsivity' spectrum for each type of photodetector which must be matched to the spectrum of the light which is to be detected.

Secondly, suppose that we are seeking to detect a light power of P watts at an optical frequency ν. This means that $P/h\nu$ photons are arriving every second. Suppose now that a fraction, η, of these produce electron hole pairs. Then there are $\eta P/h\nu$ charge carriers of each sign produced every second. If all are collected, and noting that each charge carrier only travels, on average, a distance of half the depletion width, and thus contributes, effectively, half the flow rate, the observed electric current is:

$$I = \frac{e\eta P}{h\nu} \qquad (7.17)$$

Thus the current is proportional to the optical power. This means that the electrical power is proportional to the square of the optical power. It is important, then, when specifying the signal-to-noise ratio (SNR) for a detection process, to be sure about whether the ratio is stated in terms of electrical or optical power. (This is a fairly common source of confusion in the specification of detector noise performance.)

Another very important feature is speed of response. How quickly can the above processes respond to changes in input light level? Again some simple ideas establish the basic principles. Consider the electron created in the electron-hole creation process. Each electron will continue to contribute to the current flow until it is annihilated. If the average time for recombination of an electron-hole pair in the semiconductor is τ and the transit time across the layer is τ_0, then the average number of times the electron is used in the current, clearly, is τ/τ_0. Hence the current due to the light flux is

$$i_p = \frac{e\eta P}{h\nu} \frac{\tau}{\tau_0} \qquad (7.18)$$

However, the minimum time within which the photoconductor can respond to a change in light level is τ, so the bandwidth of the measurement is given by

$$\Delta f = \frac{1}{2\pi\tau} \qquad (7.19)$$

We can define a sensitivity for the device as the current produced for a given optical power. From (7.18) this is seen to be:

$$S = \frac{i_p}{P} = \frac{e\eta\tau}{h\nu\tau_0} \qquad (7.20)$$

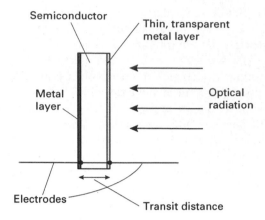

Fig. 7.26 'Side-illuminated' photoconductive detector.

This will thus increase with τ and η, but decrease with τ_0 and ν. We can increase η by increasing the width of the absorbing layer, but this will also increase τ_0 unless the geometry is changed to that shown in Fig. 7.26, known as a side-illuminated device.

The value for τ, the recombination time, will depend upon the material, but, for a large value, giving high sensitivity, we shall have to put up with a slow response, and thus a small measurement bandwidth. In fact, from (7.19) and (7.20) we note that the sensitivity-bandwidth product is independent of τ:

$$S = \frac{i_p}{P}\Delta f = \frac{e\eta}{2\pi h\nu\tau_0}$$

Hence we begin to see the kind of compromises which must be made in photodetector design.

There are two significant disadvantages possessed by photoconductive devices when compared with other types of photodetector. The first is that, with the absolute values which must be inserted in equation (7.18), high sensitivity (values of $\tau/\tau_0 \geq 10^3$, say) can only be achieved for response times ~ 50 ms. This is much too slow a response for most optoelectronic applications. The second is that the electron/hole recombination process is a random one, and thus one which introduces noise. This 'recombination noise', as it is called, is troublesome in many applications. An advantage, however, of these devices is that they can be used at long wavelengths, up to $\sim 20\,\mu$m. Since the band gap must be quite small for such low energy photons there is a problem with thermally created electron hole pairs which give rise to thermal noise. Hence such detectors usually must be cooled to, at least, liquid nitrogen temperatures (77 K).

Another advantage of these devices is that they are very cheap. For this reason they were often used in domestic cameras as CdS or CdSe cells.

Other materials which are commonly used are:

$$PbS \qquad 1-3\,\mu m \text{ wavelength}$$
$$InS \qquad 3-7\,\mu m \text{ wavelength}$$
$$HgCdTe \quad 5-14\,\mu m \text{ wavelength.}$$

7.4.2 Photodiodes

(i) *Junction photodiodes*

Consider, again, the $p-n$ junction of Fig. 7.7. When considering luminescence it was noted that the physical contact between these two types of semiconductor (i.e., p and n) led to a diffusion of majority carriers across the junction in an attempt to equalize these concentrations on either side. The result, however, was to establish an electric field across the junction, as a consequence of the charge separation. The region over which this field is established is known as the depletion region, since it must lead to a depletion of charge carriers where it is active in causing the charge separation. The $p-n$ junction comes to equilibrium with a voltage across it which will, from the discussions of Section 7.2.2(i), be given by:

$$eV_J = E_J \approx \tfrac{1}{2}E_g + \tfrac{1}{2}(E_D - E_A)$$

It will also be remembered, from previous discussions, that the equilibrium will leave the p-type material negatively charged and the n-type positively charged (Fig. 7.27). Suppose now that a photon (energy $h\nu$) is incident upon the region of semiconductor which is exposed to the junction's electric field and that, as before, the requirement $h\nu > E_g$ (the semiconductor bandgap) is satisfied. The photon, again, creates an electron-hole pair which immediately comes under the action of the electric field. The result is that the electron is swept to the positively charged n-type material and the hole to the negatively charged p-type material, thus reducing the junction voltage below its equilibrium value. This voltage reduction is easily derived. We know that, at equilibrium, the drift current due to the electric field across the junction and the diffusion current due to the concentration gradient across the junction are equal at, say, i_0. When a photon flux equivalent to an optical power, P, is incident upon the junction, the electron-hole pair which is created is acted upon by the electric field across the junction and thus increases the drift current by (7.17):

$$i_P = \frac{e\eta P}{h\nu}$$

Hence the new drift current becomes

$$i_0 + i_P$$

In order to accommodate this change, the diffusion current must also increase, to bring the system back to equilibrium. If the reduction in junction voltage brought about by the photon flux is ΔV_J then more charge carriers can cross the junction by virtue of their thermal energy kT, and the current will increase by the Boltzmann factor $\exp(e\Delta V_J/kT)$ (this being a good approximation to the Fermi-Dirac function (6.18b) on the tails of the distribution, i.e., in the respective bands). Hence the diffusion current will increase to

$$i_D = i_0 \exp\left(\frac{e\Delta V_J}{kT}\right)$$

So the new equilibrium requires that:

$$i_P + i_0 = i_0 \exp\left(\frac{e\Delta V_J}{kT}\right)$$

or

$$\Delta V_J = \frac{kT}{e} \ln\left(\frac{i_P}{i_0} + 1\right)$$

Now i_0 will be a constant for a given material, so a measurement of ΔV_J gives i_P, and thus P. However the logarithmic relationship between ΔV_J and P is non-linear, and thus this 'photovoltaic' mode of operation is not generally convenient to measure P. The mode does, nevertheless, have the advantage of simplicity (no power supply is required) and is used for some special applications.

A much more convenient arrangement is the 'photoconductive' mode, which will now be described.

We take again our $p - n$ junction and apply to it a 'reverse-bias' voltage ($\sim 10\,\text{V}$), i.e., a voltage which makes the p side negative with respect to the n side, so that charge carriers are discouraged from making the journey across the junction (Fig. 7.27(b)). This has the effect of reducing the two opposing junction currents to very low ('leakage') values and of increasing the width of the depletion layer, since majority carriers of both signs will be pulled away from the junction by the applied voltage. If, now, photons of sufficient energy are incident upon the depletion region, the electron-hole pairs created are immediately separated into their constituent charges, which are quickly drawn to the side of the junction which has opposite charge. In this case, the reverse bias has disturbed the equilibrium to the point where only very small diffusion and drift currents flow in the absence of photogenerated carriers, since the barrier between the two types of material has increased, and the depletion region has widened.

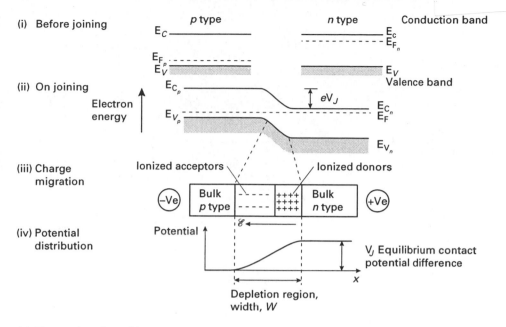

(a) The p-n junction without bias

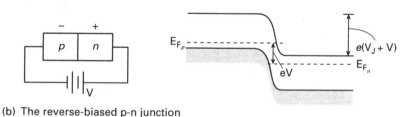

(b) The reverse-biased p-n junction

Fig. 7.27 The reverse-biased $p - n$ junction for photodetection.

Consequently, essentially the only current which flows is that due to the charges created by the incoming photons, and hence this current is directly proportional to the input optical power; this, of course, is a very convenient relationship for most types of measurement.

Another advantage of this arrangement is that the charges are quickly separated and almost immediately annihilated by the large numbers of oppositely charged carriers at the edge of the junction. This removes much of the randomness of the recombination process and leads to a much smaller 'recombination noise' than in the photovoltaic case.

The speed of response of the photoconductive photodiode depends upon the several factors but the limiting one is the time of drift of the created charges across the depletion layer (Fig. 7.28). Here, again, we meet the con-

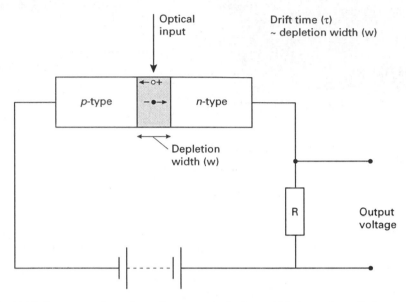

Fig. 7.28 Response time dependence on depletion width in a photodiode.

flict between sensitivity and response time. As the reverse bias increases, the drift velocity of the charges will increase until it saturates (at about $5 \times 10^4 \, \text{m s}^{-1}$ for silicon).

At this point, any increase in the bias voltage will increase the depletion region's width without any increase in drift velocity, so the response time must increase (and the bandwidth decrease). On the other hand, the larger is the depletion width the greater is the volume available to receive the incoming light and thus the greater is the number of electron pairs which can be created; hence the greater is the sensitivity. In addition, some pairs will be created outside the depletion region and, if created within a diffusion length of the region (i.e., if they are able to travel that distance, by diffusion, before recombining), the carriers will enter the field of the depletion region and then contribute to the current. This latter contribution can be seen as a fairly distinct 'tail' on the response (to a sharp pulse) of the photodiode and can be reduced by increasing the reverse bias, thereby increasing the width of the depletion region relative to the diffusion length. (Another solution is to use a 'PIN' structure, which will be described shortly).

A typical silicon photodiode structure is shown in Fig. (7.29). The $p-n$ junction is formed between the heavily-doped (p^+) p-type material and a lightly-doped n-type material, so that the depletion region extends well into the n-type region. Clearly, the p^+ region must be very thin, to allow light to penetrate well into the depletion region. The n^+ region at the bottom is to ensure a good electrical contact between the metal electrode and the n-type region.

(ii) *Performance parameters for photodiodes*

The practical performance of photoconductive photodiodes may be characterized by three features: sensitivity, speed of response (i.e., bandwidth) and noise figure. We shall deal with these in turn.

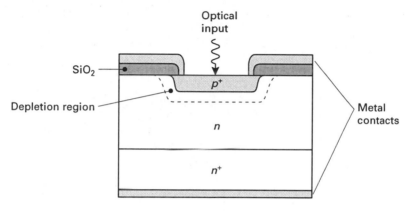

Fig. 7.29 Structure for a silicon photodiode.

The sensitivity of photodiodes normally is referred to specifically as the 'responsivity', and is defined as the photocurrent produced for a given optical power. From equation (7.17) we see that this is given by

$$R = \frac{i_p}{P} = \frac{e\eta}{h\nu}$$

with wavelength, λ, in micrometres. This can be written in the form:

$$R = \frac{\eta\lambda}{1.24} \text{ amps watt}^{-1}$$

Of course, η also will vary with wavelength, so $R(\lambda)$ will be a non-linear function of λ and will usually exhibit a broad peak around the band-gap energy. The responsivity for a typical silicon photodiode is shown in Fig. 7.30. The low value at the longer wavelengths is due to the fact that the photons have insufficient energy to create electron hole pairs; at the shorter wavelengths the light is absorbed in a thinner surface layer, so fewer pairs are created in the depletion region. A typical value for η is 0.8 at $\lambda \sim 0.85\,\mu\text{m}$ (near infrared) in silicon, so that we have for a typical value of R near to the peak value, $R \sim 0.55\,\text{A W}^{-1}$.

Moving on now to deal with speed of response and noise performance, it is necessary to consider the complete photodiode circuit. Referring to Fig. 7.31 we might ask, first, how quickly the output voltage level can respond to a change in input optical power. There are three factors which contribute to the circuit's response time: (a) the time taken by the photogenerated carriers to drift across the depletion layer, (b) the time taken by the carriers to diffuse through the respective p or n materials to the electrical contacts, and (c) the RC time constant of the associated electrical circuit.

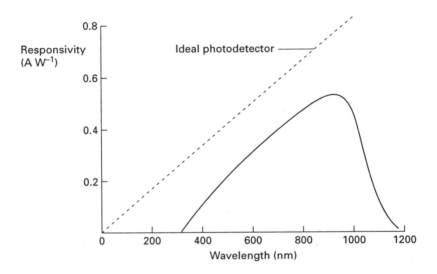

Fig. 7.30 Responsivity spectrum for a typical silicon photodiode.

Let us deal briefly with each of these in turn.

(a) The time taken for the photogenerated carriers to drift across the depletion layer obviously depends upon both their drift velocity and the width of the layer. The drift velocity can be increased by increasing the reverse bias voltage, as we have seen. (This also increases the width of the depletion layer but only as $V^{1/2}$, whereas the velocity increases as V, so that it is still advantageous to do it.) However, as mentioned in the previous section, the velocity saturates at about $5 \times 10^4 \, \mathrm{m\,s^{-1}}$ in silicon, as a result of other factors. For a typical depletion width of $\sim 2.5 \, \mu\mathrm{m}$ this value for the saturation velocity gives a response time $\sim 0.05 \, \mathrm{ns}$ and thus $\sim 20 \, \mathrm{GHz}$ bandwidth. Clearly this is a very useful value for the bandwidth.

In the case of (b), the diffusion time across the p and n materials on either side of the junction, in order to reach the electrical contacts, must also be minimized. This means that these regions must be made quite narrow, the optimum width being that which equalizes the diffusion time and the depletion layer drift time. This implies smaller thicknesses for

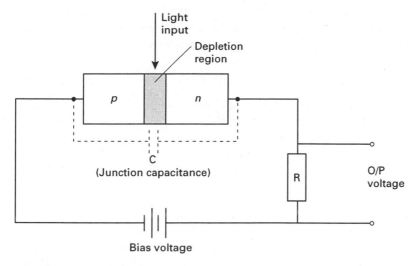

Fig. 7.31 Photodiode circuit.

these regions than the depletion width, since diffusion velocities are lower than drift velocities. Thicknesses of ~ 0.5 μm are normally used.

Finally, in the case of (c), the *RC* time constant of the associated circuit clearly has to be smaller than the required response time. The value of the resistance *R* determines the voltage produced by the photodiode current (Fig. 7.31) and therefore must be fairly large. The junction capacitance, *C*, must be kept small, but this means that the photosensitive area of the photodiode will be small, which in turn means that the sensitivity will be low because it will intercept fewer photons. We begin to see the kinds of compromise which must be made here.

A typical p-n junction capacitance is ~ 25 pF, so for a 20 GHz bandwidth, we require:

$$R < \frac{1}{2\pi fC} = 0.32\,\Omega$$

Clearly this is a very small value of resistance, meaning that the output voltage (generated by the photocurrent) will be small. A voltage amplifier will be needed in this case, and it would have to be quite a good one (large bandwidth, low noise, high gain, properly impedance-matched). A more likely value, in practice, for the load resistor is ~ 50 Ω, since this would match into standard coaxial cables (which have 50 Ω as their characteristic impedance). With this value of resistor the *RC* limitation on bandwidth is easily calculated as ~ 100 MHz. This is more than sufficient for the vast majority of applications (but these do not include high-bit-rate optical communications!).

What now of the final performance feature, the noise in photodiodes?

The source of noise primarily is due to the randomness of the photon arrivals, which was dealt with in section 4.4.4. This is called shot noise and is given by the standard shot noise expression:

$$i_n = (2eBi_D)^{1/2} \qquad (7.21)$$

where i_n is the noise current, e the electronic charge, B the circuit bandwidth and i_D the current flowing through the circuit. This latter current, i_D, will be the sum of two components, i.e.,

$$i_D = i_P + i_d$$

where i_P is the photogenerated current and i_d is the 'dark' current. This dark current is the current which flows through the circuit in the absence of any light input, and is due to the 'leakage' of thermally generated charge carriers across the junction, under the action of the reverse-bias field. This leakage current depends upon the temperature, the junction materials and the photodiode design, most notably its area: the larger the area the greater the potential for leakage.

The noise performance of the photodiode itself (as opposed to the circuit) is usually expressed in the form of the 'noise-equivalent power' (NEP), which is the optical power input which would produce a photocurrent equal to the dark current, i_d. It is clear from equations (7.17) and (7.21) that this quantity will be given by:

$$\text{NEP} = \frac{h\nu(2ei_dB)^{1/2}}{e\eta}$$

Thus the NEP is seen to vary as the square root of the dark current and of the bandwidth. It is quoted as a power 'per root hertz', in other words, for unit bandwidth. A typical value of the NEP for a silicon photodiode is $\sim 10^{-14}\,\text{W Hz}^{-1/2}$, corresponding to dark currents of $\sim 1\,\text{nA}$.

However, in order to gain a true practical appreciation of the noise performance it is again necessary to consider the complete photodetection current and to include the noise generated by the load resistor, R. The thermal (Johnson) noise power always will be $4\,kTB$, so the noise current in a resistor of value R will be

$$i_R = \left(\frac{4kTB}{R}\right)^{1/2}$$

and hence the total circuit noise current will be:

$$i_N = (2ei_dB)^{1/2} + \left(\frac{4kTB}{R}\right)^{1/2}$$

Clearly, the relative values of these two terms, for a given temperature, depend upon the values of i_d and R. For $i_d \sim 1\,\text{nA}$ the thermal noise term will dominate when $R \leq 25\,\text{M}\Omega$. We know that this normally will be the case since it is necessary to minimize R in order to achieve a respectable bandwidth.

All device and system design, in optoelectronics as in every other field, is the art of compromise. Sometimes, however, no amount of compromise will satisfy the requirements, and we must contemplate more radical measures. Three such measures will now be discussed, briefly, in respect of photodiode design.

(iii) The PIN photodiode

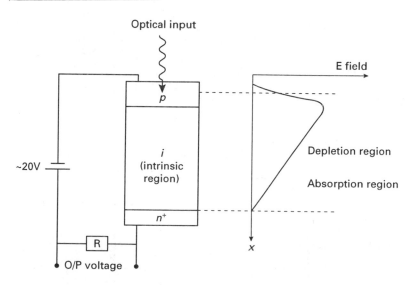

Fig. 7.32 Schematic structure for a 'pin' photodiode.

The first 'radical' measure is the PIN photodiode, as was promised earlier (section (ii)). The structure of this device is shown in Fig. 7.32. The important difference between this and the structure of the $p - n$ junction device is the existence of a large 'intrinsic' region between the two doped regions. The effect of this is to insert a semiconducting, photosensitive region within which an electric field can act to separate and mobilize the carriers from the electron hole pairs as they are produced by the incoming photons. Thus we have what is effectively a much larger depletion region, leading to a large sensitivity. With a fairly large bias voltage $(10 - 20\,\text{V})$ the drift velocity in the intrinsic region can be made high enough for a fast response, the actual response time depending on the detailed design, but response times ∼ 250 ps have been achieved. A typical PIN design is shown in Fig. 7.33(a). The design depends critically on the wavelength, which determines the penetration depth for the photons into the intrinsic region. Longer wavelength diodes (say, $> 1\,\mu$m) tend to use side illumination (Fig. 7.33(b)).

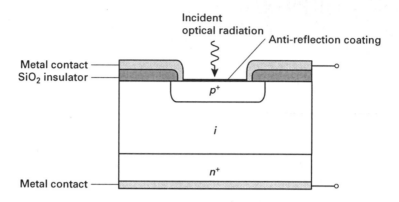

(a) Front illuminated Si pin photodiode

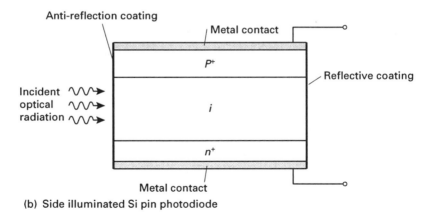

(b) Side illuminated Si pin photodiode

Fig. 7.33 'Pin' photodiode structures.

(iv) *The photomultiplier*

Photodetector diodes, like photoemitter diodes, have the advantages of compactness, ruggedness, good temperature stability and low operating voltage (10 − 20 V). However, they are of limited sensitivity and if very

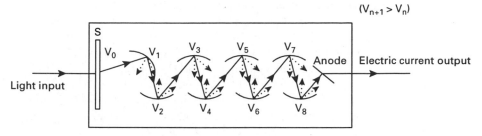

Fig. 7.34 Principle of the photomultiplier.

high sensitivities are required we might turn to a device known as a photo-multiplier. This latter device is shown schematically in Fig. 7.34. Photons impinging on a suitably chosen sensitive photocathode material, S, have sufficient energy to eject electrons from the material into a surrounding vacuum. The electrons are then accelerated to another electrode (dynode), by the electric field which exists between the two. On arriving at the second dynode they eject a larger number of secondary electrons by energy of impact. This process continues for up to about ten more stages, providing an electron multiplication factor of 10^6 to 10^9, depending on the materials and the voltages employed. Hence, quite large electrical currents can result from very small optical power inputs. Powers as low at 10^{-15} watts can be measured with these devices, the limitation being, as in all detection devices, the inherent noise level generated within the detection bandwidth. Whilst being very sensitive, photomultipliers have the disadvantages of re-quiring several kilovolts for their operation, and of using bulky vacuum tubes, which are vulnerable to damage, and suffer from ageing problems.

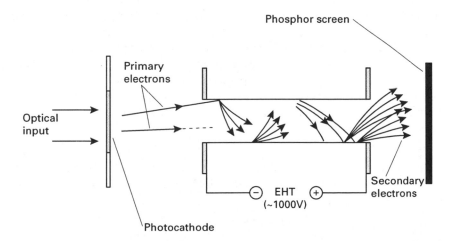

Fig. 7.35 Schematic for one channel in a microchannel plate.

A recent development of the photomultiplier idea is the microchannel plate. In this a slab of insulating material ($\sim 500\,\mu$m thick) contains a large number of small diameter ($\sim 10\,\mu$m) holes (channels) within it, normal to its faces. A high voltage ($500 - 1000$ V) is applied between its two faces (Fig. 7.35). Photons incident upon a photo-cathode material close to one of the surfaces generate electrons which enter the holes and then produce cascades of secondary electrons as they bounce along the holes, spurred on by the potential difference. The number of emerging electrons, per photon of input, can thus be very large and the device has the additional advantages of ruggedness, compactness, lower operating voltage, lower noise and increased bandwidth, since the spread of electron arrival times is reduced.

The emerging electrons in the microchannel plate device also can be directed on to a phosphor material to provide a considerably more intense version of the incoming photon stream. Such an arrangement is called an 'image intensifier' and, since the incident and output wavelengths can be different, it can be used to intensify infrared images and thus to provide 'night vision' via the infrared radiation emitted by objects, even at night, as a result of their temperature above absolute zero.

(v) The Avalanche Photodiode (APD)

A good compromise between the photodiode and the photomultiplier is the avalanche photodiode (APD). In this case a PIN junction photodiode is operated with a very high reverse bias voltage (~ 200 V). When an electron/hole pair is created by the incoming photon, the resulting charge carriers can acquire sufficient energy from the field across the junction to create further electron/hole pairs themselves, leading to an 'avalanche' multiplication process similar to that in the photomultiplier. An increase in sensitivity is thus obtained at the expense of a higher operating voltage.

The multiplication process is also quite electrically noisy, so adding to the noise level of the photo-detection. The reason for this lies in the statistically random nature of the multiplication process. If each photogenerated charge carrier produced a constant number of secondary carriers, then the only noise present would be the multiplied-up shot noise of the photogenerated current, i.e.,

$$\Delta i_s = M(2ei_P\Delta f)^{1/2}$$

where M is the multiplication factor, e the electronic charge, i_P the photogenerated current and Δf the bandwidth. Since also the final, detected, current, i_D, is given by

$$i_D = Mi_P$$

we have

$$\Delta i_s = (2eMi_D\Delta f)^{1/2}$$

However, there is excess noise associated with the multiplication process, since the number of secondaries produced by a given photogenerated carrier will vary statistically. The noise actually observed is expressed as:

$$\Delta i_s = F^{1/2}(2eMi_D\Delta f)^{1/2}$$

where F is the excess noise factor. F is a function of M but, typically, lies in the range 2-20.

On speed of response, broadly we may say that all types of photodetectors work quite readily up to optical modulation frequencies of about 1 GHz. To get beyond this figure requires special attention to materials and design, but bandwidths up to 60 GHz have been reported with carefully designed PIN photodiodes.

7.4.3 Photon counting

A valuable technique for improving the detection signal-to-noise ratio (SNR) when the signal level is very low is known as photon counting. The essence of this technique is that the arrival of a single photon at the detector will give rise to a pulse of current (in a photodiode for example) whose height lies within a known, narrow range. The randomly generated thermal noise (i_d) will be distributed over a much larger range. Thus there is an essential difference between the two types of noise, and this can be utilized.

To do this the detector feeds into a 'pulse-height discriminator' which rejects those pulses which lie outside a pre-set range. The pulses within the range are passed to a pulse counter which counts the pulses arriving in a pre-set time. Clearly, the number arriving in any given time is proportional to the optical power.

The pulse from a photodetector which receives a single photon is very small, so large intrinsic sensitivity is required. For this reason either photomultipliers or avalanche photodiodes are often used for photon counting. To improve the SNR still further the input light is often 'chopped', i.e., periodically blocked from the photodiode (or modulated at the source). This allows the photon counting system to compare the 'light-on' count rate with the 'light-off' rate. This, in turn, allows some rejection of those random signals which happen to lie within the preset range.

Clearly, the photon counting technique will only work to major advantage when the light level is low enough for there to be just one photon per count interval for most such intervals. Improvements in the SNR are typically by a factor of 2 or 3 (3 − 5 dB).

Figure 7.36 illustrates the discriminatory feature. The pulse count rate in this diagram is plotted as a function of the pulse height and it shows clearly the range over which the count rate can be seen to contain the photon contribution. The optimum pulse height acceptance range also is

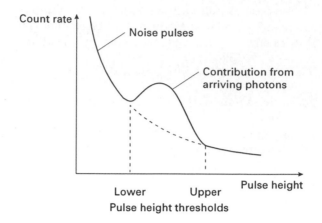

Fig. 7.36 Count rate versus pulse height for photon-counting detection.

clearly evident from such a diagram. An example of an application where photon counting is used is that of Raman backscatter in optical fibres, and this is described in section 10.8.

7.5 CONCLUSIONS

In this chapter we have learned how to construct the devices, sources, modulators and detectors, which comprise the essential, practical tools of optoelectronics. We now know how to generate light, impress information upon it, and extract the information when again required.

Having impressed information on the light it is necessary to guide the light to those places where the information is needed. To do this, optical waveguides are required. These are the subject of the next chapter.

PROBLEMS

7.1 Discuss the relative advantages and disadvantages of gas lasers, solid state lasers and semiconductor lasers. When would you expect to have to use an excimer laser (if ever)?

7.2 Describe how a medium which amplifies light can be turned into a laser oscillator and describe the spectrum of the output. What are the requirements for oscillation? What happens to the spectrum and time domain output when the laser is mode-locked? Why is a four-level system normally more efficient than a three-level system? What are the primary characteristics of an argon laser?

7.3 It is now possible to design lasers using optical fibres. A fibre doped with Nd has been found to have a small signal gain coefficient $g = 1.1 \, \mathrm{m}^{-1}$. Assuming that the refractive index of the fibre is 1.5, derive an expression for and hence calculate the Fresnel reflection at the glass/air interface for a cleave at 90° to the fibre axis. Describe the requirements to obtain laser oscillation and calculate the minimum length of fibre required to produce laser oscillation where the reflectors are the cleaved fibre ends (ignore material losses).

What is the frequency mode spacing for this length of laser? If it is possible to mode lock this laser, what would be the repetition rate and approximate pulsewidth of the output pulse when the linewidth of the laser is $10 \, \mathrm{GHz}$? State any assumptions made. Suggest a way in which this laser could be pumped.

7.4 Describe the principles underlying the semiconductor laser diode.

If the current passing into an operating semiconductor laser is $10 \, \mathrm{mA}$ and its quantum efficiency is 40% how much light power is it emitting if it operates at a wavelength of $0.85 \, \mu\mathrm{m}$?

7.5 Describe the main features of the Pockels and Kerr electro-optic effects.

A linearly polarized laser beam passes into an electro-optic medium under the influence of an electric field whose direction is transverse to the direction of propagation. If the polarization direction lies at angle ϑ to the field, and the field inserts a phase of φ between the polarization component parallel to its direction, compared with the orthogonal component, show that for maximum power transmission through a polarizing analyser placed at the output from the medium, the acceptance direction of the analyser must lie at angle α to the field direction, where:

$$\tan 2\alpha = \tan 2\vartheta \cos \varphi$$

If $\vartheta = 30°$ and $\varphi = 65°$, what fraction of the original laser power passes through the analyser?

7.6 Describe the Faraday magneto-optic effect. How can it be used for impressing information on a beam of light?

A cylindrical rod of magneto-optic material, of length L with Verdet constant V at the wavelength of light to be used, is placed wholly within an N-turn solenoid, also of length L, so that the rod and solenoid axes are parallel. A beam of laser light, with power P watts, passes through the rod along its axis. The light entering the rod is linearly polarized in the vertical direction. The light emerging from the rod is passed through a polarization analyser whose acceptance direction is vertical, and then on to a photodiode which has a sensitivity of S amps watt^{-1}.

A current ramp is applied to the solenoid, of the form:

$$i = kt$$

where i is the current at time t, and k is a constant. Assuming that optical losses are negligible for the arrangement, derive expressions for the amplitude and frequency of the AC component of the signal delivered by the photodiode.

7.7 Describe the differences between the Raman-Nath and Bragg regimes for acousto-optic modulation. A modulator operating in the Bragg regime consists of a slab of lead molybdate (PbMO$_4$) with thickness 50 mm and refractive index 2.3 at 633 nm wavelength.

An acoustic wave with frequency 80 MHz is launched into the slab parallel with its faces; its velocity in the material is 3.75×10^3 m s^{-1}. At what angle must a 633 nm He-Ne laser beam be launched into the slab in order to suffer maximum deflection? What is the magnitude of the deflection and what is the frequency shift suffered by the deflected beam. Is there any amplitude change?

How would you design an optical switch based on this operation?

7.8 Explain how a $p - n$ junction diode may be used to detect optical radiation. Distinguish between the photoconductive and photovoltaic modes of operation.

A photodiode is to be used, without applied reverse bias, to detect optical radiation with vacuum wavelength $0.8\,\mu$m and incident powers in the range $10\,\mu$W to 1 mW. The photodiode has a 70% quantum efficiency and a dark reverse saturation current of 1 nA.

Calculate the value of the load resistance needed to ensure shot noise limited detection at room temperature. Why cannot linear operation be achieved if the photodetector is directly connected to this load? Explain how a transimpedance amplifier may be used to overcome this problem.

7.9 Compare the construction, principles of operation and characteristics of photomultiplier tube and avalanche photodiode detectors. What are the relative advantages and disadvantages of each type of detector?

An analogue optical link is to be set up between two points 2 km apart. The required link bandwidth and signal-to-noise ratio are 100 MHz and 60 dB respectively. The source can emit 0.5 mW peak optical power at $0.83\,\mu$m into the fibre, whose loss is 1.5 dB/km at this wavelength. Assuming 100% modulation depth, calculate the average photocurrent generated in a PIN detector placed at the output end of the fibre. Assume the detector quantum efficiency is 70%. Does the shot noise in the signal current allow the SNR specification to be met? What is the system margin, ignoring any other noise sources?

7.10 List and briefly describe the characteristics and origins of the main sources of noise encountered when measuring an optical signal. Give relevant formulae and suggest remedies where applicable.

A transimpedance amplifier has a $1\,M\Omega$ feedback resistor with a stray capacitance of $16\,pF$ in the feedback path. Calculate the bandwidth of the amplifier and the RMS noise current at the input when the resistor is operating at $20°C$.

A PIN diode detector is connected to this transimpedance amplifier. The detector has a risetime of $10\,ns$ and an NEP of $10^{-14}\,W\,Hz^{-1/2}$ above the $1/f$ noise corner at $\lambda = 0.8\,\mu m$ and a quantum efficiency of 0.7. What is the overall NEP of this receiver combination for an operating wavelength of $\lambda = 0.8\,\mu m$?

7.11 Why is it that a photoconductive detector gives a current gain while a photoemissive detector does not? What are the disadvantages of photoconductive detectors? Under what circumstances would you use one?

REFERENCES

[1] Kaplan, W. (1981) *Advanced Mathematics for Engineers*, Addison-Wesley, chap. 12.
[2] Ghatak, A.K. and Thyagarajan, K. (1989) *Optical Electronics*, Cambridge University Press, chap. 19.

FURTHER READING

Siegman, A. E. (1986) *Lasers*, University Science Books.
Vasil'ev, P. (1995) *Ultra-fast Diode Lasers*, Artech House.
Dennis, P. N. J. (1986) *Photodetectors*, Plenum Press.
Yariv, A. (1986) *Optical Electronics*, Holt-Saunders, 3rd edn, chaps. 7, 9 and 11.

8

Optical waveguides

8.1 INTRODUCTION

The basic principles of optical waveguiding were introduced in section 2.8. We saw there that waves are guided when they are constrained to lie within a channel between two other media, the refractive index of the channel material being slightly higher than those of the other media, so that the light can 'bounce' along the channel by means of a series of total internal reflections (TIRs) at the boundaries between media. The case considered in section 2.8, and shown again in Fig. 8.1, is that where a channel of refractive index n_1 lies between two slabs, each with refractive index n_2 $(n_1 > n_2)$; this is the easiest arrangement to analyse mathematically, yet it illustrates all the important principles.

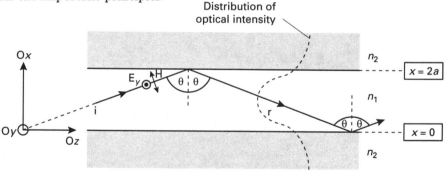

Fig. 8.1 Optical slab waveguide.

The other important point which was made earlier is that, in order to progress down the guide indefinitely, the waves from the successive boundaries must interfere constructively, forming what is essentially a continuous, stable, interference pattern down the guiding channel. If the interference is not fully constructive, the wave will eventually 'self-destruct', owing to the out-of-phase cancellations (although, clearly, if the phasings are almost correct, the wave might persist for a considerable distance, attenuating only slowly). The condition which must be imposed for constructive interference defines for us the guided wave parameters, in particular those angles of bounce which can give rise to the 'modes' of the waveguide, i.e., the various patterns of constructive interference which are allowed by the restrictions (boundary conditions) of the guide geometry.

The ideas involved in waveguiding are thus quite simple. In order to make use of them we need, as always, a proper mathematical description, so we shall in this chapter develop this description.

8.2 THE PLANAR WAVEGUIDE

For convenience and completeness it will be necessary, first, to review the arguments and results presented in section 2.8 for the slab waveguide.

Hence, let us study again the symmetrical slab waveguide shown in Fig. 8.1. The guiding channel consists here of a slab of material of refractive index n_1 surrounded by two outer slabs, each of refractive index n_2. From equation (2.18) we have that the resultant electric field for light which is linearly polarized in a direction perpendicular to the plane of incidence (the so-called transverse electronic (TE) mode as shown in Fig. 2.13) is given by the sum of the upwards and downwards propagating rays:

$$E_T = E_i + E_r$$

where

$$E_i = E_0 \exp(i\omega t - ikn_1 x \cos\vartheta - ikn_1 z \sin\vartheta)$$

(i.e., a wave travelling in the xz plane at angle ϑ to the slab boundaries which lie parallel to the yz plane) and

$$E_r = E_0 \exp(i\omega t + ikn_1 x \cos\vartheta - ikn_1 z \sin\vartheta + i\delta_s)$$

which is the wave resulting from the reflection at the boundary, and differs from E_i in two respects: it is now travelling in the negative direction of Ox, hence the change of sign in the x term, and there has been a change of phase at the reflection, hence the $i\delta_s$ term. We must also remember that δ_s depends on the angle, ϑ, the polarization of the wave and, of course, n_1 and n_2. Hence

$$E_T = E_i + E_r$$

$$= 2E_0 \cos(kn_1 x \cos\vartheta + \tfrac{1}{2}\delta_s) \exp(i\omega t - ikn_1 z \sin\vartheta + i\tfrac{1}{2}\delta_s) \qquad (8.1a)$$

which is a wave propagating in the Oz direction, but with amplitude varying in the Ox direction according to $2E_0 \cos(kn_1 x \cos\vartheta + \tfrac{1}{2}\delta_s)$ (see Fig. 8.1).

The symmetry of the arrangement tells us that the intensity (*square* of the electric field) of the wave must be the same at the two boundaries and thus that it is the same at $x = 0$ as at $x = 2a$. Hence:

$$\cos^2(\tfrac{1}{2}\delta_s) = \cos^2(kn_1 2a \cos\vartheta + \tfrac{1}{2}\delta_s)$$

which implies, following the arguments given in section 2.8, that:

$$2akn_1 \cos \vartheta + \delta_s = m\pi \tag{8.1b}$$

where m is an integer. This is our 'transverse resonance condition' and it is a condition on ϑ (remember that δ_s also depends on ϑ) which defines a number of allowed values for ϑ (corresponding to the various integer values of m) which in turn define our discete, allowable modes (or interference patterns) of propagation.

Now the wavenumber, $k = 2\pi/\lambda$, for the free space propagation of the wave has suffered a number of modifications. First, the wavelength of the light is smaller in the medium than in free space (the frequency remains the same, but the velocity is reduced by a factor $n_{1,2}$), so we can conveniently define

$$\beta_1 = n_1 k$$

$$\beta_2 = n_2 k$$

as the wavenumbers in the guiding and outer slabs, respectively. Secondly, however, if we choose to interpret equation (8.1a) as one describing a wave propagating in the Oz direction with amplitude modulated in the Ox direction, it is convenient to resolve the wavenumber in the guiding medium into components along Oz and Ox, i.e.,

along Oz:

$$\beta = n_1 k \sin \vartheta \tag{8.2a}$$

along Ox

$$q = n_1 k \cos \vartheta \tag{8.2b}$$

Of these two components β is clearly the more important, since it is the effective wavenumber for the propagation down the guide. In fact equation (8.1a) can now be written:

$$E_T = 2E_0 \cos(qx + \tfrac{1}{2}\delta_s) \exp i(\omega t - \beta z + i\tfrac{1}{2}\delta_s)$$

What can be said about the velocity of the wave down the guide? Clearly the phase velocity is given by:

$$c_p = \frac{\omega}{\beta}$$

However, from section 4.3 we know that this is not the end of the story, for the velocity with which optical energy propagates down the guide is given by the group velocity, which, in this case is given by

$$c_g = \frac{d\omega}{d\beta}$$

What, then, is the dependence of ω upon β?

To answer this, let us start with equation (8.2a), i.e.,

$$\beta = n_1 k \sin \vartheta$$

The first thing to note is that, for all real ϑ, this requires:

$$\beta \leq n_1 k$$

Also, since the TIR condition requires that

$$\sin \vartheta \geq \frac{n_2}{n_1}$$

it follows that:

$$\beta = n_1 k \sin \vartheta \geq n_2 k$$

Hence we have

$$n_1 k \geq \beta \geq n_2 k$$

or

$$\beta_1 \geq \beta \geq \beta_2$$

In other words, the wavenumber describing the propagation along the guide axis always lies between the wavenumbers for the guiding medium (β_1) and the outer medium (β_2). This we might have expected from the physics, since the propagation lies partly in the guide and partly in the outer medium (evanescent wave). We shall be returning to this point later.

Remember that our present concern is about how β varies with ω between these two limits, so how else does equation (8.2a) help?

Clearly, the relation

$$k = \frac{\omega}{c_0}$$

where c_0 is the free space velocity, gives one dependence of β on ω, but what about $\sin \vartheta$? For a given value of m (i.e., a given mode) the transverse resonant condition (8.1b) provides the dependence of ϑ on k. However, this is quite complex since, as we know, δ_s is a quite complex function of ϑ. Hence in order to proceed further this dependence must be considered.

The expressions for the phase changes which occur under TIR at a given angle were derived in section 2.6 and are restated here:

$$\tan \tfrac{1}{2}\delta_s = \frac{\left(n_1^2 \sin^2 \vartheta - n_2^2\right)^{1/2}}{n_1 \cos \vartheta}$$

for the case where the electric field is perpendicular to the plane of incidence and

$$\tan \tfrac{1}{2}\delta_p = \frac{n_1 \left(n_1^2 \sin^2 \vartheta - n_2^2\right)^{1/2}}{n_2^2 \cos \vartheta}$$

for the case where it lies in the plane of incidence.

Note also that:

$$\tan \tfrac{1}{2}\delta_p = \frac{n_1^2}{n_2^2} \tan \tfrac{1}{2}\delta_s$$

Finally, let us define, for convenience, a parameter, p, where

$$p^2 = \beta^2 - n_2^2 k^2 = k^2 (n_1^2 \sin^2 \vartheta - n_2^2) \qquad (8.3)$$

The physical significance of p will soon become clear.

We now discover that we can cast our 'transverse resonance' condition (8.1b) into the form

$$\tan\left(aq - \frac{1}{2}m\pi\right) = \frac{p}{q} \quad (E_\perp) \qquad (8.4a)$$

for the perpendicular polarization and

$$\tan\left(aq - \frac{1}{2}m\pi\right) = \frac{n_1^2}{n_2^2}\frac{p}{q} \quad (E_{||}) \qquad (8.4b)$$

for the parallel polarization.

The conventional waveguide notation designates these two cases as 'transverse electric (TE)' for $E_\perp$ and 'transverse magnetic (TM)' for $E_{||}$. The terms refer, of course, to the direction of the stated fields with respect to the plane of incidence of the ray.

We can use equations (8.4) to characterize the modes for any given slab geometry. The solutions of the equations can be separated into odd and even types according to whether m is odd or even. For odd m we have

$$tan(aq - \tfrac{1}{2}m_{odd}\pi) = \cot aq \qquad (8.5a)$$

and for even m

$$tan(aq - \tfrac{1}{2}m_{even}\pi) = \tan aq \qquad (8.5b)$$

Taking m to be even we may then write equation (8.4a), for example, in the form:

$$aq \tan aq = ap \quad (E_\perp) \qquad (8.6)$$

Now from the definitions of p and q it is clear that:

$$a^2 p^2 + a^2 q^2 = a^2 k^2 (n_1^2 - n_2^2) \qquad (8.7)$$

Taking rectangular axes ap, aq this latter relation between p and q translates into a circle of radius $ak \left(n_1^2 - n_2^2\right)^{1/2}$ (Fig. 8.2). If, on the same axes, we also plot the function $aq \tan aq$, then equation (8.6) is satisfied at all points of intersection between the two functions (Fig. 8.2). (A similar set

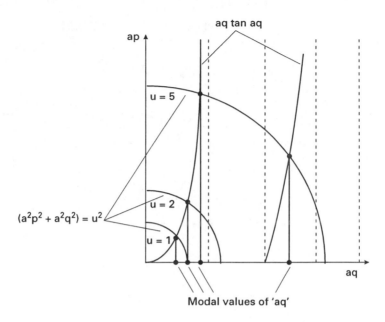

Fig. 8.2 Graphical solution of the modal equation for the slab waveguide.

of solutions clearly can be found for odd m.) These points, therefore, provide the values of ϑ which correspond to the allowed modes of the guide. Having determined a value for ϑ for a given k, β can be determined from:

$$\beta = n_1 k \sin \vartheta$$

and hence β can be determined as a function of k (for a given m) for the TE modes.

Now, finally, with

$$k = \frac{\omega}{c}$$

we have the relationship between β and ω which we have been seeking. For obvious reasons these are called dispersion curves, and are important determinants of waveguide behaviour. They are drawn either as β versus k or as ω versus β. The three lowest order modes for a typical slab waveguide are shown in Fig. 8.3(a) using the latter representation. Clearly, this is the more convenient form for determining the group velocity $d\omega/d\beta$ by simple differentiation (Fig. 8.3(b)). Dispersion will be considered separately in section 8.6.2.

A final point of great importance should be made. As k decreases, so the quantity

$$a^2 p^2 + a^2 q^2 = a^2 k^2 (n_1^2 - n_2^2)$$

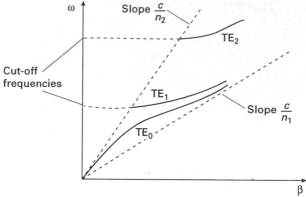

(a) Dispersion diagram for slab waveguide

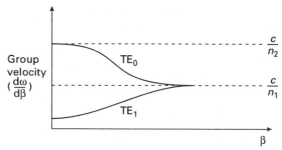

(b) Variation of group velocity with wavenumber

Fig. 8.3 Dispersion and group velocity for a slab waveguide.

decreases and the various modes are sequentially 'cut-off' as the circle (Fig. 8.2) reduces in radius. This is also apparent in Fig. 8.3(a) since a reduction in k corresponds, of course, to a reduction in ω. Clearly the number of possible modes depends upon the waveguide parameters a, n_1 and n_2. However, it is also clear that there will always be at least one mode, since the circle will always intersect the tan curve at one point, even for a vanishingly small circle radius. If there is only one solution, then Fig. 8.2 shows that the radius of the circle must be less than $\frac{1}{2}\pi$, i.e.,

$$ak(n_1^2 - n_2^2)^{1/2} < \tfrac{1}{2}\pi$$

or

$$\frac{2\pi a}{\lambda}(n_1^2 - n_2^2)^{1/2} < 1.57 \tag{8.8}$$

This quantity is another important waveguide parameter, for this and many other reasons. It is given the symbol V and is called the 'normalized frequency', or, quite often, simply the 'V number'. Thus,

$$V = \frac{2\pi a}{\lambda}(n_1^2 - n_2^2)^{1/2}$$

Equation (8.8) is thus the single-mode condition for this symmetrical slab waveguide. It represents an important case, since the existence of just one mode in a waveguide simplifies considerably the behaviour of radiation within it, and thus facilitates its use in, for example, the transmission of information along it. Physically, equation (8.8) is stating the condition under which it is impossible for constructive interference to occur for any ray other than that which (almost) travels along the guide axis.

Clearly, a very similar analysis can be performed for the TM modes, using equation (8.4(b)).

Look again now at Fig. 8.1. It is clear that there are waves travelling in the outer media with amplitudes falling off the farther we go from the central channel. This matter was dealt with in section 2.6, where it was seen that this was a direct result of the necessity for fields (and their derivatives) to be continuous across the media boundaries. We know from equation (8.1a) that the field amplitude in the central channel varies as:

$$E_x = 2E_0 \cos(kn_1 x \cos \vartheta + \tfrac{1}{2}\delta_s)$$

How does the field in the outer slabs vary? The answer to this question was given in section 2.6 when dealing with the TIR phenomenon. It was seen there that the evanescent field in the second medium, when TIR occurred, fell off in amplitude according to

$$E_x = E_a \exp\left(-\frac{2\pi x}{\lambda_2}\sinh \gamma\right); \quad x > a$$

where:
(i) E_a is the value of the field at the boundary, i.e.,

$$E_a = 2E_0 \cos(kn_1 a \cos \vartheta + \tfrac{1}{2}\delta_s)$$

(ii) λ_2 is the wavelength in the second medium, and is equal to λ/n_2
(iii) $2\pi\dfrac{\sinh \gamma}{\lambda_2} = k(n_1^2 \sin^2 \vartheta - n_2^2)^{1/2}$ and this can now be identified with p from equation (8.3).
Hence:

$$E_x = E_a \exp(-px); \quad x > a$$

and we see that p is just the exponential decay constant for the amplitude of the evanescent wave (Fig. 8.4) and, from (8.3), we note that $p \sim 0.1k$.

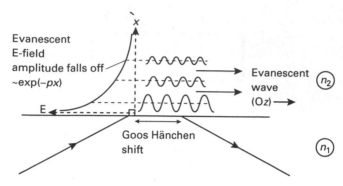

Fig. 8.4 Evanescent wave decay.

(It is a fact of any physical analysis that all parameters of mathematical importance will always have a simple physical meaning.)

So the evanescent waves are waves which propagate in the outer media parallel with the boundary but with amplitude falling off exponentially with distance from the boundary.

These evanescent waves are very important. First, if the total propagation is not to be disturbed the thickness of each outer slab must be great enough for the evanescent wave to have negligible amplitude at its outer boundary: the wave falls off as $\sim \exp(-x/\lambda)$, so at $x \sim 20\lambda$ it normally will be quite negligible ($\sim 10^{-9}$). At optical wavelengths, then, the slabs should have a thickness $\geq 20\,\mu\mathrm{m}$.

Secondly, since energy is travelling (in the Oz direction!) in the outer media, the evanescent wave properties will influence the core propagation, in respect, for example, of loss and dispersion. We shall consider these aspects in more detail in sections 8.5 and 8.6.

8.3 INTEGRATED OPTICS

Planar waveguides find interesting application in integrated optics. In this, waves are guided by planar channels and are processed in a variety of ways. An example is shown in Fig. 8.5. This is an electro-optic modulator, following the basic principles described in section 7.3.1. However, the electric field is acting on a waveguide which, in this case, is a channel (such as we have just been considering) surrounded by 'outer slabs' called here a 'substrate'. The electric field is imposed by means of the two substrate electrodes, and the interaction path is under close control, as a result of the waveguiding. The material of which both the substrate and the waveguide are made should, in this case, clearly be an electro-optic material, such as lithium tantalate ($LiTaO_3$). The central waveguiding channel may be constructed by diffusing ions into it (under careful control); an example of a suitable ion is niobium (Nb), which will thus increase the refractive index of the 'diffused' region and allow total internal reflection to occur at its boundaries with the 'raw' LiTaO$_3$. Many other functions are possible using suitable materials, geometries and field influences. It is possible to fabricate couplers, amplifiers, polarizers, filters, etc., all within a planar 'integrated' geometry.

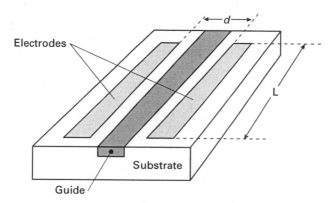

Fig. 8.5 An integrated-optical electro-optic phase modulator.

One of the advantages of this integrated optics technology is that the structures can be produced to high manufactured tolerance by 'mass production' methods, allowing them to be produced cheaply if required numbers are large, as is likely to be the case in optical telecommunications, for example.

A fairly recent, but potentially very powerful, development is that of the 'Optoelectronic Integrated Circuit' (OEIC) which combines optical waveguide functions with electronic ones such as optical source control, photodetection and signal processing, again on a single, planar, readily-manufacturable 'chip'.

Note, finally, that in Fig. 8.5 the 'upper' slab (air) has a different refractive index from the lower one (substrate). This is thus an example of an asymmetrical planar waveguide, the analysis of which is more complex than the symmetrical one which we have considered. However, the basic principles are the same; the mathematics is just more cumbersome, and is covered in many other texts (see, for example [1]).

8.4 CYLINDRICAL WAVEGUIDES

Let us now consider the cylindrical dielectric structure shown in Fig. 8.6. This is the geometry of the optical fibre, the central region being known as the 'core' and the outer region as the 'cladding'. In this case the same

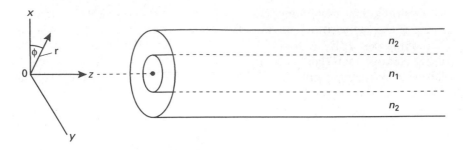

Fig. 8.6 Cylindrical waveguide geometry.

basic principles apply as for the dielectric slab, but the circular, rather than planar, symmetry complicates the mathematics. We use, for convenience, cylindrical co-ordinates (r, φ, z) as defined in Fig. 8.6. This allows us to cast Maxwell's wave equation (see Appendix I) for the dielectric structure into the form:

$$\nabla^2 E = \frac{1}{r} \frac{\partial}{\partial r} \left(r \frac{\partial E}{\partial r} \right) + \frac{1}{r^2} \frac{\partial^2 E}{\partial \varphi^2} + \frac{\partial^2 E}{\partial z^2} = \mu \varepsilon \frac{\partial^2 E}{\partial t^2}$$

If we try a solution for E in which all variables are separable, we write:

$$E = E_r(r) E_\varphi(\varphi) E_z(z) E_t(t)$$

and can immediately, from the known physics, take it that:

$$E_z(z) E_t(t) = \exp[i(\beta z - \omega t)]$$

In other words the wave is progressing along the axis of the cylinder with wavenumber β and with angular frequency ω. It follows, of course, that its (phase) velocity of progression along the axis is given by:

$$c_p = \frac{\omega}{\beta}$$

By substitution of these expressions into the wave equation (8.9) we may rewrite it in the form:

$$\frac{\partial}{\partial r} \left(r \frac{\partial (E_r E_\varphi)}{\partial r} \right) + \frac{1}{r^2} \frac{\partial^2 (E_r E_\varphi)}{\partial \varphi^2} - \beta^2 E_r E_\varphi + \mu \varepsilon \omega^2 E_r E_\varphi = 0$$

Now if we suggest a periodic function for E_φ of the form:

$$E_\varphi = \exp(\pm i l \varphi)$$

where l is an integer, we can further reduce the equation to:

$$\frac{\partial^2 E_r}{\partial r^2} + \frac{1}{r}\frac{\partial E_r}{\partial r} + \left(n^2 k^2 - \beta^2 - \frac{l^2}{r^2}\right) E_r = 0$$

This is a form of Bessel's equation, and its solutions are Bessel functions (see any advanced mathematical text, e.g., [2]). If we use the same substitutions as for the previous planar case, i.e.,

$$n_1^2 k^2 - \beta^2 = q^2$$
$$\beta^2 - n_2^2 k^2 = p^2$$

we find for $r \leq a$ (core)

$$\frac{\partial^2 E_r}{\partial r^2} + \frac{1}{r}\frac{\partial E_r}{\partial r} + \left(q^2 - \frac{l^2}{r^2}\right) E_r = 0$$

and for $r > a$ (cladding)

$$\frac{\partial^2 E_r}{\partial r^2} + \frac{1}{r}\frac{\partial E_r}{\partial r} + \left(p^2 + \frac{l^2}{r^2}\right) E_r = 0$$

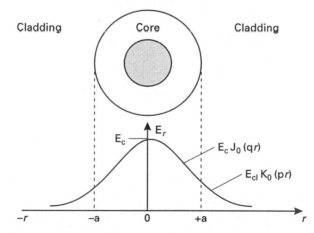

(a) Lowest order solution of the cylindrical wave equation ($l = 0$)

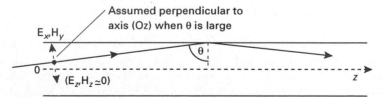

(b) The geometry of the weakly-guiding approximation

Fig. 8.7 Solution for the cylindrical waveguide equation and the weakly-guiding approximation.

Solutions of these equations are (see Fig. 8.7(a)):

$$E_r = E_c J_l(qr); \quad r \le a$$

$$E_r = E_{cl} K_l(pr); \quad r > a$$

where J_l is a 'Bessel function of the first kind' and K_l is a 'modified Bessel function of the second kind' (sometimes known as a 'modified Hankel function'). The two functions must clearly be continuous at $r = a$, and we have for our full 'trial' solution in the core

$$E = E_c J_l(qr) \exp(\pm il\varphi) \exp i(\beta z - \omega t)$$

and a similar one for the cladding

$$E = E_{cl} J_l(pr) \exp(\pm il\varphi) \exp i(\beta z - \omega t)$$

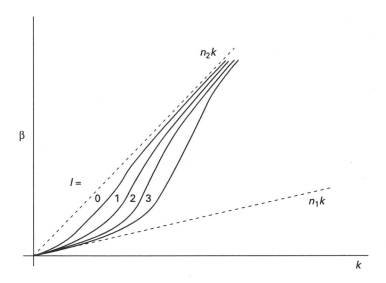

Fig. 8.8 Dispersion curves for the cylindrical waveguide.

Again we can determine the allowable values for p, q and β by imposing the boundary conditions at $r = a$ [5]. The result is a relationship which provides the β versus k, or 'dispersion' curves, shown in Fig. 8.8. The mathematical manipulations are tedious, but are somewhat eased by using the so-called 'weakly guiding' approximation. This makes use of the fact that if $n_1 \sim n_2$, then the ray's angle of incidence on the boundary must be very large, if TIR is to occur. The ray must bounce down the core almost at grazing incidence. This means that the wave is very nearly a transverse wave, with very small z components. By neglecting the longitudinal components H_z, E_z, a considerable simplification of the mathematics results (Fig. 8.7(b)). Since the wave is, to a first approximation, transverse, it can be resolved conveniently into two linearly polarized components, just as for free space propagation. The modes are thus dubbed 'linearly polarized (LP)' modes, and the notation which describes the profile's intensity distribution is the 'LP' notation.

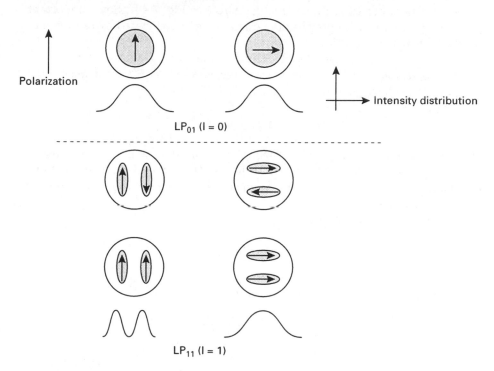

Fig. 8.9 Some low order modes of the clyindrical waveguide (with weakly-guiding labels).

8.5 OPTICAL FIBRES

The cylindrical geometry relates directly, of course, to the optical fibre. The latter has just the geometry we have been considering and, for a typical fibre:

$$\frac{n_1 - n_2}{n_1} \sim 0.01$$

so that the weakly guiding approximation is valid. Some of the low-order 'LP modes' of intensity distribution are shown in Fig. 8.9, together with their polarizations, and values for the azimuthal integer, 1. There are, then, two possible linearly polarized optical fibre modes. For the cylindrical geometry the 'single-mode condition' is (analogously to equation (8.8) for the planar case):

$$V = \frac{2\pi a}{\lambda}(n_1^2 - n_2^2)^{1/2} < 2.405$$

The figure 2.405 derives from the value of the argument for which the lowest order Bessel function, J_0, has its first zero (see Fig. 7.19). Some important practical features of optical-fibre design can be appreciated by reversion to geometrical (ray) optics.

Let us consider, first, the problem of launching light into the fibre. Referring to Fig. 8.10(a), we have for a ray incident on the front face of the fibre at angle ϑ_0, and with refracted angle ϑ_1:

$$n_0 \sin \vartheta_0 = n_1 \sin \vartheta_1$$

where n_0 and n_1 are the refractive indices of air and the fibre core material, respectively. If the angle at which the ray then strikes the core/cladding boundary is ϑ_T, then, for TIR, we must have: $\sin \vartheta_T > n_2/n_1$ where n_2 is the cladding index.

Since $\vartheta_T = \frac{1}{2}\pi - \vartheta_1$ the inequality is equivalent to

$$\cos \vartheta_1 > \frac{n_2}{n_1}$$

so from the Snell's law expression above,

$$\cos \vartheta_1 = \left(1 - \frac{n_0^2 \sin^2 \vartheta_0}{n_1^2}\right)^{1/2}$$

or

$$n_0 \sin \vartheta_0 < (n_1^2 - n_2^2)^{1/2}$$

The quantity on the RHS of this inequality is known as the numerical aperture (NA) of the fibre. It is a specification of the 'acceptance' cone of light, this being a cone of apex half-angle ϑ_0. Clearly, a large refractive index difference between core and cladding is necessary for a large acceptance angle; for a typical fibre, $\vartheta_0 \sim 10°$.

The discrete values of reflection angle which are allowed by the transverse resonance condition (within the TIR condition) can be represented by the ray propagations shown in Fig. 8.10(b). This makes clear that for a large number of allowable rays (i.e., modes) the TIR angle should be large, implying a large NA. However it is also clear, geometrically, that the rays will progress down the guide at velocities which depend on their angles of reflection: the smaller the angle, the smaller the velocity. This leads to large 'modal dispersion' at large NA since, if the launched light energy is distributed among many modes, the differing velocities will lead to varying times of arrival of the energy components at the far end of the fibre. This is undesirable in, for example, communications applications, since it will lead to a limitation on the communications bandwidth. In a digital system, a pulse cannot be allowed to spread into the pulses before or after it. For greatest bandwidth only one mode should be allowed, and this requires a small NA. Thus a balance must be struck between good signal level (large NA) and large signal bandwidth (small NA). We shall return to this topic in section 8.6.

A fibre design which attempts to attain a better-balanced position between these is shown in Fig. 8.10(c). This fibre is known as graded-index (GI) fibre and it possesses a core refractive index profile which falls off parabolically (approximately) from its peak value on the axis. This profile constitutes, effectively, a continuous convex lens, which allows large acceptance angle while limiting the number of allowable modes to a relatively small value. GI fibre is used widely in short and medium distance communications systems. For trunk systems single-mode fibre is invariably used, however. This ensures that the modal dispersion is entirely absent, thus removing this limitation on bandwidth. Single-mode fibre possesses a communications bandwidth which is an order of magnitude greater than that of multimode fibre. However, it is not without its problems. It is time now to deal with the communications application for optical fibre in a more coherent fashion.

8.6 OPTICAL FIBRES FOR COMMUNICATIONS

In discussing the properties of optical fibres several references have been made to communications systems, for this is their most important application. Indeed, it is arguably the most important application area for the whole of opto-electronics at the present time.

The basic arrangement for an optical-fibre communications system is shown in Fig. 8.11. A laser source provides light which is modulated by the information required to be transmitted, the information being in the form of an electrical signal which is applied to an optical modulator. This light is then launched into an optical fibre which guides it to its destination. At the destination the light emerges from the fibre and falls on to a photodetector which converts it into an electrical signal. This electrical signal will be a close reproduction of that which was used to modulate the laser source:

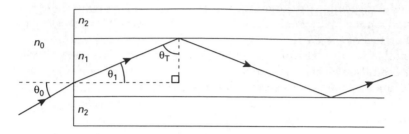

(a) Acceptance angle for an optical fibre

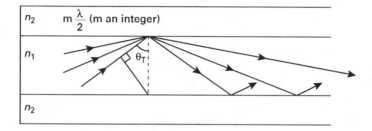

(b) Ray representations of fibre modes

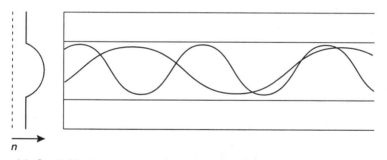

(c) Graded index ray paths

Fig. 8.10 Ray propagations in optical fibres.

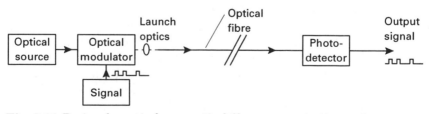

Fig. 8.11 Basic schematic for an optical-fibre communication system.

the closer it is, the better is the communications link.

The primary advantage of such an optical arrangement is the enormous communications bandwidth which it offers, for bandwidth is equivalent to information-carrying capacity. The reason for this is that the frequency of the light is so much greater than that of the more conventional carrier signals such as radio and microwave transmissions: light in the visible range has a frequency of 5×10^{14} Hz compared with microwaves at $\sim 10^{10}$ Hz and radio $\sim 10^{8}$ Hz. The higher is the frequency of the carrier, the smaller is the relative effect of a given modulation bandwidth, for any modulation signal will spread the carrier signal over a band at least equal to the modulation bandwidth. Hence a 1 GHz (10^9 Hz) modulation bandwidth will spread a microwave carrier by 10%, but an optical carrier by only 2×10^{-4}% (using the above figures). The smaller perturbation of the carrier frequency means that the properties of all the components in the communications system are substantially constant over the transmission bandwidth, and this applies especially to the transmission medium. For, if the medium acts differently for different frequencies over the modulation bandwidth, the information becomes distorted, and the communications link performance degrades. Hence, at optical frequencies, using optical-fibre waveguides, very high-bandwidth systems can be expected. Most long-distance communication systems presently are digital, which means that the signal information exists in the form of a series of pulses which encode the information as a series of 'yes' or 'no' answers to the question 'is a pulse present in a particular time slot?' (Digital representation is discussed in more detail in section 10.5.) The advantage of this is that the detection system has only to answer 'yes' or 'no' to this simple question and not to decide on the precise level of the signal over a range, as is the case for analogue systems. Digital systems are thus very robust in terms of signal level, the only requirement being that the level should be above a certain threshold, but they do require more bandwidth than analogue systems. Optical-fibre communications systems readily provide this. However, even optical fibres both attenuate and distort the transmitted signals to some extent. It is necessary to understand the processes which lead to attenuation and distortion in fibres in order to get the best from them for communications purposes. These are the subjects for the next two sections.

8.6.1 Optical-fibre attenuation

The mechanisms responsible for the attenuation of light propagation in materials have been dealt with in section 4.2. We saw there that they are due to absorption and scattering by the atoms of the optical medium.

Most high-grade optical fibres are fabricated from amorphous silica by drawing a thin fibre strand from a melt (Fig. 8.12). The block of the material which is melted is called the 'preform' and it is carefully constructed to have the required scaled-up geometry of the fibre. The core is given greater refractive index than the cladding by doping with materials such as beryllium or germanium. The geometry is preserved in the drawing process. Many tens of kilometres of fibre can be drawn from a single preform. (For details of fibre fabrication processes see, for example, [3]).

The absorption spectrum of silica was shown in Fig. 4.3. The two peaks shown are in fact harmonics ('overtones') of a fundamental vibration at $2.8\,\mu m$ which is due to the stretching of the O-H bond. These peaks are troublesome since they exist in a region of the spectrum which is potentially very useful for optical communications: there are good LED and laser sources in the region (GaAs) and it is also, as we shall see in the next section, a region of low dispersion, which means that its information-carrying capacity is very large.

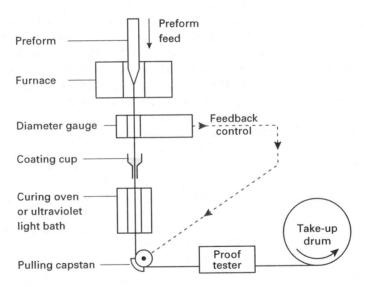

Fig. 8.12 Schematic of a fibre-pulling rig.

Optical-fibre communications technology really began when, in the mid-1960s, it was realized [4] that the loss in silica 'glass' was due to impurities which were removable by known processes: these impurities were mostly metallic ions such as $Fe^{3+}, Mn^{2+}, Ni^{2+}, Co^{2+}$, etc. Having removed these, the problem of the O-H resonance remained: this was the result of residual water in the structure, and it proved very difficult to remove. However, by the mid-1970s a concentrated attack on this material problem had given rise to silica of such purity that the secondary 'water' peaks were hardly noticeable below $1.2\,\mu m$ (Fig. 8.13). The attenuation which remained was due almost entirely to Rayleigh scattering ($\sim 1/\lambda^4$), which is a fundamental property of the amorphous silica material structure and cannot be reduced

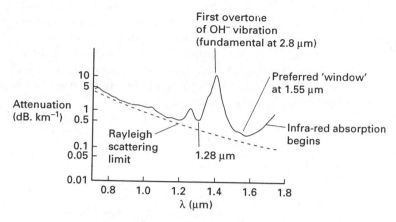

Fig. 8.13 Absorption spectrum for a silica fibre.

substantially. Clearly, under these conditions, the larger the optical wavelength (the smaller the frequency), the smaller will be the attenuation, and the better will be the communications link. This remains true until we reach wavelengths in excess of $\sim 1.55\,\mu m$ when other resonances such as Si-O (fundamental material), Be-O and Ge-O (core dopant) start to give rise to absorption again. Thus $1.55\,\mu m$ clearly will be a good wavelength to use for communications. Losses as low as $0.2\,\text{dB Km}^{-1}$ can be achieved there. However, there are considerations other than just attenuation to consider in the choice of the working communications wavelength. One of these is the availability of suitable sources. We saw in section 6.2.4 that the difficulty of providing lasers rises as ν^3, where ν is the optical frequency. Another problem is that of the dispersion. We shall now take a closer look at this last feature.

8.6.2 Optical-fibre dispersion

We met optical material dispersion in section 4.3. It is a consequence of the variation of refractive index with optical frequency, and it has its origins in the same atomic absorption processes which give rise to the absorptive component of the attenuation spectrum.

Clearly, any optical energy propagating in a material medium will comprise a range of wavelengths. It is not possible to devise a source of radiation which has zero spectral width. Consequently, in the face of optical dispersion in the medium, different parts of the propagating energy will travel at different velocities; and if that energy is carrying information (i.e., it has been modulated in some way) that information will become distorted by the velocity differences. The further it travels, the greater will be the distortion; the greater the wavelength spread, the greater will be the distortion; the greater the dispersion power of the medium, again the greater will be the distortion. For good communications we need, therefore, to choose our sources, wavelengths and materials very carefully, and in order to make these choices we must understand the processes involved.

In optical fibres and in all other optical waveguides there are three types of dispersion: modal dispersion (in multimode guides only), material dispersion (which we already know something about) and waveguide dispersion (a consequence of the guide's geometry).

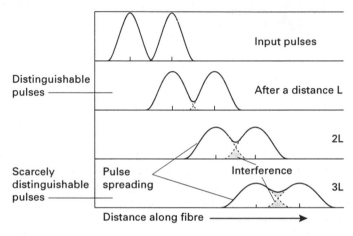

Fig. 8.14 Effect of fibre dispersion on a pulsed input optical signal.

The effect of dispersion in a waveguide is to limit its communications-carrying capacity, i.e., its bandwidth. This is seen most readily by considering a digital communications system, that is, one which transmits information by means of a stream of pulses (Fig. 8.14). (The presence of a pulse indicates a '1', the absence of one indicates a '0' and the stream thus comprises a digital coding of the information to be transmitted.) A stream of clear, distinct pulses is launched into the fibre (for example) by modulating a laser source. As the pulses propagate down the fibre, the spread of optical wavelengths of which they are comprised will be acted upon by the dispersive effects in the fibre, and the result will be a broadening of the pulses (Fig. 8.14). When the broadening has become so great that it is no longer possible to distinguish between two successive pulses, the communications link fails.

Clearly, for a given dispersive power, the broadening will increase linearly with distance.

Hence

$$\frac{\Delta \tau}{L} = \text{constant}$$

where $\Delta \tau$ is the broadening (in time) of the pulse over a fibre length L. Now the 'bit-rate', effectively the digital bandwidth, which the fibre length

L can carry will be $1/\Delta\tau$, since the spacing between pulses, $\Delta\tau$, will be closed up by the dispersion when the broadening is equal to $\Delta\tau$. Hence we have:

$$BL = \text{constant} \qquad\qquad (8.10)$$

where B is the allowable bit rate in pulses (or 'bits', i.e., '*binary digits*') per second. It gives a good idea of the capacity of modern optical communications systems to know that this bit rate is usually quoted in megabits/second ($\mathrm{Mb\,s^{-1}}$) or Gigabits/second ($\mathrm{Gb\,s^{-1}}$).

The result of dispersion is thus to impose a 'bandwidth × distance' limitation: the greater the distance the smaller is the bandwidth which can be transmitted for a given fibre, and vice versa.

Let us look now at the particular causes of dispersion in optical fibres.

(i) *Modal dispersion*

Modal dispersion was introduced briefly in section 8.5. This dispersion exists only in multimode fibres, since it results from the differing velocities of the range of modes supported by the fibre. This is not a material dispersion but results from the fibre's structure. Optical energy is launched into the fibre and will be launched into many, perhaps all, of the modes supported by the fibre. The effect of modal dispersion clearly will depend on how the propagating energy is distributed among the possible modes, and this will vary along the fibre as the energy redistributes itself according to local conditions (e.g., bends, joints, etc.). In order to get a 'feel' for its order of magnitude, however, we can very easily calculate the difference in time of flight, over a given distance, between the fastest and slowest modes supported by the fibre.

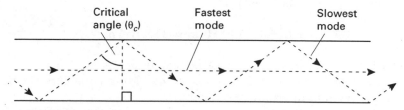

Fig. 8.15 Modal dispersion.

The fastest mode will be that which travels (almost) straight down the fibre, along the axis (Fig. 8.15). This will have the velocity of the unbounded core medium, c_0/n_1.

The slowest mode will be that which is represented by a ray which is incident on the core/cladding boundary at the TIR angle (for any greater angle the ray will not be guided). Clearly (Fig. 8.15) this ray travels at velocity $(c_0/n_1)\sin\vartheta_c$, where ϑ_c is the critical angle.

Since we have

$$\sin\vartheta_c = \frac{n_2}{n_1}$$

it is easily seen that the two times of flight along a distance L of fibre are:

$\tau_f = t_{min}$

$\tau_s = t_{max}$

$$\tau_f = L\frac{n_1}{c_0}; \quad \tau_s = \frac{Ln_1}{c_0\sin\vartheta_c} = \frac{Ln_1^2}{n_2}$$

Hence

$$\Delta\tau = \tau_s - \tau_f = \frac{L}{c_0}\frac{n_1}{n_2}(n_1 - n_2)$$

And since

$$n_1 \approx n_2 \quad \left(\frac{n_1 - n_2}{n_1} \sim 0.01\right)$$

then

$$\Delta\tau \approx \frac{L}{c_0}\Delta n \tag{8.11}$$

where Δn is the difference in refractive index between core and cladding. Equation (8.11) is a clear specific example of the general equation (8.10), for we have, from (8.10):

$$\frac{\Delta\tau}{L} = \frac{\Delta n}{c_0}; \quad B = \frac{1}{\Delta\tau}$$

Hence

$$BL = \frac{c_0}{\Delta n}$$

The RHS is thus a constant for a given fibre. Typically, with a refractive index difference of ~ 0.01 we have:

$$BL = 3 \times 10^{10}\,\text{Hz}\,\text{m} = 30\,\text{MHz}\,\text{km}$$

Hence, for such a fibre, only a 30-MHz bandwidth is available over a 1-km length; only 3 MHz over 10 km, and so on. Multimode fibre clearly is seriously limited in its bandwidth capability.

It is instructive also to relate B to the amount of optical power which can be launched into the fibre from a given source. From section 8.5 we know that the numerical aperture (NA) is given by:

$$\text{NA} = (n_1^2 - n_2^2)^{1/2} = [(n_1 - n_2)(n_1 + n_2)]^{1/2}$$

or

$$(\text{NA})^2 = (\Delta n)(2n_1)$$

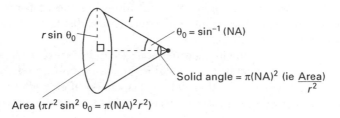

$r \sin \theta_0$

r

$\theta_0 = \sin^{-1}(\mathrm{NA})$

Solid angle $= \pi(\mathrm{NA})^2$ (ie $\dfrac{\text{Area}}{r^2}$)

Area ($\pi r^2 \sin^2 \theta_0 = \pi(\mathrm{NA})^2 r^2$)

Fig. 8.16 Solid angle for fibre's light acceptance

since

$$\frac{(n_1 - n_2)}{n_1} \sim 0.01$$

Substituting for Δn in (8.11) we find

$$\Delta \tau = \frac{L}{2n_1 c_0}(\mathrm{NA})^2$$

Now the numerical aperture is a measure of the ease with which the fibre will accept light from a source. We can see from Fig. 8.16 that the 'solid angle' of acceptance is just $\pi(\mathrm{NA})^2$. Hence the greater the value of $(\mathrm{NA})^2$, the greater will be the launched power. If we also assume that the noise on the received signal is independent of fibre length (a fair assumption since almost all the noise will be shot and thermal noise generated in the receiver), then it follows that the detection signal-to-noise ratio (SNR) is proportional to the launched power (for a given fibre length) and thus to $(\mathrm{NA})^2$. Hence

$$\mathrm{SNR} \sim (\mathrm{NA})^2$$

$$\frac{1}{\Delta \tau} \sim B \sim \frac{2n_1 c_0}{L(\mathrm{NA})^2}$$

and

$$\mathrm{SNR} \times B \sim \frac{2n_1 c_0}{L}$$

Thus, for a given fibre length, the product of SNR and bandwidth is also a constant. Increasing the NA (for example) may increase the power into the fibre, but this is at the expense of a reduced bandwidth, owing to the increased modal dispersion which results from the greater NA. Such relationships are generally true in communications systems but are especially easy to appreciate for multimode optical-fibre links. These relationships allow a glimpse of the kinds of compromise which must be faced by optical-fibre communications systems designers. In order to minimize multimode dispersion, and thus maximize the bandwidth for a given fibre length, it is clear that the number of modes must be minimized. The absolute minimum number that we can have is one: a monomode fibre. It is for this reason that monomode (or 'single' mode) fibres are the preferred medium for optical-fibre communications: only quite short distance ($< 1\,\mathrm{km}$) links now employ multimode fibres.

However, in monomode fibres other sources of dispersion become important. These are overwhelmed by modal dispersion in multimode fibres, but when this is removed, it is these other dispersion effects which limit the bandwidth of the communications system. These other effects will now be considered.

(ii) *Material dispersion*

Material dispersion was covered in some detail in section 4.3. It is due, as was detailed there, to the fact that the refractive index of any optical material will vary with wavelength, owing to the structure of the atomic resonances.

It was noted that the variation of refractive index with wavelength gave rise to a group velocity, different from the phase velocity, given by:

$$c_g = \frac{d\omega}{dk}$$

which, with the aid of the relation

$$\omega = \frac{c_0}{n}k$$

where n is the refractive index of the material, could be translated into

$$c_g = \frac{c_0}{n}\left(1 - \frac{k}{n}\frac{dn}{dk}\right)$$

where c_g is the velocity with which the optical energy travels.

Clearly, c_g will not be a constant across the source spectrum unless dn/dk is constant across it, i.e., unless n is a linear function of wavelength. If it is not, then different portions of the source spectrum travel at different velocities and this will result in, among other effects, the broadening of an optical pulse.

The time taken for energy to travel a distance L when its group velocity is c_g is

$$\tau = \frac{L}{c_g}$$

Hence the spread of times for a source spectrum of width $\delta\omega$ will be:

$$\Delta\tau = L\frac{d\left(c_g^{-1}\right)}{d\omega}\delta\omega = L\frac{d^2k}{d\omega^2}\delta\omega$$

Since $k = 2\pi/\lambda = n\omega/c_0$, this can also be expressed as:

$$\Delta\tau = \frac{L}{c_0}\lambda \left(\frac{\mathrm{d}^2 n}{\mathrm{d}\lambda^2}\right)_\lambda \delta\lambda$$

where $\left(\mathrm{d}^2 n/\mathrm{d}\lambda^2\right)_\lambda$ is the value of $\left(\mathrm{d}^2 n/\mathrm{d}\lambda^2\right)$ at the wavelength λ. Again it can be noted that, with $B \sim 1/\Delta\tau$

$$BL = \frac{c_0}{\lambda \left(\frac{\mathrm{d}^2 n}{\mathrm{d}\lambda^2}\right)\delta\lambda}$$

and is a constant for a given source and material.

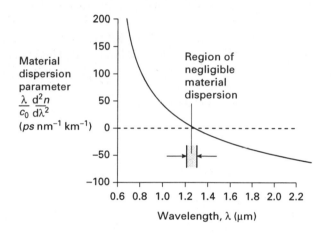

Fig. 8.17 The material dispersion 'zero' for silica.

It is clear that, for maximum bandwidth × distance product, $d^2 n/d\lambda^2$, λ and $\delta\lambda$ must be as small as possible. Now, clearly $(\lambda/c_0)(d^2 n/d\lambda^2)$ (at the wavelength λ) is a characteristic of the material. Its variation with λ is shown in Fig. 8.17 for silica.

From this it can be seen that $d^2 n/d\lambda^2 = 0$ at $\lambda \sim 1.28\,\mu$m.

It follows that this is the preferred wavelength for maximum bandwidth in a silica fibre. It is extremely fortuitous that this wavelength also corresponds to a minimum in the absorption spectrum for silica (see Fig. 8.13), thus also giving low attenuation at this wavelength. It was the combination of these two factors which led to the rapid progress of monomode optical-fibre communications technology in the 1980s.

In order to take maximum advantage of the bandwidth capability at this wavelength it is necessary, of course, also to minimize $\delta\lambda$, the optical source's spectral width. This requirement has led to the development of relatively higher power ($\sim 100\,$mW) narrow band ($\sim 0.1\,$nm) semiconductor laser sources and, more recently, optical-fibre lasers.

Of course, $d^2n/d\lambda^2$ will not be zero over the whole width of the source. The best we can do is ensure that it is zero at or around the centre of the spectrum, λ_0 say.

Let us insert some typical numbers into our equations, to get a feel for practicalities.

Let us assume that the light source is a semiconductor laser with a wavelength spread of $\sim 1\,$nm. From Fig. 8.17 we can see that the mean material dispersion for a range of 1 nm around the zero point ($1.28\,\mu$m) is $\sim 3\,$ps nm^{-1}km^{-1}.

Hence $\Delta\tau \approx 3 \times 10^{-12}$km^{-1} or $BL \approx 330\,$Gbs^{-1}km.

Therefore, for a link of length 100 km, for example, we shall have a bandwidth of $3.3\,$Gbs^{-1}. This is a respectable bandwidth but it should be possible to do better over this distance. The most obvious approach for improvement is to reduce still further the spectral width of the source, to $\sim 0.1\,$nm, say, where $33\,$Gbs^{-1} will be available over the same distance. However, before settling for this there is yet another source of dispersion to worry about in regard to monomode fibres. This is known as waveguide dispersion and will now be considered.

(iii) *Waveguide dispersion*

When considering planar waveguides in section 8.2 and cylindrical waveguides in section 8.4 the relationships between ω and β (Figs 8.3 and 8.8) were seen to be non-linear for any given mode and, in particular, for the lowest order mode. This lowest order mode is, of course, the only mode propagating in a monomode fibre. The non-linearity derives from the necessity to satisfy the guide's boundary condition at the different optical frequencies, and especially from the fact that the angle at which rays strike the boundary varies with frequency, which, in turn, leads to a different phase change on TIR; this latter is a complex function of the angle of incidence (section 2.6).

From the discussion in the last section, (ii), and in section 4.3 on pulse spreading it was noted (equation 4.11) that the spread of arrival times, after a distance L, for a pulse whose energy is spread over a spectral width $\delta\omega$, is:

$$\Delta\tau = L\frac{d^2k}{d\omega^2}\delta\omega$$

Clearly, for the present case, the guide wavenumber β must be substituted for k to give:

$$\Delta\tau = L\frac{d^2\beta}{d\omega^2}\delta\omega$$

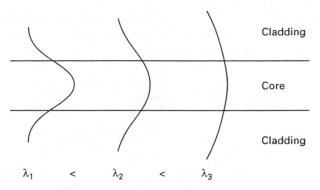

(a) Variation with wavelength of power distribution in a fibre

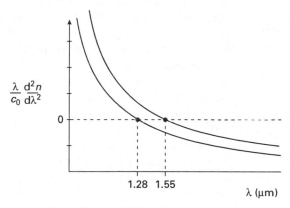

(b) Dispersion shifting for silica fibre

Fig. 8.18 Waveguide dispersion.

Hence, in the case of waveguide dispersion, $\Delta\tau$ can be calculated from the β v. ω curves (the dispersion curves!) for any given mode and, in particular, for the lowest mode in the monomode case.

The physical origins of this form of dispersion lie in two factors. First, as the frequency varies for a given mode, the change of angle of incidence at the boundary means that the ray progresses down the guide at a different velocity (the velocity varies as $\sin\vartheta$, in fact). Secondly, as the frequency varies and the angle at the boundary varies, the penetration of the evanescent wave into the second medium (cladding) varies, in accordance with equation (2.17). Hence the relative amounts of power in the guiding channel (e.g., core) and in the outer medium's evanescent field will vary with frequency. Since the refractive indices for the two media are different, it is unsurprising that this leads to a variation of the guided wave's effective refractive index with frequency.

It transpires, in the case of cylindrical waveguides [5] that the ratio of the power in the core to that in the cladding for the lowest order mode is given, as a first approximation, by:

$$\frac{P_{core}}{P_{cladding}} = V^2 \tag{8.12}$$

where

$$V = \frac{2\pi a}{\lambda}(n_1^2 - n_2^2)^{1/2}$$

Hence, as λ rises, the power in the core increasingly transfers to the cladding. This, again, can readily be justified qualitatively from the physics: as λ rises, the angle of incidence (ϑ) decreases (i.e., the ray becomes more steeply inclined to the axis), as required by equation (8.2a). As ϑ decreases, the penetration of the evanescent wave increases (equation 2.17). Hence the cladding medium has greater effect at higher wavelengths (lower β) and the guide refractive index tends towards that of the cladding, n_2 (Fig. 8.18a). Clearly, all these arguments are reversed as the wavelength decreases. In particular, a decreasing wavelength causes the refractive index to tend towards that of the core; indeed, for very small wavelengths, very much less than the core diameter, the wave is, to first order, unaware of the boundary between the media, and propagates as if it were doing so in an unrestricted core medium, of refractive index n_1.

This waveguide dispersion can, in fact, be very useful, for it can be arranged to oppose the material dispersion. The waveguide dispersion depends upon the fibre geometry (equation (8.12), so, by choosing the geometry appropriately, the dispersion minimum for the fibre can be shifted from its value defined by material dispersion to another convenient (but quite close) wavelength (Fig. 8.18b). Fibre which has been 'adjusted' in this way is called 'dispersion-shifted' fibre, and this stratagem is used to move the wavelength of the dispersion minimum from its 'unrestricted' value of 1.28 μm to 1.55 μm, for example, where the attenuation due to absorption is lower. Unrepeatered trunk telecommunications systems of several hundred kilometre lengths can be installed using such fibre.

8.7 POLARIZATION-HOLDING WAVEGUIDES

In section 3.9 there was a brief discussion of polarization-holding optical-fibre waveguides. We are now in a position to understand these better.

First, consider again the fibre with stress-induced birefringence (two refractive indices). The stress, via the strain-optic effect, has produced a different refractive index for light linearly polarized parallel with the stress direction when compared with that linearly polarized perpendicular to that direction (Fig. 8.19). The difference in refractive indices implies, of course,

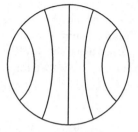

(a) Stress distribution in a polarization-holding fibre

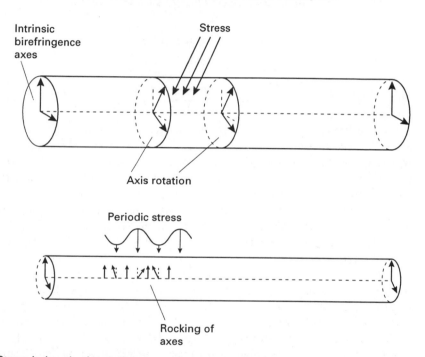

Intrinsic
birefringence
axes

Stress

Axis rotation

Periodic stress

Rocking of
axes

(b) Stress-induced axis rotation

Fig. 8.19 Light coupling in a polarization-holding fibre.

a difference in velocity between the two polarization states which are, as
we know, the two polarization eigenmodes of the guide. This velocity difference means that light of one linear polarization launched into one of the
eigenmodes is locked into that state unless the fibre is specially perturbed.
The reason for this is that any random couplings from one eigenmode to
the other will not, in general, be in phase, owing to the velocity difference
between the components. Hence they will tend to interfere destructively.
However, coupling can occur if a coupling perturbation (e.g., an extra ex-

ternal stress which locally rotates the axes, Fig. 8.19(b)) has a spatial period equal to the distance over which the two eigenmodes come into phase. What is this distance? Suppose that the difference in refractive index is Δn; this quantity is also known as the birefringence in this case, and is sometimes given the symbol B. The phase lag due to Δn over a distance b is given by

$$\varphi_b = \frac{2\pi}{\lambda}\Delta n\ b$$

at wavelength λ. If the two eigenmodes are in phase at a given point in the fibre they will be also in phase after a distance b provided that:

$$\varphi_b = \frac{2\pi}{\lambda}\Delta n\ b = 2\pi$$

i.e.,

$$b = \frac{\lambda}{\Delta n} \tag{8.13}$$

The quantity b is known as the 'beat length' and is an important characterizing parameter for birefringent fibres. Note that it is proportional to the wavelength of the light. Now, clearly, a perturbation with spatial period equal to b will always couple in-phase components from one eigenmode to the other, so constructive interference will occur, and the coupling is strong. For an arbitrary perturbation function the coupling will depend upon the amplitude of the Fourier component with the beat length period. The smaller the beat length in a given fibre, the less likely is a given perturbation in practice to contain a large amplitude Fourier component with that small a period. For example, a single weight resting on the fibre would need to apply its stress over a distance of less than one beat length to possess such a component. Consequently, small beat length (i.e., high birefringence) optical fibres will hold a given launched eigenmode very well, and can be use to convey linearly polarized light from one place to another. Such fibres are called 'high-birefringence' or, often, more simply 'hi-bi' fibres, and are used for polarization control in a variety of applications. They are sometimes also referred to as 'polarization holding' fibres in appropriate applications. A typical hi-bi fibre will have a beat length of $\sim 2\,\mathrm{mm}$ at a wavelength of 850 nm. From equation (8.13) this gives:

$$\Delta n \sim 4.25 \times 10^{-4}$$

With $n_1 \approx n_2 \sim 1.47$ this is a refractive index difference of only $\sim 0.03\%$, but it is enough to provide strong polarization properties. The basic reason for this is that the short optical wavelengths are operating over much longer optical paths, so that the phase effects quickly accumulate, according to $\varphi(2\pi/\lambda)\Delta n\ l$.

The polarization properties of hi-bi optical fibres are very sensitive to external perturbations of all kinds, and it is for this reason that they are widely used in optical-fibre sensors. Some examples of this use are given in Chapter 10.

Consider now another type of birefringent fibre, the elliptically-cored fibre shown in Fig. 8.20. In this case the birefringence is a result of the waveguide geometry, and our recently-acquired understanding of waveguiding action allows us some penetration into the mechanism (the full mathematical description is quite complex and can be studied in [6]). Consider a ray bouncing across the maximum dimension (Fig. 8.20(b)) with linear polarization state lying perpendicular to the plane of incidence. Roughly speaking, the condition for constructive interference corresponds to equation (8.1b) with the major axis equal to a, i.e., $2akn_1 \cos \vartheta + \delta_s = m\pi$.

Hence there is defined a value for ϑ and for δ_s corresponding to that dimension and that polarization state. These parameters, in turn, will define a set of ω/β dispersion curves, like those in Fig. 8.3. There will be a different set for the parallel polarization.

For a ray bouncing across the minimum dimension, b, there will be two more such sets, again one for each linear polarization state and, clearly, since $a \neq b$, all four sets will be different. It is now not difficult to appreciate that a combination of any two corresponding perpendicular states leads to a mode which has a different set of ω/β curves from that produced by combining two parallel states (Fig. 8.20(c)). Hence, at a given value of ω (i.e., a given optical frequency), $d\omega/d\beta$ will differ for the two linear polarization states, and we then have differing group velocities, and thus linear birefringence. It is a natural consequence of the symmetry of the ellipse that the directions of the two linear eigenmodes should correspond with those of the axes of the ellipse. It is only when bouncing between major or minor axis extremities that a ray can be confined to one plane, i.e., the plane which contains the guide axis and the ellipse axis. All other rays will criss-cross the ellipse.

It can be shown [6] that, if the ellipticity is not too high and the mode not too far from cut off, the birefringence in this case is given by:

$$B \approx 0.28 \left(\frac{a}{b} - 1\right))(n_1 - n_2)^2 \tag{8.14}$$

where n_1 is again the core refractive index and n_2 that of the cladding.

Values of a/b usually are in the region of 1.2 for these fibres. The value of $n_1 - n_2$ is limited by the necessity to maintain the single-mode condition:

$$V = \frac{2\pi a, b}{\lambda}(n_1^2 - n_2^2)^{1/2} < 2.405$$

In practice, $n_1^2 - n_2^2$ is made as high as possible, while maintaining the single-mode condition by reducing the dimensions, a and b, of the ellipse.

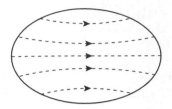

(a) Electrical field distribution for one
   eigenmode in an elliptically-cored fibre

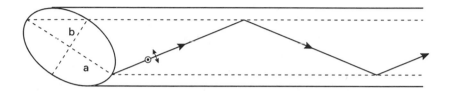

(b) Ray diagram for the eigenmode in (a)

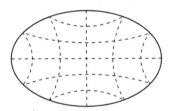

(c) E-field distribution for lowest-order
   orthogonal eigenmodes

Fig. 8.20 The elliptically-cored fibre.

However, this has the effect, of course, of reducing the core area and hence
making alignment with sources, and other sections of fibre, rather difficult
for these fibres. Another problem is that if n_1 is raised, by too much core
doping, the fibre attenuation will rise.

With differences in refractive index between core and cladding (usually
by doping the core with germanium) of 4%, equation (8.13) shows that
beat lengths of the order of 1 mm, at ~ 850 nm, can be obtained with such
fibre.

One important advantage possessed by elliptically-cored hi-bi fibre over
the stress-induced variety is that the former has a lower temperature coefficient of birefringence (by a factor ~ 5), owing to the fact that it is geometrically induced, and does not suffer from the temperature dependence
of the strain-optic coefficient, as does the stress-induced birefringence.

As was noted in section 3.9, birefringence in optical fibres has disadvantages in optical telecommunications in that it introduces yet another form of dispersion, and hence bandwidth limitation, unless steps are taken to reduce it to acceptable levels, such as spinning the preform from which the fibre is drawn, so as to average out any cross-sectional anisotropies in the waveguide.

Furthermore, linear birefringence is not the only birefringent form which can be present (again as was noted in section 3.9). Any waveguide might possess polarization eigenmodes of any elliptical form, including the linear and circular ones as special cases. These eigenmodes (by definition) will possess different velocities, thus providing, in general, elliptical birefringence in the guide. The effects of such a waveguide on any, arbitrary, input polarization state can be determined analytically by resolving the input state into the two eigenmode components (always possible) and then recombining them at the guide's output with their relative phase equal to that inserted by the guide, as a consequence of the velocity difference of the eigenmodes. In practice, linear and circular birefringences are the types most commonly encountered, and optical fibres of both these varieties are available commercially.

The control which the waveguide allows over the direction and localization of the propagating light is further enhanced by this control over its polarization state, and there are many practical examples in optoelectronics where this control is used to considerable advantage. Some of these applications are described in Chapter 10.

8.8 CONCLUSIONS

Optical waveguiding is of primary importance to the optoelectronic designer. With its aid it is possible to confine light and to direct it to where it is needed, over short, medium and long distances.

Furthermore, with the advantage of confinement, it is possible to control the interaction of light with other influences, such as electric, magnetic or acoustic fields, which may be needed to impress information upon it. Control also can be exerted over its intensity distribution, its polarization state and its non-linear behaviour; this last one is the subject of the next chapter.

In short, optical waveguiding is crucial to the control of light. For the designers of devices and systems this control is essential.

PROBLEMS

8.1 Define what is meant by the term 'total internal reflection' in the context of a dielectric waveguide.

A plane wave travelling in air impinges on the end of a slab guide with core index $n_1 = 1.5$ and cladding index $n_2 = 1.46$. What is the maximum angle of incidence of the incoming wave with respect to the axial plane of the guide for this wave to be guided?

Show that the propagation constant of guided modes, β, must satisfy the inequality

$$\boxed{k_0 n_2 \leq \beta \leq k_0 n_1}$$

where

$$k_0 = \frac{\omega}{c_0}$$

Draw the two transverse forms of a lower order mode when the propagation constant is in the middle, and at the limits, of the range given above.

8.2 A dielectric slab waveguide is made from two materials of respective refractive index 1.51 and 1.50. What thickness should the core be in order for the first higher order mode cut-off wavelength to be $1\,\mu m$?

Sketch the field distribution of the fundamental and the first higher order mode for the above case.

The effective index of the fundamental TE mode is 1.505 at an operating wavelength of $1\,\mu$ m. What is the normalized amplitude of the electric field at the core boundary?

8.3 Compare the principal characteristics of typical single-mode fibres for telecommunications with those of multimode fibres for short distance links.

What is meant by the term 'normalized frequency' of an optical fibre? For a fibre with a core index of 1.50 and a cladding index of 1.47, what is the maximum core radius allowable for single-mode operation with $1.3\,\mu m$ sources?

If a fibre is manufactured with a core radius ten times greater than this value but with the same refractive indices and this is used with the same sources, approximately how many modes will the fibre support and what will be the group delay (in ns) between the highest and lowest order modes for a link length of 100 metres?

8.4 Describe the main types of dispersion which can occur in optical fibres, including types relevant to multimode and monomode systems. Use diagrams where appropriate. For the single-mode case, discuss the relative importance of the various dispersion components in controlling the total dispersion.

A step index, multimode fibre with a core of refractive index 1.47 and cladding of refractive index 1.46 has an attenuation of $1\,dB/km$ at $850\,nm$. The fibre is used to connect a source at this wavelength to a detector

5 km away. The desired optical signal-to-noise ratio of the link is 40 dB. If the power delivered from the source into the fibre is 0.1 mW and the receiver has a noise-equivalent power of $5 \times 10^{-13}\,\mathrm{W\,Hz^{-1/2}}$, what limits the maximum bandwidth of the link and what is the order of magnitude of this bandwidth?

8.5 Describe the role played by glassy fibres in optical communications and list some of the advantages associated with their use. Calculate the maximum permissible total loss in dB/km in order not to exceed an error rate of 10^{-9} (200 photons per bit); assume an operating wavelength $\lambda = 900\,\mathrm{nm}$, transmission rate of 100 Mbit/s, fibre length of 10 km, laser input power of 1 mW and operating margins of 8 dB.

8.6 Discuss what is meant by an optical fibre with high linear birefringence. Derive the relationship between birefringence and beat length in such a fibre.

Why are such fibres sometimes used as polarization-holding waveguides and what is the source of the polarization-holding capability? How can such fibres be fabricated?

REFERENCES

[1] Syms, R. and Cozens, J. (1992) *Optical Guided Waves and Devices*, McGraw-Hill, chap. 6.

[2] Kaplan, W. (1981) *Advanced Mathematics for Engineers*, Addison-Wesley, chap. 12.

[3] Senior, J. M. (1992) *Optical Fibre Communications: Principles and Practice*, Prentice Hall, 2nd edn, chap. 4.

[4] Kao, K. C. and Hockham, G. A. (1966), *Dielectric fibre surface waveguides for optical frequencies*, Proc. IEE, 113 (7), 1151-8.

[5] Adams, M. J. (1981) *An Introduction to Optical Waveguides*, John Wiley, chap. 7.

[6] Dyott, R. B. (1995) *Elliptical Fiber Waveguides*, Artech House.

FURTHER READING

References [1], [5] and [6] above plus:

Marz, R. (1995) *Integrated Optics: Design and Modelling*, Artech House.

Najafi, S. I. (1992) *Introduction to Glass Integrated Optics*, Artech House.

Midwinter, J. E. (1978) *Optical Fibres for Transmission*, John Wiley.

9

Nonlinear optics

9.1 GENERAL INTRODUCTION

In all of the various discussions concerning the propagation of light in material media so far, we have been dealing with linear processes. By this we mean that a light beam of a certain optical frequency which enters a given medium will leave the medium with the same frequency, although the amplitude and phase of the wave will, in general, be altered.

The fundamental physical reason for this linearity lies in the way in which the wave propagates through a material medium and this was considered in some detail in section 4.2. It was seen there that the effect of the electric field of the optical wave on the medium was to set the electrons of the atoms (of which the medium is composed) into forced oscillation; these oscillating electrons then radiated secondary wavelets (since all accelerating electrons radiate) and the secondary wavelets combined with each other and with the original (primary) wave, to form a resultant wave. Now the important point here is that all the forced electrons oscillate at the same frequency (but differing phase, in general) as the primary, driving wave, and thus we have the sum of waves all of the same frequency, but with different amplitudes and phases.

If two such sinusoids are added together:

$$A_T = a_1 \sin(\omega t + \varphi_1) + a_2 \sin(\omega t + \varphi_2)$$

and we have, from simple trigonometry:

$$A_T = a_T \sin(\omega t + \varphi_T)$$

where

$$a_T^2 = a_1^2 + a_2^2 + 2a_1 a_2 \cos(\varphi_1 - \varphi_2)$$

and

$$\tan \varphi_T = \frac{a_1 \sin \varphi_1 + a_2 \sin \varphi_2}{a_1 \cos \varphi_1 + a_2 \cos \varphi_2}$$

In other words, the resultant is a sinusoid of the same frequency but of different amplitude and phase. It follows, then, that no matter how many more such waves are added, the resultant will always be a wave of the same frequency, i.e.,

$$A_T = \sum_{n=0}^{N} a_n \sin(\omega t + \varphi_n) = \alpha \sin(\omega t + \beta)$$

where α and β are expressible in terms of the a_n and φ_n.

It follows, further, that if there are two primary input waves, each will have the effect described above independently of the other, for each of the driving forces will act independently and the two will add to produce a vector resultant. We call this the 'principle of superposition' for linear systems since the resultant effect of the two (or more) actions is just the sum of the effects of each one acting on its own. This has to be the case whilst the displacements of the electrons from their equilibrium positions in the atoms varies linearly with the force of the optical electric fields. Thus, if we pass into a medium, along the same path, two light waves, of angular frequencies ω_1 and ω_2, emerging from the medium will be two light waves (and only two) with those same frequencies, but with different amplitudes and phases from the input waves.

Suppose now, however, that the displacement of the electrons is *not* linear with the driving force. Suppose, for example, that the displacement is so large that the electron is coming close to the point of breaking free from the atom altogether. We are now in a nonlinear regime. Strange things happen here. For example, a given optical frequency input into the medium may give rise to waves of several different frequencies at the output. Two frequencies ω_1 and ω_2 passing in may lead to, among others, sum and difference frequencies $\omega_1 \pm \omega_2$ coming out.

The fundamental reason for this is that the driving sinusoid has caused the atomic electrons to oscillate non-sinusoidally (Fig. 9.1). Our knowledge of Fourier analysis tells us that any periodic non-sinusoidal function contains, in addition to the fundamental component, components at harmonic frequencies, i.e., integral multiples of the fundamental frequency.

This is a fascinating regime. All kinds of interesting new optical phenomena occur here. As might be expected, some are desirable, some are not. Some are valuable in new applications, some just comprise sources of noise. But to use them to advantage, and to minimize their effects when they are a nuisance, we must, of course, understand them better. This we shall now try to do.

9.2 NONLINEAR OPTICS AND OPTICAL FIBRES

Let us begin by summarizing the conditions which give rise to optical nonlinearity.

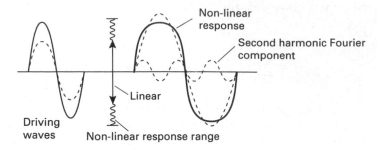

Fig. 9.1 Nonlinear response to a sinusoidal drive.

In a semiclassical description of light propagation in dielectric media, the optical electric field drives the atomic/molecular oscillators of which the material is composed, and these oscillators become secondary radiators of the field; the primary and secondary fields then combine vectorially to form the resultant wave. The phase of this wave (being different from that of its primary) determines a velocity of light different from that of free space, and its amplitude determines a scattering/absorption coefficient for the material.

Nonlinear behaviour occurs when the secondary oscillators are driven beyond the linear response; as a result, the oscillations become non-sinusoidal. Fourier theory dictates that, under these conditions, frequencies other than that of the primary wave will be generated (Fig. 9.1).

The fields necessary to do this depend upon the structure of the material, since it is this which dictates the allowable range of sinusoidal oscillation at given frequencies. Clearly, it is easier to generate large amplitudes of oscillation when the optical frequencies are close to natural resonances, and one expects (and obtains) enhanced nonlinearity there. The electric field required to produce nonlinearity in material therefore varies widely, from $\sim 10^6\,\mathrm{Vm}^{-1}$ up to $\sim 10^{11}\,\mathrm{Vm}^{-1}$, the latter being comparable with the atomic electric field. Even the lower of these figures, however, corresponds to an optical intensity of $\sim 10^9\,\mathrm{Wm}^{-2}$, which is only achievable practically with laser sources. It is for this reason that the study of nonlinear optics only really began with the invention of the laser, in 1960.

The magnitude of any given nonlinear effect will depend upon the optical intensity, the optical path over which the intensity can be maintained, and the size of the coefficient which characterizes the effect.

In bulk media the magnitude of any nonlinearity is limited by diffraction effects. For a beam of power P watts and wavelength λ focused to a spot of radius r, the intensity, $P/\pi r^2$, can be maintained (to within a factor of ~ 2) over a distance $\sim r^2/\lambda$ (Rayleigh distance), beyond which diffraction will rapidly reduce it. Hence the product of intensity and distance is $\sim P/\pi\lambda$, independent of r, and of propagation length (Fig. 9.2(a)).

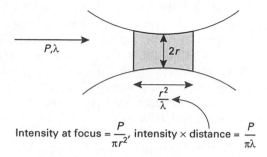

Intensity at focus $= \dfrac{P}{\pi r^2}$, intensity × distance $= \dfrac{P}{\pi \lambda}$

(a) The Rayleigh distance for free-space focusing

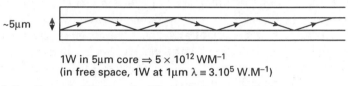

1W in 5μm core $\Rightarrow 5 \times 10^{12}$ WM^{-1}
(in free space, 1W at 1μm $\lambda \equiv 3.10^5$ W.M^{-1})

(b) Non-linear facility in optical fibres

Fig. 9.2 The intensity-distance product for non-linearity.

However, in an optical fibre the waveguiding properties, in a small diameter core, serve to maintain a high intensity over lengths of up to several kilometres (Fig. 9.2(b)). This simple fact allows magnitudes of nonlinearities, in fibres, which are many orders greater than in bulk materials. Further, for maximum overall effect, the various components' effects per elemental propagation distance must add coherently over the total path. This implies a requirement for phase coherence throughout the path which, in turn, implies a single propagation mode: monomode rather than multimode fibres must, in general, be used.

9.3 THE FORMALISM OF NONLINEAR OPTICS

In section 4.2 we assumed a linear relationship between the electric polarization (P) of a medium and the electric field (E) of an optical wave propagating in it, by taking:

$$\chi = \frac{P}{E}$$

(for convenience the constant ε_0 has been absorbed into χ; this only implies a change of units) where χ is the volume susceptibility of the medium, and is assumed constant. The underlying assumption for this is that the separation of atomic positive and negative charges is proportional to the imposed field, leading to a dipole moment per unit volume (P) which is proportional to the field.

However, it is clear that the linearity of this relationship cannot persist for ever-increasing strengths of field. Any resonant physical system must eventually be torn apart by a sufficiently strong perturbing force and, well before such a catastrophe occurs, we expect the separation of oscillating components to vary nonlinearly with the force. In the case of an atomic system under the influence of the electric field of an optical wave, we can allow for this nonlinear behaviour by writing the electric polarization of the medium in the more general form:

$$P(E) = \chi_1 E + \chi_2 E^2 + \chi_3 E^3 + \dots \chi_j E^j + \dots \tag{9.1}$$

The value of χ_j (often written $\chi^{(j)}$) decreases rapidly with increasing j for most materials. Also the importance of the j-th term, compared with the first, varies as $(\chi_j/\chi_1)E^{(j-1)}$, and so depends strongly on E. In practice, only the first three terms are of any great importance, and then only for laser-like intensities, with their large electric fields. It is not until one is dealing with power densities of $\sim 10^9\,\mathrm{Wm^{-2}}$, and fields $\sim 10^6\,\mathrm{Vm^{-1}}$, that $\chi_2 E^2$ becomes comparable with $\chi_1 E$.

Let us now consider the refractive index of the medium. In section 4.2 it was noted that:

$$\varepsilon = 1 + \chi, \quad n^2 = \varepsilon$$

Hence

$$n = (1+\chi)^{1/2} = \left(1 + \frac{P}{E}\right)^{1/2}$$

i.e.,

$$n = (1 + \chi_1 + \chi_2 E + \dots \chi_j E^{j-1} + \dots)^{1/2} \tag{9.2}$$

Hence we note that the refractive index has become dependent on E. The optical wave, in this nonlinear regime, is altering its own propagation conditions as it travels. This is a central feature of nonlinear optics.

9.4 SECOND HARMONIC GENERATION AND PHASE MATCHING

Probably the most straightforward consequence of nonlinear optical behaviour in a medium is that of the generation of the second harmonic of a fundamental optical frequency. To appreciate this mathematically, let us assume that the electric polarization of an optical medium is quite satisfactorily described by the first two terms of equation (9.1), i.e.,:

$$P(E) = \chi_1 E + \chi_2 E^2 \tag{9.3}$$

Before proceeding, there is a quite important point to make about equation (9.3).

Let us consider the effect of a change in sign of E. The two values of the field, $\pm E$, will correspond to two values of P:

$$P(+E) = \chi_1 E + \chi_2 E^2$$

$$P(-E) = -\chi_1 E + \chi_2 E^2$$

These two values clearly have different absolute magnitudes. Now if a medium is isotropic (as is the amorphous silica of which optical fibre is made) there can be no directionality in the medium and thus the matter of the sign of E, i.e., whether the electric field points up or down, cannot be of any physical relevance and cannot possibly have any measurable physical effect. In particular, it cannot possibly affect the value of the electric polarization (which is, of course, readily measurable). We should expect that changing the sign of E will merely change the sign of P, but that the magnitude of P will be exactly the same: the electrons will be displaced by the same amount in the opposite direction, all directions being equivalent. Clearly this can only be so if $\chi_2 = 0$. The same argument extended to higher order terms evidently leads us to the conclusion that all even-order terms *must* be zero for amorphous (isotropic) materials, i.e., $\chi_{2m} = 0$. This is a point to remember. The corollary of this argument is, of course, that in order to retain any even order terms the medium must exhibit some anisotropy. It must, for example, have a crystalline structure without a centre of symmetry. It follows that equation (9.3) refers to such a medium.

Suppose now that we represent the electric field of an optical wave entering such a crystalline medium by:

$$E = E_0 \cos \omega t$$

Then substituting into equation (9.3) we find

$$P(E) = \chi_1 E_0 \cos \omega t + \tfrac{1}{2}\chi_2 E_0^2 + \tfrac{1}{2}\chi_2 E_0^2 \cos 2\omega t$$

The last term, the second harmonic term at twice the original frequency, is clearly in evidence. Fundamentally, it is due to the fact that it is easier to polarize the medium in one direction than in the opposite direction, as a result of the crystal asymmetry. A kind of 'rectification' occurs.

Now the propagation of the wave through the crystal is the result of adding the original wave to the secondary wavelets from the oscillating dipoles which it induces. These oscillating dipoles are represented by P, as we saw in section 4.2. Thus $\partial^2 P/\partial t^2$ leads to e/m waves, since radiated power is proportional to the acceleration of charges, and waves at all of P's frequencies will propagate through the crystal.

Suppose now that an attempt is made to generate a second harmonic over a length L of crystal. At each point along the path of the input wave a second harmonic component will be generated. But, since the crystal medium will almost certainly be dispersive, the fundamental and second harmonic components will travel at different velocities. Hence the successive portions of second harmonic component generated by the fundamental will not, in general, be in phase with each other, and thus will not interfere constructively. Hence, the efficiency of the generation will depend upon the velocity difference between the waves. We met a very similar phenomenon when dealing with coupling in hi-bi fibres, in sections 3.9 and 8.7.

A rigorous treatment of this process requires a manipulation involving Maxwell's equations, and is given in Appendix VIII, but a semi-analytical treatment which retains a firm grasp of the physics will be given here.

Suppose that the amplitude of the fundamental (driving) wave between distances z and $z+dz$ along the optical path in the crystal is $e \cos(\omega t - kz)$. Then from equation (9.1), there will be a component of electric polarization (dipole moment per unit volume) of the form: $\chi_2 e^2 \cos^2(\omega t - kz)$ giving a time-varying second harmonic term $\frac{1}{2}\chi_2 e^2 \cos 2(\omega t - kz)$, as before. Consider, then, a slab, in the medium, of unit cross-section, and thickness dz (Fig. 9.3).

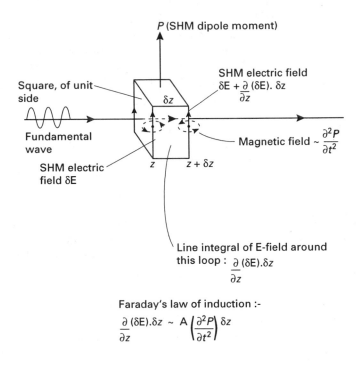

Fig. 9.3 Infinitesimals for second harmonic generation along the path of the fundamental.

For this slab the dipole moment will be:

$$P = \tfrac{1}{2}\chi_2 e^2 \cos 2(\omega t - kz)\mathrm{d}z \tag{9.4}$$

Now a time-varying dipole moment represents a movement of charge, and therefore an electric current. This current will create a magnetic field. A time-varying magnetic field (second derivative $\partial^2 P/\partial t^2$ of the dipole moment) will generate a voltage around any loop through which it threads (Faraday's law of electromagnetic induction). From Fig. 9.3 it can be seen that, if δE is the elemental component of the second harmonic electric field generated by the changing dipole moment in the thin slab, then this voltage is proportional to $\partial(\delta E)/\partial z$. Hence, we have

$$\frac{\partial(\delta E)}{\partial z} = A\frac{\partial^2 P}{\partial t^2}$$

where A is a constant.

Hence, from 9.4:

$$\frac{\partial(\delta E)}{\partial z} = -A.2\omega^2\chi_2 e^2 \cos 2(\omega t - kz)\mathrm{d}z$$

Integrating this w.r.t. z gives:

$$\delta E = A\frac{\omega^2}{k}\chi_2 e^2 \sin 2(\omega t - kz)\mathrm{d}z$$

and, with $\omega/k = c$, we have:

$$\delta E = Ac\omega\chi_2 e^2 \sin 2(\omega t - kz)\mathrm{d}z$$

as the element of the second harmonic electric field generated by the slab between z and $z+\mathrm{d}z$. But the second harmonic component now propagates with wavenumber k_s, say (since the crystal will have a different refractive index at frequency 2ω, compared with that at ω), so when this component emerges from the crystal after a further distance $L-z$, it will have become:

$$\delta E_L = Ac\omega\chi_2 e^2 \sin[2\omega t - 2kz - k_s(L-z)]\mathrm{d}z$$

Hence the total electric field amplitude generated over the length L of crystal will be, on emergence:

$$E_L(2\omega) = \int_0^L Ac\omega\chi_2 e^2 \sin[2\omega t - 2kz - k_s(L-z)]\mathrm{d}z$$

Performing this integration gives:

$$E_L(2\omega) = Ac\chi_2 e^2 L\omega \frac{\sin\left(k - \frac{1}{2}k_s\right)L}{\left(k - \frac{1}{2}k_s\right)L} \sin[2\omega t - (2k + k_s)L]$$

The intensity of the emerging second harmonic will be proportional to the square of amplitude of this, i.e.,

$$I_L(2\omega) = B\chi_2^2 e^4 L^2 \omega^2 \left[\frac{\sin\left(k - \frac{1}{2}k_s\right)L}{\left(k - \frac{1}{2}k_s\right)L}\right]^2$$

where B is another constant. Now the intensity of the fundamental wave is proportional to e^2, so the intensity of the second harmonic is proportional to the square of the intensity of the fundamental, i.e.,

$$I_L(2\omega) = B'\chi_2^2 I_L^2(\omega)L^2\omega^2 \left[\frac{\sin\left(k - \frac{1}{2}k_s\right)L}{\left(k - \frac{1}{2}k_s\right)L}\right]^2 \tag{9.5}$$

where B' is yet another constant. [Appendix VIII derives equation (9.5) rigorously.] From this we can define an efficiency η_{SHG} for the second harmonic generation process as:

$$\eta_{\text{SHG}} = \frac{I_L(2\omega)}{I_L(\omega)}$$

Note that η_{SHG} varies as the square of the fundamental frequency and of the length of the crystal; note also that it increases linearly with the power of the fundamental.

From equation (9.5) it is clear that, for maximum intensity, we require that the sinc^2 function has its maximum value, i.e., that:

$$k_s = 2k_f$$

This is the *phase matching condition* for second harmonic generation. Now the velocities of the fundamental and the second harmonic are given by:

$$c_f = \frac{\omega}{k_f}, \quad c_s = \frac{2\omega}{k_s}$$

These are equal when $k_s = 2k_f$, so the phase-matching condition is equivalent to a requirement that the two velocities are equal. This is to be expected, since it means that the fundamental generates, at each point in the material, second harmonic components which will interfere constructively. We encountered the same argument when dealing with coupling between polarization eigenmodes in a hi-bi fibre, in sections 3.9 and 8.7.

The phase-match condition usually can be satisfied by choosing the optical path to lie in a particular direction within the crystal. It has already been noted that the material must be anisotropic for second harmonic generation to occur; it will also, therefore, exhibit birefringence (section 3.3). One way of solving the phase-matching problem, therefore, is to arrange that the velocity difference resulting from birefringence is cancelled by that resulting from material dispersion. In a crystal with normal dispersion the refractive index of both the eigenmodes (i.e., both the ordinary and extraordinary rays) increases with frequency. Suppose we consider the specific example of quartz, which is a positive uniaxial crystal (see section 3.3). This means that the principal refractive index for the extraordinary ray is greater than that for the ordinary ray, i.e.,

$$n_e > n_0$$

Since quartz is also normally dispersive, it follows that:

$$n_e^{(2\omega)} > n_o^{(\omega)}$$

$$n_o^{(2\omega)} > n_o^{(\omega)}$$

Hence the index ellipsoids for the two frequencies are as shown in Fig. 9.4(a). Now it will be remembered from section 3.3 that the refractive indices for the 'o' and 'e' rays for any given direction in the crystal are given by the major and minor axes of the ellipse in which the plane normal to the direction, and passing through the centre of the index ellipsoid, intersects the surface of the ellipsoid. The geometry (Fig. 9.4(a)) thus makes it clear that a direction can be found [1] for which

$$n_o^{(2\omega)}(\vartheta_m) = n_e^{(\omega)}(\vartheta_m)$$

so SHG phase matching occurs provided that

$$n_o^{(2\omega)} < n_e^{(\omega)}$$

The above is indeed true for quartz over the optical range. Simple trigonometry allows ϑ_m to be determined in terms of the principal refractive indices as:

$$\sin^2 \vartheta_m = \frac{\left(n_e^{(\omega)}\right)^{-2} - \left(n_e^{(2\omega)}\right)^{-2}}{\left(n_o^{(2\omega)}\right)^{-2} - \left(n_e^{(2\omega)}\right)^{-2}}$$

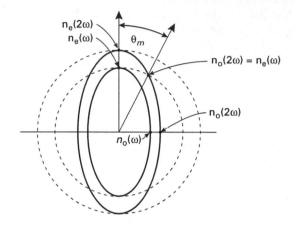

(a) Phase matching with the birefringence index ellipsoids

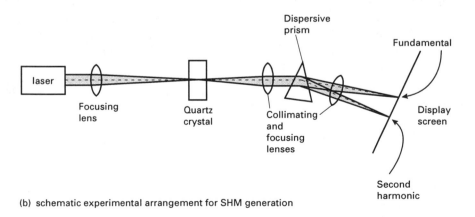

(b) schematic experimental arrangement for SHM generation

Fig. 9.4 Conditions for second harmonic generation in quartz.

Hence ϑ_m is the angle at which phase matching occurs. It also follows from this that, for second harmonic generation in this case, the wave at the fundamental frequency must be launched at angle ϑ_m with respect to the crystal axis and *must have the 'extraordinary' polarization*; and that the second harmonic component *will appear in the same direction and will have the 'ordinary' polarization*, i.e., the two waves are collinear and have orthogonal linear polarizations! Clearly, other crystal-direction and polarization arrangements also are possible in other crystals.

The required conditions can be satisfied in many crystals but quartz is an especially good one owing to its physical robustness, its ready obtainability with good optical quality and its high optical power-handling capacity.

Provided that the input light propagates along the chosen axis, the conversion efficiency ($\omega \rightarrow 2\omega$) is a maximum compared with any other path (per unit length) through the crystal. Care must be taken, however, to minimize the divergence of the beam (so that most of the energy travels in the chosen direction) and to ensure that the temperature remains constant (since the birefringence of the crystal will be temperature dependent).

The particle picture of the second harmonic generation process is viewed as an annihilation of two photons at the fundamental frequency, and the creation of one photon at the second harmonic frequency. This pair of processes is necessary in order to conserve energy i.e.,

$$2h\nu_f = h(2\nu_f) = h\nu_s$$

The phase-matching condition is then equivalent to conservation of momentum. The momentum of a photon wave number k is given by:

$$p = \frac{h}{2\pi}k$$

and thus conservation requires that:

$$k_s = 2k_f$$

as in the wave treatment.

Quantum processes which have no need to dispose of excess momentum are again the most probable, and thus this represents the condition for maximum conversion efficiency in the particle picture.

The primary practical importance of second harmonic generation is that it allows laser light to be produced at the higher frequencies, into the blue and ultraviolet, where conditions are not intrinsically favourable for laser action, as was noted earlier (section 6.2.4). In this context we note again, from equation (9.5), that the efficiency of the generation increases as the square of the fundamental frequency, which is of assistance in producing these higher frequencies.

9.5 OPTICAL MIXING

Optical mixing is a process closely related to second harmonic generation. If, instead of propagating just one laser wave through the same nonlinear crystal, we superimpose two (at different optical frequencies) simultaneously along the same direction, then we shall generate sum and difference frequencies, i.e.,

$$E = E_1 \cos \omega_1 t + E_2 \cos \omega_2 t$$

and thus again using equation (9.3):

$$P(E) = \chi_1(E_1 \cos \omega_1 t + E_2 \cos \omega_2 t) + \chi_2(E_1 \cos \omega_1 t + E_2 \cos \omega_2 t)^2$$

This expression for $P(E)$ is seen to contain the term

$$2\chi_2 E_1 E_2 \cos \omega_1 t \cos \omega_2 t = \chi_2 E_1 E_2 \cos(\omega_1 + \omega_2)t + \chi_2 E_1 E_2 \cos(\omega_1 - \omega_2)t$$

giving the required sum and difference frequency terms. Again, for efficient generation of these components, we must ensure that they are phase matched. For example, to generate the sum frequency efficiently we require that:

$$k_1 + k_2 = k_{(1+2)}$$

which is equivalent to

$$\omega_1 n_1 + \omega_2 n_2 = (\omega_1 + \omega_2)n_{(1+2)}$$

where the n represent the refractive indices at the suffix frequencies. The condition again is satisfied by choosing an appropriate direction relative to the crystal axes.

This mixing process is particularly useful in the reverse sense. If a suitable crystal is placed in a Fabry-Perot cavity which possesses a resonance at ω_1, say, and is 'pumped' by laser radiation at $\omega_{(1+2)}$, then the latter generates both ω_1 and ω_2. This process is called parametric oscillation: ω_1 is called the signal frequency and ω_2 the idler frequency. It is a useful method for 'down conversion' of an optical frequency, i.e., conversion from a higher to a lower value.

The importance of phase matching in nonlinear optics cannot be overstressed. If waves at frequencies different from the fundamental are to be generated efficiently they must be produced with the correct relative phase to allow constructive interference, and this, as we have seen, means that velocities must be equal to allow phase matching to occur. This feature dominates the practical application of nonlinear optics.

9.6 INTENSITY-DEPENDENT REFRACTIVE INDEX

It was noted in section 9.4 that all the even-order terms in the expression (9.1) for the non-linear susceptibility (χ) are zero for an amorphous (i.e., isotropic) medium. This means, of course, that, in an optical fibre, made from amorphous silica, we can expect that $\chi_{(2m)} = 0$, so it will not be possible to generate a second harmonic according to the principles outlined in section 9.4. (However, second harmonic generation has been observed in fibres [2] for reasons which took some time to understand!) It is possible to generate a third harmonic, however, since to a good approximation the electric polarization in the fibre can be expressed by:

$$P(E) = \chi_1 E + \chi_3 E^3 \tag{9.6}$$

Clearly, though, if we wish to generate the third harmonic efficiently we must again phase match it with the fundamental, and this means that somehow we must arrange for the two relevant velocities to be equal, i.e., $c_\omega = c_{3\omega}$. This is very difficult to achieve in practice, although it has been done.

There is, however, a more important application of equation (9.6) in amorphous media. From section 4.2 we know that the effective refractive index in this case can be written:

$$n_e = (1 + \chi_1 + \chi_3 E^2)^{1/2}$$

and, if $\chi_1, \chi_3 E^2 \ll 1$,

$$n_e \approx 1 + \tfrac{1}{2}\chi_1 + \tfrac{1}{2}\chi_3 E^2$$

Hence:

$$n_e = n_o + \tfrac{1}{2}\chi_3 E^2 \tag{9.7a}$$

where n_o is the 'normal', linear refractive index of the medium. But we know that the intensity (power/unit area) of the light is proportional to E^2, so that we can write:

$$n_e = n_o + n_2 I \tag{9.7b}$$

where n_2 is a constant for the medium. Equation (9.7b) is very important and has a number of practical consequences. We can see immediately that it means that the refractive index of the medium depends upon the intensity of the propagating light: the light is influencing its own velocity as it travels.

In order to fix ideas to some extent let us consider some numbers for silica. For amorphous silica $n_2 \sim 3.2 \times 10^{-20}\,\mathrm{m^2 W^{-1}}$, which means that a 1% change in refractive index (readily observable) will occur for an intensity $\sim 5 \times 10^{17}\,\mathrm{Wm^{-2}}$. For a fibre with a core diameter $\sim 5\,\mu\mathrm{m}$ this requires an optical power level of 10 MW. Peak power levels of this magnitude are readily obtainable, for short durations, with modern lasers.

It is interesting to note that this phenomenon is another aspect of the electro-optic effect which was described in section 3.9. Clearly the refractive

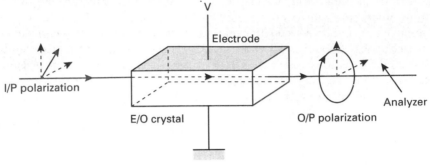

(a) 'Normal' electro-optic Kerr effect

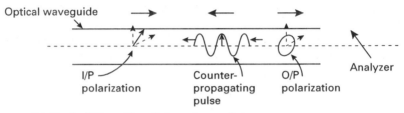

(b) 'Optical' Kerr effect: light acting on light

Fig. 9.5 'Normal' and 'optical' Kerr effects.

index of the medium is being altered by an electric field. This will now be considered in more detail.

9.6.1 Optical Kerr effect

The normal electro-optic Kerr effect was considered in section 3.9. It is an effect whereby an electric field imposed on a medium induces a linear birefringence with slow axis parallel with the field (Fig. 9.5(a)). The value of the induced birefringence is proportional to the square of the electric field. In the optical Kerr effect the electric field involved is that of an optical wave, and thus the birefringence probed by one wave may be that produced by another (Fig. 9.5(b)).

The phase difference introduced by an electric field E over an optical path L is given by:

$$\Delta\varphi = \frac{2\pi}{\lambda}\Delta n L$$

where $\Delta n = KE^2$ [see equation (7.1b)], K being the Kerr constant.

Now from (9.7a) and (9.7b) we have:

$$\Delta n = n_2 I = \tfrac{1}{2}\chi_3 E^2 = KE^2 \qquad (9.8)$$

From our discussions of elementary electromagnetics in section 2.4 we derived equation (2.6):

$$I = c\varepsilon E^2$$

Hence we have, from (9.8):

$$K = n_2 c\varepsilon = \tfrac{1}{2}\chi_3$$

showing that the electro-optic effect, whether the result of an optical or an external electric field, is a nonlinear phenomenon, depending on χ_3. Using similar arguments it can easily be shown that the electro-optic Pockels effect (section 7.3.1) also is a nonlinear effect, depending on χ_2. (Remember that the Pockels effect can only occur in anisotropic media, so that χ_2 will be non-zero).

The optical Kerr effect has several other interesting consequences. One of these is self-phase modulation, which is the next topic for consideration.

9.6.2 Self-phase modulation (SPM)

The fact that refractive index can be dependent on optical intensity clearly has implications for the phase of the wave propagating in nonlinear medium. We have:

$$\varphi = \frac{2\pi}{\lambda} nL$$

Hence for $n = n_o + n_2 I$,

$$\varphi = \frac{2\pi L}{\lambda}(n_o + n_2 I)$$

Suppose now that the intensity is a time-dependent function $I(t)$. It follows that φ also will be time dependent, and, since:

$$\omega = \frac{\mathrm{d}\varphi}{\mathrm{d}t}$$

the frequency spectrum will be changed by this effect, which is known as *self-phase modulation* (SPM).

In a dispersive medium a change in the spectrum of a temporally varying function (e.g. a pulse) will change the shape of the function. For example, pulse broadening or *pulse compression* can be obtained under appropriate circumstances. To see this, consider a Gaussian pulse (Fig. 9.6(a)). The Gaussian shape modulates an optical carrier of frequency ω_0, say, and the new instantaneous frequency becomes

$$\omega' = \omega_0 + \frac{\mathrm{d}\varphi}{\mathrm{d}t}$$

If the pulse is propagating in the Oz direction:

$$\varphi = -\frac{2\pi z}{\lambda}(n_o + n_2 I) \tag{9.9a}$$

and we have

$$\omega' = \omega_0 - \frac{2\pi z}{\lambda} n_2 \frac{dI}{dt} \tag{9.9b}$$

At the leading edge of the pulse $dI/dt > 0$, hence

$$\omega' = \omega - \omega_I(t)$$

At the trailing edge

$$\frac{dI}{dt} < 0$$

and

$$\omega' = \omega + \omega_I(t)$$

Hence the pulse is now 'chirped', i.e., the frequency varies across the pulse. Fig. 9.6(b) shows an example of this effect.

Suppose, for example, a pulse from a mode-locked Argon laser, initial width 180 ps, is passed down 100 m of optical fibre. As a result of self-phase modulation the frequency spectrum is changed by the propagation. Fig. 9.6(c) shows how the spectrum varies as the initial peak power of the pulse is varied. The peak power will lead to a peak phase change, according to equation (9.9a) and this phase change is shown for each of the spectra. It can be seen that the initial spectrum ($\Delta\varphi = 0$) is due just to the modulation of the optical sinusoid (Fourier spectrum of a Gaussian pulse) and, as the value of $\Delta\varphi$ increases, the first effect is a broadening. At $\Delta\varphi = 1.5\pi$ the spectrum has split into two clear peaks, corresponding to the frequency shifts at the back and front edges of the pulse. The spectra then develop multiple peaks.

It is important to realize that this does not necessarily change the shape of the pulse envelope, just the optical frequency within it. However, if the medium through which the pulse is passing is dispersive, the pulse shape will change. This is an interesting possibility and it will be considered further in section 9.9.

9.7 FOUR-PHOTON MIXING (FPM)

In sections 9.4 and 9.5 we saw how two photons of certain frequencies could be 'mixed' to generate photons at different frequencies in the processes of second harmonic generation, and of sum and difference frequency generation. These processes were 'mediated' (as we say) by χ_2. In section 9.6

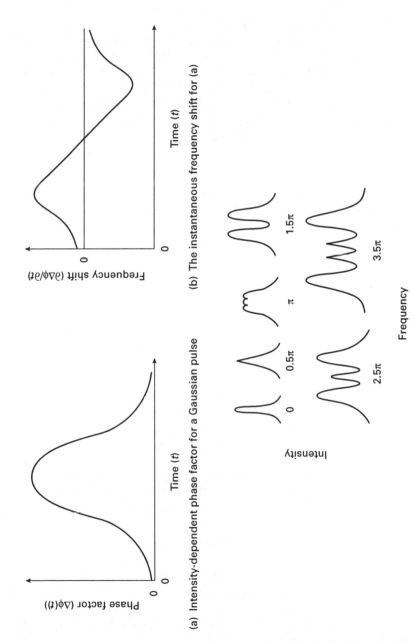

(a) Intensity-dependent phase factor for a Gaussian pulse

(b) The instantaneous frequency shift for (a)

(c) Frequency spectra for (a) designated by maximum phase shift, at peak

Fig. 9.6 Self-phase modulation for a Gaussian pulse [6].

we saw how the electro-optic effect in amorphous media (Kerr effect) was mediated by χ_3. Can χ_3 be used to generate new frequencies? If it can, it would be very convenient if it could do so in amorphous silica because then it could be used with the high intensity, long path lengths associated with optical fibres, and efficient generations could be expected. The two-photon mixing processes considered in the preceding sections relied upon the mixing of two fields, and thus on two photons, in the squared-field second term of equation (9.1): $\chi_2 E^2$. If χ_3 is to be used via the third term, $\chi_3 E^3$, we naturally expect, therefore, that three photons will be involved.

Let us consider three optical waves of frequencies ω_p, ω_s and ω_a and further suppose (for reasons which will soon become clear) that these are related by:

$$2\omega_p = \omega_s + \omega_a \qquad (9.10)$$

(Note that four photons are involved here).

It is clear that, under this condition, the term $\chi_3 E_{\omega_p}^2 E_{\omega_a}$ will generate a frequency:

$$2\omega_p - \omega_a = \omega_s$$

and that the term $\chi_3 E_{\omega_p}^2 E_{\omega_s}$ will generate a frequency:

$$2\omega_p - \omega_s = \omega_a$$

Hence ω_s and ω_s are continuously generated by each other, with the assistance of ω_p and χ_3. However, as we know very well now, this can only take place efficiently if there is phase matching, i.e., if the velocities of ω_p, ω_s and ω_a are all the same.

Interestingly, this can be achieved using high-linear-birefringence (hi-bi, section 8.7) fibre. Remember that the two linearly polarized eigenmodes in such a fibre have different velocities. Also remember that there is material and waveguide dispersion to take into account. The result is that, provided that the 'pump' frequency (ω_p) is chosen correctly in relation to the dispersion characteristic and launched as a linearly polarized wave with its polarization direction aligned with the fibre's slow axis, then, as in the case of second harmonic generation in a crystal (section 9.4), the combination of the velocity difference in the other polarization eigenmode (fast axis) and the dispersion in the fibre (material and waveguide) can allow two other frequencies, ω_s and ω_a, to have the same velocity in the fast mode as ω_p has in the slow mode, and also to satisfy (9.10). ω_s is called the 'Stokes' frequency and ω_a the anti-Stokes frequency when $\omega_a > \omega_s$. This process clearly involves four photons ($\omega_p, \omega_p, \omega_a, \omega_s$) and hence the name *four-photon mixing* (FPM). (It is sometimes also referred to as 'three-wave mixing', for obvious reasons.)

The process is analogous to that known as parametric down-conversion in microwaves, where it is used to produce a down-converted frequency

(ω_s) known as the 'signal' and an (unwanted) up-converted frequency (ω_a) known as the 'idler'. An optical four-photon mixing frequency spectrum generated in hi-bi fibre is shown in Fig. 9.7.

Four-photon mixing has a number of uses. An especially valuable one is that of an optical amplifier. If a pump is injected at ω_p, it will provide gain for signals injected (in the orthogonal polarization of course) at ω_s or ω_a. The gain can be controlled by injecting signals at ω_s and ω_a simultaneously, and then varying their relative phase. The pump will provide more gain to the component which is the more closely phase matched (Fig. 9.8).

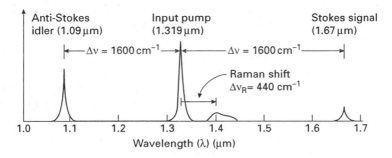

Fig. 9.7 Four-photon mixing spectrum in hi-bi fibre [7].

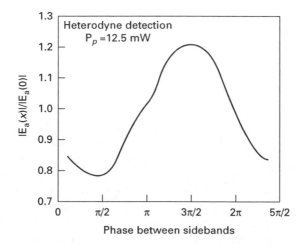

Fig. 9.8 Dependence of parametric gain on phase matching in four-photon mixing [8].

Another useful application is that of determining which is the 'fast' and which the 'slow' axis of a hi-bi fibre. Only when the pump is injected into the slow axis will FPM occur. This determination is surprisingly difficult by any other method. By measuring accurately the frequencies ω_a and ω_s, variations in birefringence can be tracked, implying possibilities for use in optical-fibre sensing of any external influences which affect the birefringence (e.g. temperature, stress).

Finally the effects of FPM can be unwanted, also. In optical-fibre telecommunications the generation of frequencies other than that of the input signal, via capricious birefringence effects, can lead to cross-talk in multichannel systems (e.g. wavelength-division-multiplexed (WDM) systems).

9.8 PARAMETRIC AND INELASTIC PROCESSES

In the discussion of nonlinear optical processes so far we have considered effects due to the nonlinear susceptibility, χ. Now as explained in section 4.2, χ is a measure of the ease with which an imposed electric field can separate the centres of positive and negative electric charge, and this separation is almost entirely due to the movement of the electron charge distributions, since electrons are so much more mobile than the positive nuclei. Hence it follows that the non-linear processes considered so far are the result of the near-instantaneous responses of atomic electrons to fields which push them beyond their linear displacements. These processes are referred to as 'parametric' nonlinear effects, since they effectively rely on the parameter χ.

However, there is another class of nonlinear optical effects in materials, the so-called 'inelastic' class, and it concerns the 'inelastic' scattering of primary, propagating radiation. The word inelastic refers to the fact that the optical energy is not simply redistributed into other optical waves via the 'mediation' effects of the atomic electrons but, in this case, is converted into other forms of energy: heat energy or acoustic energy, for example.

The two best-known inelastic scattering effects are the Raman effect and the Brillouin (pronounced 'Breelooah' after the famous Frenchman) effect. Classically, these two effects are broadly explicable in terms of Doppler shifts. When light is incident upon a moving atom or molecule, the light which is scattered will be Doppler shifted in frequency. If the scatterer is moving away from the incident light, either as a result of bulk movement of the material or as a result of electron oscillation within the molecule, then the frequency of the scattered light will be downshifted; if it is moving towards the scatterer, it will be upshifted. Downshifted, or lower frequency, scattered light is called Stokes radiation; upshifted light is called anti-Stokes radiation. (Hence the designations for ω_s and ω_a in the case of FPM, previous section.)

It should be emphasized, however, that only one photon in $\sim 10^6$ takes part in a frequency-shift Stokes or anti-Stokes process. The vast majority simply re-radiate at the same frequency to give rise to the Rayleigh scattering already discussed in section 4.2.

When the frequency shift is due to motion resulting from molecular vibration or rotations the phenomenon is referred to as the Raman effect; when it is due to bulk motions of large numbers of molecules, as when a sound wave is passing through the material, it is called the Brillouin effect.

Of course all such motions are quantized at the molecular level. Transitions can only take place between discrete energy levels, and scattering occurs between photons and photons in Raman scattering, and between photons and 'phonons' (quantized units of acoustic energy) in Brillouin scattering.

We shall deal firstly with the Raman effect.

9.8.1 Raman scattering

When an intense laser beam of angular frequency ω_L is incident upon a material the radiation scattered from the medium contains frequencies higher and lower than ω_L. As ω_L is varied, the spectrum of frequencies moves along with ω_L (Fig. 9.9). It is, in other words, the difference between ω_L and the spectrum of frequencies which the medium scatters which is characteristic of the medium.

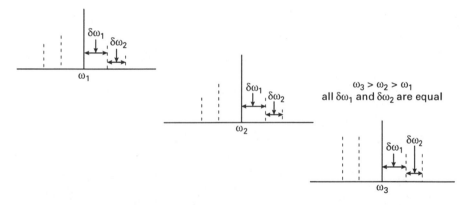

Fig. 9.9 Raman spectra at different pump frequencies.

These difference frequencies are just the vibrational and rotational modes of the material's molecular structure, and thus Raman spectroscopy is a powerful means by which this structure can be examined. For a given (quantified) vibrational frequency ω_v we have:

$$\omega_s = \omega_L - \omega_v \tag{9.11a}$$

where ω_s is the Stokes (downshifted) frequency and

$$\omega_a = \omega_L + \omega_v \tag{9.11b}$$

where ω_a is the anti-Stokes (upshifted) frequency.

It is often useful to begin with a classical (i.e., non-quantum) explanation of a physical effect (if possible!) since this provides our inadequate thought processes with 'pictures' which make us feel more comfortable, but also, and probably more importantly, gives us a better idea of which other physical quantities might influence the effect. The classical explanation of the Raman effect resides in the notion of a variable susceptibility for molecules. The normal definition of susceptibility is as given in section 4.2:

$$\chi = \frac{P}{E}$$

(where the ε_0 is again absorbed into χ for convenience). However, we know from equation (9.1) that P/E, which is a measure of the ease with which positive and negative charges are separated by an electric field, is not a constant, but can be expressed as a power series in the electric field. The fundamental reason for this is, of course, that the force needed further to separate the charges depends, to some extent, on the actual separation. Clearly, if the simple case of positive and negative point charges $+q$ and $-q$ is considered, the force between them is given by:

$$F = \frac{Cq^2}{r^2}$$

(where C is a constant), so the force required for further separation must be greater than this, and hence this force will vary as $1/r^2$.

In the classical picture of a vibrating molecule the distance between the centres of positive and negative electric charge varies sinusoidally, with small amplitude, about a mean value, so that we can expect the volume susceptibility, χ, also to vary sinusoidally, with the same frequency, according to:

$$\chi = \chi_0 + \chi_0' \sin \omega_m t$$

say, with $\chi_0' \ll \chi_0$

If now, with this molecular oscillation occurring, an optical electric field of the form:

$$E = E_0 \sin \omega_e t$$

is incident upon the molecule, the electric polarization will become:

$$P = (\chi_0 + \chi_0' \sin \omega_m t) E_0 \sin \omega_e t$$

i.e.,

$$P = \chi_0 E_0 \sin \omega_e t + \tfrac{1}{2} \chi_0' E_0 \cos(\omega_e - \omega_m) - \tfrac{1}{2} \chi_0' E_0 \cos(\omega_e + \omega_m)$$

The first term in this expression represents normal Rayleigh scattering from the molecule. It is the dominant term, and leads to scattered radiation at the same frequency as the incident optical wave. The other two, much smaller, terms give downshifted and upshifted components of P. We know from section 9.4 (and Appendix V) that a time-varying electric polarization generates radiation at the frequency of variation, so the shifted P-terms lead to the Raman radiation at frequencies $\omega_e - \omega_m$ (Stokes) and $\omega_e + \omega_m$ (anti-Stokes).

This simple, classical picture is useful quantitatively in indicating that the scattered radiation will have an intensity proportional to the intensity of the incident optical wave, and to the square of the differential susceptibility, χ_0'.

However, for a proper quantification of the Raman effect it is necessary to return to the quantum description, and we begin this by considering some of the consequences of equations (9.11a) and (9.11b). In (9.11a) a laser photon has interacted with the atom to give rise to a lower frequency Stokes photon. To conserve energy we must have:

$$h\omega_L = h\omega_s + h\omega_v$$

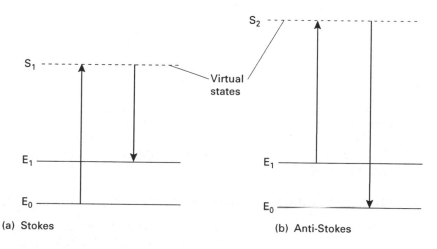

Fig. 9.10 Raman energy-level transitions.

Hence the emission of the Stokes photon must be accompanied by the excitation of the molecule to an excited vibrational state. But how can this be if $h\omega_L/2\pi$ does not correspond to a molecular transition? To answer this we have to invoke the notion of *virtual* energy levels. A molecule can exist in a virtual level which differs from a real level of the system by an energy $\Delta\varepsilon$, but only for a time $\Delta\tau$, where: $\Delta\varepsilon\,\Delta\tau \sim h$, the quantum constant.

This is another manifestation of the quantum uncertainty principle. In this case the Stokes emission can be explained by means of the diagram shown in Fig. 9.10(a). The laser photon raises the molecule from the ground state (E_0) to a short-lived virtual state (S_1) which then decays to an excited state. The anti-Stokes case is shown in Fig. 9.10(b). The laser photon now raises an already-excited molecule to an even higher virtual state (S_2) which then decays to the ground state. We now have a satisfactory explanation of the phenomenon in quantum terms.

Of course, in thermal equilibrium, the ratio of the number of molecules in the excited state, E_1, to the number in the ground state, E_0, is given by the Boltzmann factor:

$$\frac{n_1}{n_0} = \exp\left(-\frac{E_1 - E_0}{kT}\right) = \exp\left(-\frac{\hbar\omega_v}{2\pi kT}\right) \tag{9.12}$$

and will be very much less than unity. Consequently, the anti-Stokes radiation will be at a much lower level than the Stokes radiation. This is partly offset by the larger scattering efficiency at the higher frequency: the efficiency increases as the fourth power of the frequency (Appendix V). Hence the ratio of anti-Stokes to Stokes radiation levels when the medium is in thermal equilibrium is given by:

$$\frac{I_a}{I_s} = \frac{\omega_a^4}{\omega_s^4} \exp\left(-\frac{\hbar\omega_v}{2\pi kT}\right) \tag{9.13}$$

where

$$\omega_v = \omega_L - \omega_s = \omega_a - \omega_L$$

The same phase-matching conditions apply as always. For efficient generation of the Stokes radiation, for example, we must have the velocity of the laser beam equal to that of the Stokes-frequency component and this must be arranged, if dealing with a crystal medium, by choice of a suitable direction, as before.

9.8.2 Stimulated Raman scattering (SRS)

The Raman scattering process we have been considering up to now can be called 'spontaneous' scattering, since it depends, for the generation of Stokes and anti-Stokes radiation, upon the spontaneous decay of the molecule from its virtual states down to the lower states. This will be the case when the relative density of photons at the Stokes and anti-Stokes frequencies is close to the equilibrium levels given by equation (9.12), since one way of regarding a spontaneous decay is that it is 'stimulated' to decay by the photons present within the correct energy range in equilibrium. (And even for an excited, isolated atom there are 'vacuum fluctuations' which perform the same task: it is never possible to say that energy within a given volume, and frequency range, is zero, for this would mean that the energy was known exactly, and there will always be uncertainty of knowledge $\Delta\epsilon$, where $\Delta\epsilon\Delta\tau \sim h$).

Suppose now that there are many Stokes photons, caused by a particularly intense laser pumping beam. Then these Stokes photons will stimulate other Stokes transitions from the virtual level to the excited level; this will cause an increase in the excited level population which will then increase the anti-Stokes radiation, which will itself become stimulated. The whole system becomes self-driving and self-sustaining. This is stimulated Raman scattering (SRS). It occurs quite readily in optical fibres, for all the reasons already explained. Hence, we should now study the Raman effect in optical fibres.

9.8.3 The Raman effect in optical fibres

Optical fibres, as we know, allow long, high intensity, optical path interaction lengths, and thus we would expect the Raman effect, and especially the stimulated Raman effect, to be relatively easy to observe within them. This is indeed the case.

The fibre is made from fused silica, an amorphous medium. Consequently, there is a large variety of vibrational and rotational frequencies among all the varying-strength and varying-orientation chemical bondings. Hence there is a broad spectrum of Raman radiation (Fig. 9.11).

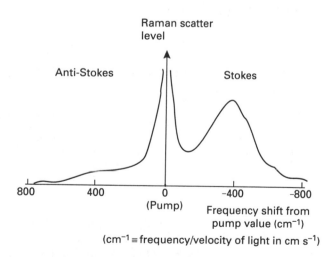

Fig. 9.11 Raman spectrum for silica.

The fact of the broad spectrum eases the phase-matching restrictions. This is because each of the scatterings occurs with a large measure of phase independence (since the structure is random) and thus the resultant intensity is simply the sum of the separate scattered intensities from the individual centres and this is true for any direction. (The theory of random scatterers was developed in section 4.2.) Hence the Raman scattering process in amorphous silica can quite readily be observed in multimode fibres (in addition to single-mode fibres) since there is no phase-matching condition to satisfy and therefore no requirement for phase coherence from the pump wave: the random contributions from all the modes can add incoherently just as can the contributions from the random scattering centres. The extra advantage of being able to use a multimode fibre is that it is easier to launch into it large levels of pump power, and therefore to generate high levels of Raman radiation.

When the pump laser signal is very intense, as has been noted, it is possible to produce stimulated Raman radiation in the fibre. In this case the Stokes photons are so plentiful that new Stokes photons are more likely to be generated from other Stokes photons, by stimulated emission, than by spontaneous decay following pump excitation. In this case, then, the rate of increase of Stokes radiation is proportional to both the pump laser photon density n_L and the Stokes photon density, n_s. In fact, taking the fibre axis as Oz, we can express the rate at which the Stokes photon density increases as the pump propagates along Oz by a differential equation of the form:

$$\frac{\mathrm{d}n_s}{\mathrm{d}z} = An_L(n_s + 1) \tag{9.14}$$

where A is a constant for the material, the '1' is necessary in the third factor $(n_s + 1)$ in order to allow the process to begin at the front end, when $n_s = 0$, from the stimulation. The single photon (per unit volume) which the '1' represents will be generated spontaneously.

Of course, it is very quickly the case that $n_s \gg 1$, so that equation (9.14) becomes, for most of the fibre

$$\frac{\mathrm{d}n_s}{\mathrm{d}z} = An_L n_s$$

the solution of which is:

$$n_s = n_s(z_0) \exp(An_L z) \tag{9.15a}$$

where $n_s(z_0)$ is the value of the Stokes density at that value of z (i.e., z_0) where $n_s \gg 1$ first can be regarded as valid.

Equation (9.15a) is more conveniently expressed in the form:

$$I_s(z) = I_z(z_0) \exp(gI_L z) \tag{9.15b}$$

where I_s is now the Stokes intensity, I_L is the pump intensity and g is called the 'Raman gain'. Since g must be positive (from the structure of the exponent in (9.15a)), it is clear that the Stokes photon density rises exponentially with distance along the fibre axis. Further, if a beam of radiation at the Stokes frequency is injected into the fibre at the same time as the pump, then this light acts to cause stimulated Stokes radiation and hence is amplified by the pump. In this case:

$$I_s(z) = I_s(0) \exp(gI_L z)$$

where $I_s(0)$ is now the intensity of the injected signal. Hence we have a Raman amplifier. Moreover, since there are no phase conditions to satisfy, pump and signal beams can even propagate in opposite directions, allowing easy separation of the two components.

Finally, if the Stokes radiation is allowed to build up sufficiently, over a long length of fibre, it can itself act as a pump which generates second-order Stokes radiation, at a frequency now $2\nu_v$ lower than the original pump. This second-order radiation can then generate a third-order Raman signal, etc. To date, five orders of Stokes radiation have been observed in an optical fibre (Fig. 9.12).

Such Stokes Raman sources are very useful multiple-laser-line sources which have been used (among other things) to measure monomode fibre dispersion characteristics.

9.8.4 Practical applications of the Raman effect

It will be useful to summarize some of the uses and consequences of the Raman effect (especially the effect in optical fibres) and, at the same time, to fix ideas by providing some numbers.

Spontaneous Raman scattering will always occur to some extent when an intense optical beam is passed through a material. It provides valuable information on the molecular structure of the material. The spontaneous Raman effect is also used in distributed sensing (see section 10.8).

Stimulated emission will occur when there is more chance of a given virtual Stokes excited state being stimulated to decay, than of decaying spontaneously. Stimulated emission will be the dominant propagation when its intensity exceeds that of the pump. For a typical monomode fibre this occurs for a laser pump power $\sim 5\,\mathrm{W}$. This means that the power-handling capacity of fibres is quite severely limited by Raman processes. Above $\sim 5\,\mathrm{W}$, the propagation breaks up into a number (at least three) of frequency components. There are also implications for cross-talk in multichannel optical-fibre telecommunications.

The Raman spectrum shown in Fig. 9.11 has a spectral width $\sim 40\,\mathrm{nm}$, emphasizing the lack of coherence and the broad gain bandwidth for use in optical amplification. Remember also that it means that amplification can occur in both forward and backward directions with respect to the pump propagation. The broad bandwidth is a consequence of the large variety

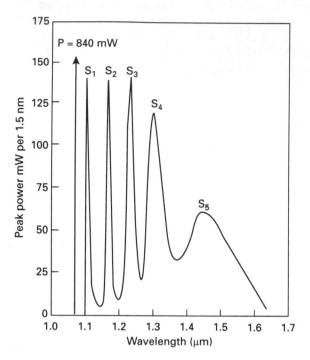

Fig. 9.12 Spectrum of multiple-order Stokes emission [9].

of rotational-vibrational energy transitions in an amorphous material. The Raman cross-section, and thus the gain, can be enhanced by the use of suitable dopants in the fibre: GeO_2 is a well-known one, and it may be remembered that this dopant is also used to increase the core refractive index in elliptically-cored hi-bi fibres. These latter fibres are thus very useful for Raman applications.

Gains of up to 45 dB have been obtained with fibre Raman lasers. The maximum gain is obtained when both pump and signal have the same (linear) polarization, thus indicating another advantage for the use of elliptically-cored hi-bi fibres.

By increasing the fibre's input pump power to $\sim 1\,kW$, up to five orders of Stokes radiation can be generated, as explained in the preceding section. Higher orders than five are broadened into a quasi-continuum by the effects of self-phase modulation and four-photon mixing discussed in Sections 9.6.2 and 9.7. By placing the fibre within a Fabry-Perot cavity, as for the fibre laser, a fibre Raman laser can be constructed. By tuning the cavity length, this laser can be tuned over $\sim 30\,nm$ of the Raman spectral width (Fig. 9.13). This tunability is extremely useful, in a source which is so readily compatible with other optical fibres, for a range of diagnostic procedures

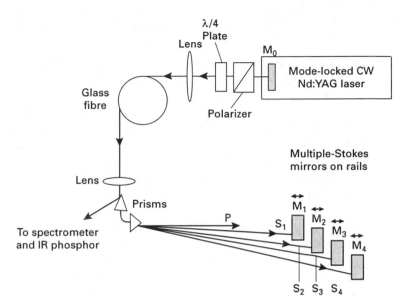

Fig. 9.13 Schematic for a fibre-Raman laser [10].

in optical-fibre technology.

9.8.5 Brillouin scattering

Brillouin scattering is the result of an interaction between a pump laser and the bulk motions of the medium in which it is propagating. The bulk motions effectively comprise an acoustic wave in the medium. The phenomenon can thus be regarded essentially as a Bragg-type scattering from the propagating acoustic wave in the medium (see section 7.3.3).

In quantum mechanical terms the effect can be explained in just the same way as was the Raman effect. It is essentially the same effect, the only difference being that the excitation energy is not now due to molecular vibration/rotation but to the bulk motion. The bulk motion must be quantized, and the relevant quanta are called 'phonons' in this case. The rest of the explanation is the same.

However, we are in a different regime and the values of the physical quantities are very different. The strength of the interaction is much greater, the bandwidth is much narrower and, since the medium moves, acoustically, as a coherent whole, phase-matching conditions are now important, even in amorphous media.

In the case of Brillouin scattering, the various phenomena are probably, initially, best understood in terms of the classical interaction between the optical pump (laser) and the acoustic wave in the medium. This is the approach we shall adopt.

If an acoustic wave propagates in a medium, the variations in pressure give rise to variations in strain which, via the strain-optic effect, give rise to corresponding variations in refractive index in the medium. We met these effects before when dealing with acousto-optic modulation (section 7.3.3). Sound waves will always be present in a medium at a temperature above absolute zero, since the molecules are in thermal motion, and the consequent dynamic interaction will couple energy into the natural vibrational modes of the structure. Hence a propagating optical wave (a pump laser) will be scattered from these refractive index variations (photon-phonon scattering). These effects will comprise spontaneous scattering and, since the acoustic waves are of low energy, this will be at a very low level.

As the power of the pump laser is increased, however, some of its power may be backscattered from an acoustic wave travelling along the same path, either forwards or backwards. The velocity of the acoustic wave will, of course, have a definite value, characteristic of the medium.

Now since the laser radiation is being backscattered from what is, essentially, a moving Bragg diffraction grating, there will be a Doppler shifting of the backscattered optical radiation to either above or below the pump's frequency (Fig. 9.14.)

This Doppler-shifted wave now interferes with the forward-propagating laser radiation to produce an optical standing wave in the medium which, owing to the Doppler frequency difference, moves through the medium, at just the acoustic velocity (this will be proved, analytically, shortly).

The standing wave so produced will consist of large electric fields at the anti-nodes and small fields at the nodes (Fig. 9.14).

Now whenever an electric field is applied to a medium, there will be a consequent mechanical strain on the medium. This is a result of the fact that the field will perturb the inter-molecular forces which hold the medium together, and will thus cause the medium to expand or contract. This phenomenon is known as electrostriction, and, as might be expected, its magnitude varies enormously from material to material.

The result of electrostriction in this case is to generate an acoustic wave in sympathy with the optical standing wave. Hence the backscattered wave has generated a moving acoustic diffraction grating from which further backscattering can occur. Then the pump wave and the Doppler-shifted scattered waves combine to produce diffraction gratings which move forwards and backwards at the acoustic velocity; each of the three-wave interactions is stable; the forward acoustic wave producing the Stokes backscattered signal, and the backward wave the anti-Stokes signal. The complete self-sustaining system comprises the stimulated Brillouin scattering (SBS) phenomenon. As always, for an understanding of the phenomenon sufficient to be able to use it, it is necessary to quantify the above ideas.

The Doppler frequency shift from an acoustic wave moving at velocity

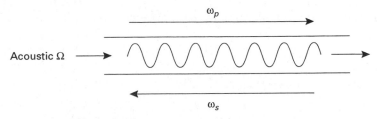

(a) Stokes scatter: $\omega_s = \omega_p - \Omega$

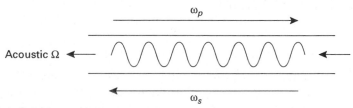

(a) Anti-Stokes scatter: $\omega_a = \omega_p + \Omega$

Fig. 9.14 Basic Brillouin scatter processes in optical fibre.

v is given by:

$$\frac{\delta\omega}{\omega_p} = \frac{2v}{c} \tag{9.16}$$

where $\delta\omega$ is the angular frequency shift, ω_p the angular frequency of the optical pump and c the velocity of light in the medium.

For the sum of forward (pump) and backward (scattered) optical waves at frequencies ω_p and ω_B, respectively, we may write

$$S = E_p \cos(\omega_p t - k_p z) + E_B \cos(\omega_B t + k_B z)$$

which gives, on manipulation:

$$S = (E_p - E_B) \cos(\omega_p - k_p z) +$$

$$E_B[\cos(\omega_p t - k_p z) + \cos(\omega_B t + k_B z)]$$

i.e.,

$$S = (E_p - E_B) \cos(\omega_p t - k_p z) +$$

$$2E_B \cos\tfrac{1}{2}[(\omega_p + \omega_B)t + (k_B - k_p)z] \cos\tfrac{1}{2}[(\omega_p - \omega_B)t + (k_B + k_p)z]$$

This expression represents a wave (first term) travelling in direction Oz, plus a standing wave (second term) whose amplitude is varying as

$$\cos\tfrac{1}{2}[(\omega_p - \omega_B)t + (k_B + k_p)z]$$

This comprises an envelope which moves with velocity

$$v_e = \frac{\omega_p - \omega_B}{k_p + k_B}$$

but $k_p = \omega_p/c$ and $k_B = \omega_B/c$, so that

$$k_p + k_B = \frac{\omega_p + \omega_B}{c} = \frac{2\omega_p + \delta\omega}{c} \approx \frac{2\omega_p}{c}$$

since $\delta\omega/\omega_p$ is very small.

Thus, $k_p + k_B \approx 2k_p$ and

$$v_e = \frac{\delta\omega}{2k_p} = v$$

(from (9.16)), and hence the standing wave moves at the acoustic velocity.

Further, the standing wave will have a distance between successive anti-nodes of $\frac{1}{2}\lambda_p$, the same, to first order, for both directions of propagation, again because $\delta\omega/\omega_p \ll 1$.. Hence this (i.e., $\frac{1}{2}\lambda_p$) will be the acoustic wavelength.

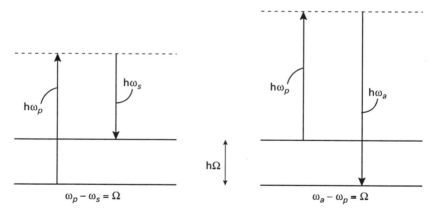

Fig. 9.15 Frequency relations for Brillouin scattering.

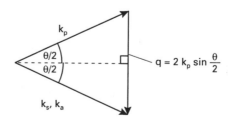

Fig. 9.16 Momentum conservation for Brillouin scattering.

Let us now look at the matching conditions. Energy conservation requires that (Fig. 9.15):

$$\delta\omega = \omega_p - \omega_s = \omega_a - \omega_p = \Omega$$

where Ω is the acoustic angular frequency and p, a, s refer to the pump, anti-Stokes and Stokes components, respectively. Momentum conservation requires that:

$$q = k_p - k_s = k_a - k_p$$

where q is the acoustic wavenumber. If ϑ is the angle between the pump's propagation direction and that of either of the two scattered waves, we have (Fig. 9.16) with $k_a \approx k_s \approx k_p$:

$$q = 2k_p \sin \tfrac{1}{2}\vartheta$$

Now for an optical fibre, to first order, we have only two possible values of ϑ, i.e., 0 or π. If $\vartheta = 0, q = 0$ and there can be no acoustic wave. If $\vartheta = \pi, q = 2k_p$ and the acoustic wave has wavelength $\tfrac{1}{2}\lambda_p$, which is the case we considered qualitatively. Thus the acoustic wavelength is half the optical pump wavelength. The magnitude of the Stokes or anti-Stokes frequency shift will be given by:

$$\delta f = \frac{\Omega}{2\pi} = \frac{2v}{\lambda_p} = \frac{2nv}{\lambda_0}$$

where λ_0 is the free-space optical wavelength, and n is the refractive index of the fibre medium. (This result also could have been obtained from the expression the Doppler shift.) For silica at $\lambda_0 = 1.5\,\mu m$ we find that $\delta f \sim 11.5\,\text{GHz}$. Hence the frequency shift in the Brillouin effect is some two orders of magnitude smaller than for the Raman effect.

An important feature of Brillouin backscatter in fibres is the optical bandwidth with which it is associated. This bandwidth is determined by the decay time of the acoustic wave, which is about 5 ns at these frequencies. Hence for coherent scatter from an acoustic wave, the optical wave also must itself remain coherent for $\sim 5\,\text{ns}$, which implies an optical bandwidth $\sim 100\,\text{MHz}$, for optimum Brillouin effect. This is a narrow bandwidth, even for laser sources, so in general only a small fraction of the light from a given source will suffer Brillouin backscatter. This explains why Raman scattering usually dominates: it is able to utilize a much greater fraction of light emitted from conventional laser sources. However, for low dispersion and, especially, for coherent optical communications systems (see section 10.10.4) very small optical bandwidths are used, much less than 100 MHz, and Brillouin scattering then becomes the limiting factor on the level of transmitted power. A threshold for SBS has been measured [3] at only

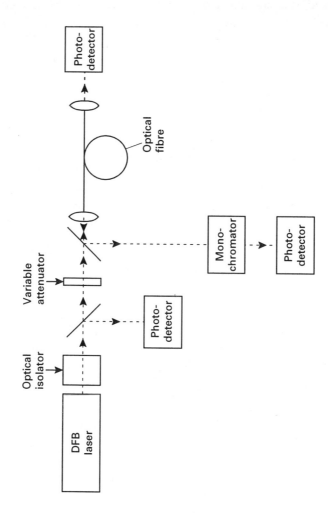

Fig. 9.17 Schematic arrangement for observing stimulated Brillouin scatter in an optical fibre [3].

5 mW for a 13.6 km length of fibre, with a 1.6 MHz optical bandwidth from a distributed feedback (DFB) laser (Fig. 9.17). SBS can be suppressed by applying a phase modulation to the source light. This effectively destroys the phase coherence necessary for constructive backscatter, and thus allows higher power levels to be transmitted. This has considerable advantages for coherent optical communications systems [3] (Figs 9.18(a) and (b)).

Brillouin amplifiers and oscillators (Fig. 9.19) can be constructed much as for the Raman effect. The significant advantages which they possess over those for the Raman effect are that the frequency differences are much smaller (~ 10 GHz) and thus allow electronic, rather than optical, filtering and detection techniques, and that the bandwidth of the oscillation and amplification can be very narrow (< 25 MHz). This latter advantage renders Brillouin amplifiers very suitable for amplification of the optical carrier, as opposed to the modulation sidebands, in coherent optical communications systems.

9.9 SOLITONS

No account of nonlinear optics even at the 'essential' level would be complete without a brief discussion of optical solitons. A soliton is a 'solitary wave', a wave pulse which propagates, even in the face of group velocity dispersion (GVD) in the medium, over long distances without change of form. It thus has enormous potential for application to long-distance optical-fibre digital communications.

The soliton is not limited to optics. It comprises a particular set of solutions of the nonlinear Schrodinger wave equation and is a possibility whenever non-linear wave motion occurs in a dispersive medium. It was first observed in 1834 (before any theory was worked out and certainly long before Schrodinger formulated his wave equation) as a large amplitude water wave propagating along the Union canal which connects Edinburgh and Glasgow, in Scotland. John Scott Russell, a Scottish civil engineer, was exercising his horse alongside the canal in the summer of 1834 when he noticed a wave pulse which had been generated by the sudden halting of a horse-drawn barge. This wave travelled without change of shape or size, and Russell followed it on his horse for two kilometres, noting how puzzlingly stable it was. This turned out to be a water-wave soliton, but solitons were subsequently observed in many other branches of physics, wherever wave motion can occur, in fact, for all restorative systems are capable of being drive into nonlinearity.

The detailed mathematical analysis of this phenomenon is complex, but the basic ideas are relatively straightforward; moreover they follow from ideas which we have already discussed. In section 4.3 we studied group velocity dispersion (GVD) and earlier in this chapter (section 9.6.2) we

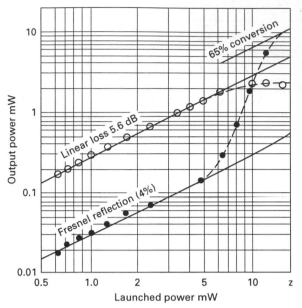

(a) Transmitted and reflected power in SBS: O Transmitted
● Reflected

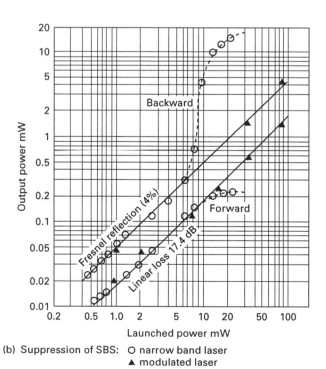

(b) Suppression of SBS: O narrow band laser
▲ modulated laser

Fig. 9.18 Power dependence and suppression of SBS in an optical fibre [3].

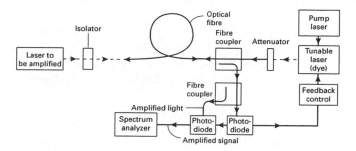

Fig. 9.19 Schematic for a stimulated Brillouin amplifier.

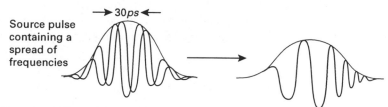

Source pulse containing a spread of frequencies

→ 30ps ←

(a) Negative GVD: high frequencies have greater velocity than low frequencies and the pulse is broadened

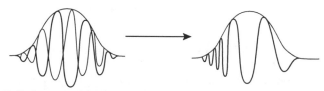

(b) Self phase modulation: high frequencies are 'chirped' to the trailing edge of the pulse

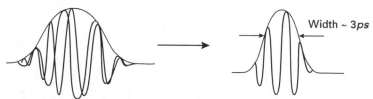

Width ~ 3ps

(c) Self phase modulation together with negative GVD: the effects in (a) and (b) balance to generate a stable 'soliton'

Fig. 9.20 Essentials of soliton formation.

studied self-phase modulation (SPM). These two ideas must be put together in order to understand solitons.

Let us, as in section 9.6.2, take an optical wave pulse with a Gaussian intensity envelope, and pass it into a dispersive medium. Since the source of the pulse will have a non-zero spectral width the GVD will act on this pulse to broaden it: the positive GVD will cause the lower frequencies to arrive at the output first, and the negative GVD the higher frequencies first. Clearly, in both cases the pulse is broadened. If self-phase modulation (SPM) also is present, however, we know that, from the consequences of equation (9.9b), higher frequencies are produced at the trailing edge of the pulse, while lower ones are produced at the leading edge. If this effect happens in the presence of negative GVD, then the trailing edge of the pulse will tend to catch up with leading edge: the pulse will be compressed. It is easy to see that under certain circumstances this compressive effect might exactly balance the spreading effect due to the source spectral width, and the pulse width can then remain constant throughout the propagation.

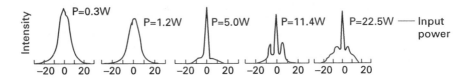

Fig. 9.21 Measured solitons emerging from an optical fibre [11].

This balance can indeed be struck and the result is a soliton. Fig. 9.20 illustrates this process.

Solitons have been observed in optical fibres (Fig. 9.21) and could lead to optical communications systems of phenomenal bandwidth × distance products, perhaps as high as 10,000 GHz km. However, the theory shows solitons to be unstable in lossy media. In addition they tend to attract each other when closer than ∼ 10 pulse widths apart. These features clearly limit their advantages in the communications area.

Their potential remains considerable, however, and the research into them undoubtedly will continue.

An interesting corollary to the discussion of the effects which give rise to solitons involves the interaction of a Gaussian pulse with a positive GVD. Fig. 9.22 shows the effect, on the spectrum of this pulse, of SPM with a *positive* GVD medium. It is seen there that the result is a much extended region of linear chirp (Fig. 9.22(c)) where the frequency varies linearly from the front edge to the back edge of the pulse. If, after this has been done, the pulse is then passed into a purely *negative* GVD medium (e.g. another fibre), the pulse can be very strongly compressed. Pulse widths as small as

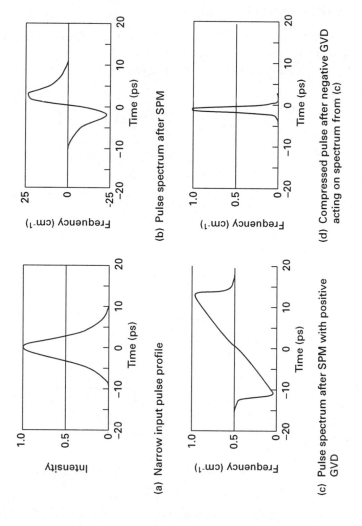

(a) Narrow input pulse profile

(b) Pulse spectrum after SPM

(c) Pulse spectrum after SPM with positive GVD

(d) Compressed pulse after negative GVD acting on spectrum from (c)

Fig. 9.22 Pulse compression using SPM with positive GVD followed by negative GVD.

8 fs (8×10^{-15} s) have been produced, using such a method, at a wavelength of 620 nm. Such pulses contain only about four optical cycles!

Pulses such as these can be used in research to study, for example, very fast molecular processes, single-atom chemical reactions and very fast switching phenomena.

9.10 PHOTOSENSITIVITY

It is possible for light to bring about changes in a fibre's refractive index which persist long after the optical stimulus has ceased. Sometimes the changes are, in fact, permanent (or semipermanent).

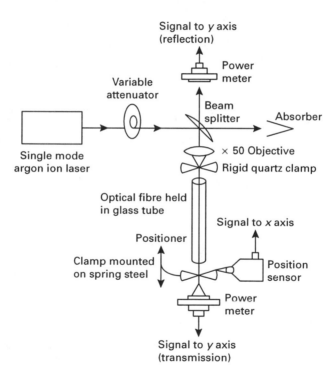

Fig. 9.23 Schematic arrangement for recording reflection gratings in an optical fibre [12].

The first such phenomenon was observed in 1978 [4]. In this, argon laser light was launched into ∼ 50 cm of a Ge-doped monomode fibre, and both the backscattered and transmitted beams were carefully monitored (Fig. 9.23). The reflected light level was observed to rise quite steeply over a period of a few minutes, with a corresponding fall in the transmitted level. Investigation revealed that a standing-wave interference pattern had

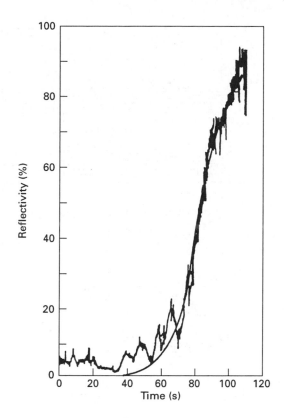

Fig. 9.24 Growth of reflectivity with time during formation of a photosensitive grating in an optical fibre [13].

been set up in the fibre, via the light Fresnel-reflected (initially) from the fibre end. This intensity pattern had written, over a period of minutes, a corresponding periodic refractive index variation in the fibre, which then acted to Bragg-reflect the light. Clearly, as the reflection increased in strength, so also did the 'visibility' of the interference pattern: positive feedback was present.

Up to 90% of the light could be backreflected in this way (Fig. 9.24). The 'grating' persists when the source light is switched off. Such 'fibre filters' have many potential uses. First, they are spectrally very narrow ($\sim 300\,\text{MHz}$ over 30 cm of fibre); secondly, they are tunable over a limited range (e.g., by stretching the fibre or changing its temperature). Interesting polarization signatures may also be written into the fibre using this technique.

A different kind of application involves the dispersion characteristic in the region of the fabricated grating's absorption line. This can allow a negative GVD to be generated at a convenient wavelength, so as to provide

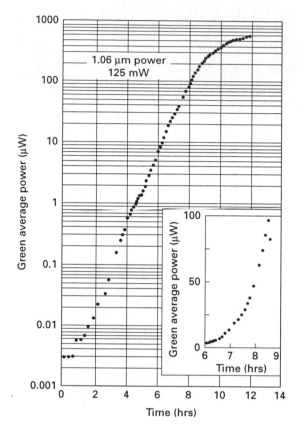

Fig. 9.25 Growth of second harmonic component in a monomode optical fibre exposed to laser radiation at 1.06 μm wavelength [5].

the possibility of pulse compression without an external delay line. It can also provide the means by which the GVD accumulated over a long optical-fibre communication path can be compensated, with consequent annulment of the pulse distortion caused by the GVD, and thus a greatly increased communications bandwidth.

Photosensitivity is also thought to be the cause of second harmonic generation (SHG), which has been observed [5] in single-mode fibre exposed to pulsed 1.06 μm radiation from a Nd-YAG laser. SHG should not be possible in an amorphous fibre (as discussed in section 9.6, $\chi_2 = 0$). However, Fig. 9.25 shows how the SHG component grew to $\sim 3\%$ of the input power over a 10-hour period. The exact mechanism for this generation is not (at the time of writing) fully understood, although it is thought to be due to a 'poling' (i.e., ordering) of the silica molecules by the optical electric field. Neither, in fact, is the general nature of the 'writing' process well understood. It is believed to be due to the germanium atom's effect on the silica lattice in producing lattice 'defects', or 'dangling bonds', which become traps for electrons (F-centres). The refractive index depends upon the occupancy of these traps, and the occupancy depends upon the input light intensity; photons can eject the electrons from their traps. However, there remain some puzzling features, and a fully quantified understanding has yet to be worked out.

9.11 CONCLUSIONS

We have seen in this chapter that nonlinear optics has its advantages and disadvantages. When it is properly under control it can be enormously useful; but on other occasions it can intrude, disturb and degrade.

The processes by which light waves produce light waves of other frequencies need very high optical electric fields and thus high peak intensities. It was for this reason that nonlinear optics only became a serious subject with the advent of the laser. Optical fibres provide a convenient means by which peak intensities can be maintained over relatively long distances, and are thus very useful media for the study and control of nonlinear optical effects.

We must also remember that in order to cause one optical frequency component to generate another, the second must be generated in phase with itself along the generation path: phase matching is an important feature of such processes.

We have noted the difference between parametric effects (optical interactions are 'mediated' by an optical material) and 'inelastic' effects (the material joins in enthusiastically). And we have seen how the effects which occur when electrons are stretched beyond the comfortable sinusoidal oscillations in their atoms or molecules can yield useful extra optical waves (Raman, Brillouin) and can influence their own propagation conditions (optical Kerr effect, SPM), and how when balancing two well-known effects (SPM, GVD) we can generate the amazing distortion-free solitons. Finally we can also use light to alter, permanently or semi-permanently, the optical properties of a medium, and thus provide the means whereby a new class of optical components, especially fibre components, can be fabricated.

There is a wealth of potential here. The exploration of possibilities for non-linear optics, especially in regard to new, natural or synthetic optical materials (e.g. organics, high T_c superconductors) etc., has not even really begun. The prospects, for example, for new storage media, fast switching of light by light, and three-dimensional television, which will be opened up in the future by such materials, are intriguing, and it could well be that nonlinear optical technology soon will become a powerful subject in its own right.

PROBLEMS

9.1 Define what is meant by the nonlinear electrical polarization of an optical medium.

The electrical polarization for a general optical medium is given by

$$P = \sum_{j=1}^{\infty} \chi_j E^j$$

Explain how can this expression be modified for amorphous (i.e., isotropic) media, and give reasons for the validity of the modification.

Show how the expression for P can lead to the generation of the second harmonic of an optical wave propagating in the medium, and discuss the conditions which affect the efficiency with which this harmonic is generated. Why is such generation (apparently) only possible in a crystalline medium?

9.2 Discuss the meaning of the term 'nonlinear optical propagation'. The electric polarization of an optical medium is given by:

$$P(E) = \chi_1 E + \chi_2 E^2$$

where E is the electric field and χ_1, χ_2 are constants. If an optical wave whose electric field is given by

$$E = E_0 \cos \omega t$$

propagates through the medium, show that the intensity of the second harmonic (at frequency 2ω), after a distance L of propagation path, is proportional to $\chi_2^2 L^2 \omega^2$

9.3 The dielectric susceptibility for an optical material is given by:

$$\chi = \chi_0 + \chi_1 E + \chi_2 E^2 + \chi_3 E^3$$

An optical wave of the form

$$E = E_0 \exp i(\omega t - kz)$$

is incident upon the material. What is the value of the electric polarization of the component which is generated at frequency 3ω within the material? What conditions must be satisfied for this component to be generated with maximum efficiency? How would you seek to satisfy these conditions in practice?

9.4 Discuss the essential differences between parametric and inelastic non-linear optical processes.

Describe in detail the *photon* description of: (i) second harmonic generation, (ii) spontaneous Raman scattering and (iii) stimulated Brillouin scattering.

9.5 A high-energy pulse of light, with wavelength 514 nm, is launched into an optical fibre at a uniform temperature of 295 K. The backscattered spontaneous Raman radiation is detected. The Stokes spectrum peaks at 495 nm and is, at that wavelength, at a power level 65 dB lower than that of the launched pulse. If fibre attenuation can be neglected, what will be the level of the anti-Stokes radiation at 533 nm, compared with the launched power? What will the level be at 300 K?

9.6 Explain the two three-wave interactions which give rise to the stimulated Brillouin scattering phenomenon in optical fibres, including the part played by electrostriction.

If the backscattered optical frequency is shifted by 11.75 GHz from the forward, pump wave at 1550 nm what is the acoustic velocity in the fibre medium if its refractive index is 1.46 at the pump wavelength?

Devise a purely electronic means for measuring the amplitude of the frequency-shifted signal, using only a PIN photodetector.

9.7 Explain how optical solitons can be generated in an optical fibre. How could the same phenomena be used for pulse compression?

9.8 What is meant by photosensitivity in optical fibres? Provide a plausible explanation of the phenomenon. Assuming that your explanation is correct, how might the photosensitivity be enhanced?

To what practical uses might this phenomenon be put?

REFERENCES

[1] Nye, J. F. (1976) *Physical Properties of Crystals*, Clarendon Press, Oxford, chap. 13.

[2] Fujii, Y., Kawasaki, B. S., Hill, K. O. and Johnson, D. C. (1980) Sum-frequency light generation in optical fibres, *Opt. Lett.*, **5**, 48.

[3] Cotter, D. (1982) Suppression of stimulated Brillouin scattering during transmission of high power narrowband laser light in monomode fibre, *Electron. Lett.*, **18**, 638.

[4] Hill, K. O., Fujii, Y., Johnson D. C. and Kawasaki, B.S. (1978) Photosensitivity in optical-fibre waveguides, *Appl. Phys. Lett.*, **32**, 647.

[5] Osterberg, U. Margulis, W. (1986) Dye laser pumped by Nd-YAG laser pulses frequency doubled in a glass optical fibre, *Opt. Lett.*, **11**, 516.

[6] Stolen R. H. and Lin, C. (1978) Self-phase modulation in optical fibres, *Phys. Rev. A*, **17**, 1448-52.

[7] Lin, C. et al. (1981) Phase matching in the minimum chromatic dispersion region of SM Fibres for stimulated FPM, *Opt. Lett.*, **6**, 493.

[8] Bar-Joseph, I. *et al.* (1986) Parametric interaction of a modulated wave in an SM fibre, *Opt. Lett.*, **11**, 534.

[9] Cohen, L. G. and Lin, C. (1978) A universal fibre-optic measurement system based on a Near IR fibre Raman laser', *IEEE J. Quant. Elect.*, **QE-14**, 855.

[10] Lin, C. and French, W. G. (1979) A near IR fibre Raman oscillator, *Appl. Phys. Lett.*, **34**, 10.

[11] Mollenauer, L.F., Stolen, R. H. and Gordon, J. P. (1980) Experimental observation of picosecond narrowing and solitons in optical fibres', *Phys. Rev. Lett.*, **45**, 1095.

[12] Kawasaki, B. S., Hill, K.O., Johnson D.C. and Fujii, Y. (1978) Narrow-band Bragg reflectors in optical fibres, *Opt. Lett.*, **3**, 66.

[13] Bures, J., Lapierre, J. and Pascale, D. (1980) Photosensitivity in optical fibres: a model for growth of an interference filter', *Appl. Phys. Lett.*, **37**, 860.

FURTHER READING

Agrawal, G. P. (1989) *Nonlinear Fiber Optics*, Academic Press.

Boyd, R. W. (1992) *Nonlinear Optics*, Academic Press.

Guenther, R. D. (1990) *Modern Optics*, John Wiley and Sons, chap. 15.

10

Optoelectronics in action

10.1 INTRODUCTION

In the preceding chapters we have discussed many optoelectronic phenomena and many ideas for making use of them. With all of this material well appreciated (hopefully!) and largely understood, it will certainly be valuable to view the ideas in action. For this reason, the present chapter will describe a number of devices and systems, in current usage, which make use of the ideas and methods that have been encountered throughout the various earlier discussions. This will serve two purposes: first, it will help to cement the ideas, and place them more firmly into a practical context; secondly, it will provide an insight into the power and utility of practical optoelectronics and offer a glimpse into what developments might be possible in the future. The following (final) chapter will attempt to peer more speculatively, but purposefully, into what the future holds in store.

In order to satisfy our present purpose of illustrating the ideas in action, I have chosen a sub-set of devices and systems which cover as broad as possible a range of the ideas previously discussed in the book, but without the encumbrance of any intimidating complexity. (The aim of this book has always been to concentrate on the essentials.) With these practical ideas firmly in place the more advanced developments will follow quite comfortably. Prior to each device/system description will be found a list of the optoelectronic ideas which it utilizes, and a reference to the chapter where each idea has been discussed. This will be of assistance if any revision should be found desirable.

We begin with a very useful device in pure optics but move quickly on to systems which employ a large variety of ideas.

10.2 ANTIREFLECTIVE COATINGS

Topics needed:
   (i) Multiple wave interference (Chapter 2)
   (ii) Fresnel's dielectric equations (2)
   (iii) Material dispersion (4).

Optoelectronic systems often contain many optical components such as lenses, filters, polarizers, etc., each of which represents one or more air/glass (or other optical material) interfaces. At normal incidence, Fresnel's equation (2.13c) tells us that the ratio of reflected to incident light intensities at such an interface between two media (1,2) is given by:

$$\frac{I_r}{I_i} = \left(\frac{n_1 - n_2}{n_1 + n_2}\right)^2$$

where n_1 is the refractive index of the first medium and n_2 that of the second (Fig. 10.1), at the wavelength of interest. This means that for an air/glass interface, with $n_1 \sim 1$ and $n_2 \sim 1.5$, we have that:

$$\frac{I_r}{I_i} \sim 0.04$$

so that $\sim 4\%$ of the light is lost at each interface. For a camera lens, for example, with two such interfaces, the loss is $\sim 8\%$; in low light level conditions this could be a quite serious disadvantage.

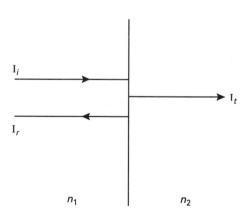

Fig. 10.1 Geometry for reflection at the boundary between dielectric media.

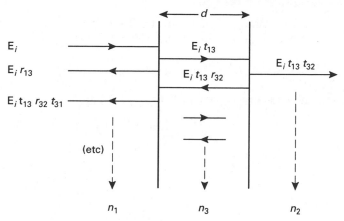

Fig. 10.2 Geometry and nomenclature for reflections at boundaries between three dielectric media.

Consider now the arrangement shown in Fig. 10.2. Here a thin layer of a third dielectric material has been interposed between the two original media. Suppose that this interposed layer has refractive index n_3 and width d. Suppose now that a (monochromatic) wave is travelling in medium 1 towards the thin layer. This wave will experience a reflection at the 1,3 interface and another reflection at the 3,2 interface. If these two reflections can be made to interfere destructively, i.e., to cancel each other, then the reflection from the front (1,3) interface will be eliminated. Let us calculate the conditions for this to occur. In section 2.9 the topic of multiple reflections of diminishing amplitude was covered, and these ideas are needed again here.

Suppose that the incident optical electric field is E_i. This will suffer a reflection at the 1,3 interface, so the reflected field may be written:

$$\frac{E_r}{E_i} = r_{13} = \frac{n_1 - n_3}{n_1 + n_3} \tag{10.1}$$

from Fresnel's equation (2.13a).

The wave transmitted will be

$$\frac{E_t}{E_i} = t_{13} = \frac{2n_1}{n_1 + n_3}$$

and this will be reflected from the 3,2 interface and will reappear at the front face with a phase lag of φ, where

$$\varphi = \frac{2\pi}{\lambda_3} 2d$$

(where λ_3 is the optical wavelength in material 3), having performed a double passage across the slab of thickness d. This wave will be partially transmitted and partially reflected at the 3,1 interface. The transmitted wave amplitude which accompanies this reflection will be:

$$E'_r = E_i t_{13} r_{32} t_{31} \exp(i\varphi)$$

where, from the Fresnel equations (2.13 a,b) we have:

$$t_{13} = \frac{2n_1}{n_1 + n_3}$$
$$r_{32} = \frac{n_3 - n_2}{n_3 + n_2} \qquad (10.2)$$
$$t_{31} = \frac{2n_3}{n_1 + n_3}$$

The reflected component will execute another double passage of layer 3 via another reflection at the 3,2 interface (see Fig. 10.2) and will give rise to a wave reflected into material 1, with amplitude

$$E''_r = E_i t_{13} r_{32} r_{31} r_{32} t_{31} \exp(2i\varphi)$$

It is clear that this process will continue, with the n-th wave reflected back into medium 1 given by:

$$E_r^{(n)} = E_i t_{13} r_{32} t_{31} \left(r_{32} r_{31}\right)^n \exp(in\varphi)$$

Hence the total wave amplitude reflected back into medium 1 will be:

$$E_R = r_{13} E_i + \sum_{n=1}^{\infty} E_i t_{13} r_{32} t_{31} \left(r_{32} r_{31}\right)^{(n-1)} \exp(in\varphi)$$

The summation term is a simple geometrical series, so we find:

$$\frac{E_r}{E_i} = r_{13} + \frac{t_{13} r_{32} t_{31}}{1 - r_{32} r_{31} \exp(i\varphi)} \qquad (10.3)$$

Suppose now that we arrange for the width d of layer 3 to be an odd integer times $\frac{1}{4}\lambda_3$. In this case the light will suffer a phase delay of π for one double passage of the layer, so successive reflections within the layer will tend to cancel out. The condition on φ thus becomes:

$$\varphi = \frac{2\pi}{\lambda_3} 2(2m+1)\frac{\lambda_3}{4} = (2m+1)\pi$$

where m is a positive integer.

Hence

$$\exp(i\varphi) = -1$$

and equation (10.3) becomes:

$$\frac{E_R}{E_i} = r_{13} + \frac{t_{13}r_{32}t_{31}}{1 + r_{32}r_{31}}$$

Substituting now for the r, t coefficients from (10.2) we find that

$$\frac{E_R}{E_i} = \frac{n_1 n_2 - n_3^2}{n_1 n_2 + n_3^2} \tag{10.4}$$

It is clear from this equation that the reflected wave can be eliminated by choosing:

$$n_3 = \sqrt{n_1 n_2} \tag{10.5}$$

In practice there are few materials which have both the right refractive index [according to (10.5)] and the appropriate physical properties to act as an anti-reflection or 'blooming' layer. For the glass/air interface we have

$$n_3 \approx \sqrt{1 \times 1.5} = 1.22$$

Either cryolite (sodium aluminium fluoride) at $n_3 = 1.35$ or magnesium fluoride at $n_3 = 1.38$ are used for their hardness and good adhesion to glass.

The 'blooming' layer is evaporated on to the surface of the glass lens, under interferometric control. The process is halted when the reflected wave disappears, at the chosen wavelength, which, for visible optics such as camera lenses, will be chosen as the centre of the visible range, i.e., at $\sim 550\,\text{nm}$.

Of course, equation (10.5) will be satisfied strictly for only one wavelength since the materials will be dispersive, and hence their refractive indices will vary with wavelength.

The removal of the greenish reflected light around 550 nm means that bloomed lenses have a bluish-red appearance to the eye. The overall reflectivity for the whole visible spectrum can be reduced from $\sim 4\%$ to $\sim 1\%$ by use of antireflection coatings and, as stressed earlier, this is a much-used device for many sensitive optical systems such as cameras, telescopes, microscopes and narrowband optical filters.

Some simple ideas in wave interference have yielded an extremely valuable practical result in this case.

10.3 OPTICAL-FIBRE CURRENT MEASUREMENT

Topics needed:
(i) Faraday magneto-optic effect (Chapter 3)
(ii) Polarization optics (Jones matrices) (3)

(iii) Polarization properties of optical fibres (3)
(iv) Photodetectors (7).

Having dealt, in the preceding section, with an optical device, we turn now, by contrast, to an optoelectronic system. By studying this we shall begin to see the advantages in bringing together several ideas from both optics and electronics.

A prevalent problem in the electricity supply industry (ESI) throughout the world is that of the measurement of the electric current produced by generating stations of all types. The current generated is distributed to customers of the supply company on a grid of high voltage lines (since the higher the voltage the smaller is the current for a given transfer of power and thus the smaller are the resistive losses). This current must be measured accurately. This measurement is necessary to ensure that the customer is charged the correct amount for the energy which is used; but it is also necessary for proper control of the grid network, and for rapid indication of fault conditions, to allow corrective action.

The current usually is measured at distribution junctions known as switching substations, and it must be done on high voltage bus-bars, the voltage level being several hundred kilovolts.

Conventional current transformers, using primary and secondary induction windings, are fairly widespread for this measurement purpose but they require costly high voltage insulation between the two windings, and they inevitably suffer from saturation effects, owing to ferromagnetic hysteresis, and from relatively low bandwidth, owing to the large winding inductances. Consider the alternative, optoelectronic approach to high-voltage current measurement illustrated in Fig. 10.3.

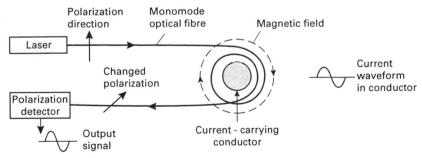

Fig. 10.3 Schematic for optical-fibre current measurement.

A length of monomode optical fibre (a good insulator) is coiled several times around a current-carrying bar. Linearly polarized light from a semiconductor laser is launched into this fibre coil. Whilst propagating through the coil of fibre the light comes under the influence of the magnetic field due to the current in the bar. The field lines around such a current are cylindrical in form, so they lie parallel with the fibre axis as it coils around the bar. Hence the linearly polarized light comes under the action of a longitudinal magnetic field as it propagates; it thus experiences the Faraday magneto-optic effect, and the direction of polarization is rotated by an amount proportional to the magnetic field, and to the path length along which the field acts. This, as was emphasized in section 3.9(b), effectively is a magnetic-field-induced circular birefringence.

Now the total rotation of the polarization direction on emergence will be proportional to the line integral of the field around the loop, i.e.,

$$\rho = NV \oint \mathbf{H}.\mathrm{dl} \qquad (10.6)$$

where V is the Verdet (magneto-optic) constant, N is the number of turns on the fibre coil, $\mathbf{H}$ is the magnetic field and l is the path length.

But from our knowledge of basic electromagnetism we know that the line integral of the magnetic field around a current-carrying conductor is just equal to the current (Ampere's circuital theorem). Hence (10.6) can be written:

$$\rho = NVI$$

where I is the current to be measured.

Thus a measurement of ρ allows us to measure I, since both N and V are known constants. This will be a very convenient means for measuring current at high voltage because:

(i) The fibre is made from a dielectric, insulating material and no costly insulation will be required between the high voltage bar and the (earthy) indication point.

(ii) There is no hysteresis effect, since no ferromagnetic materials are involved.

(iii) The magneto-optic effect is very fast (almost instantaneous) in fused silica, and thus the measurement bandwidth can be very large.

(iv) The fibre is easily coiled around a bar, and thus installation is very straightforward.

How, then, can we actually measure ρ? Suppose that the emerging linearly polarized light falls on to a linear polarizer which is set with its polarization direction parallel with that of the light's *input* polarization direction (Fig. 10.4(a)). In the absence of a magnetic field ($\rho = 0$) all the light will be passed by the polarizer (ignoring its intrinsic attenuation).

Let us assume that the electric field amplitude of the propagating light is E_0, so that an intensity proportional to E_0^2 is passed, in the absence of current. When current flows, the polarization is rotated through an angle ρ and only a field component $E_0 \cos \rho$ will now be passed by the polarizer, giving a measurable intensity proportional to $E_0^2 \cos^2 \rho$. This intensity, in principle, allows ρ to be deduced.

However, there is a more convenient way to measure ρ.

Suppose that instead of a simple polarizer, we use a Wollaston prism, with its polarization axes set at $\pm 45°$ to the input polarization direction. We now have two intensity outputs from the Wollaston prism (see Fig. 10.4(b)):

$$I_1 = KE_0^2 \cos^2\left(\tfrac{1}{4}\pi - \rho\right)$$

$$I_2 = KE_0^2 \cos^2\left(\tfrac{1}{4}\pi + \rho\right)$$

where K is the usual universal constant.

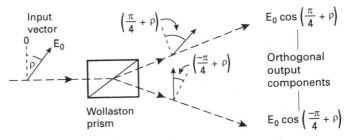

(a) Wollaston E-field components

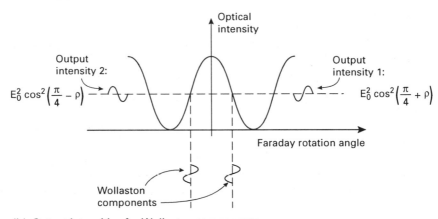

(b) Output intensities for Wollaston components

Fig. 10.4 Wollaston prism action for processing of current-measurement signal.

If we detect these two intensities separately (by measuring the optical powers falling on two separate photodiodes: remember that power = intensity $\times$ area), then we can readily arrange for the electronics to construct the function

$$S = \frac{I_1 - I_2}{I_1 + I_2} = \frac{\cos^2\left(\tfrac{1}{4}\pi - \rho\right) - \cos^2\left(\tfrac{1}{4}\pi + \rho\right)}{\cos^2\left(\tfrac{1}{4}\pi - \rho\right) + \cos^2\left(\tfrac{1}{4}\pi + \rho\right)} \tag{10.7}$$

which gives, on manipulation of the functions

$$S = \sin 2\rho$$

and, if 2ρ is small ($\ll \pi/2$):

$$S \approx 2\rho = 2NVI$$

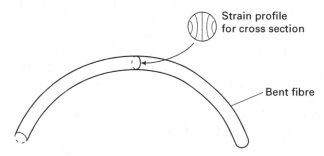

Fig. 10.5 Fibre bending strain.

Hence, S is proportional to the current I, under these conditions, and independent of the light intensity ($\sim E_0^2$), which can vary with time. What then are the problems with this method for measuring current? The first is that the fibre must be bent around the bar. If a fibre is bent, linear birefringence is introduced since the fibre is strained asymmetrically (Fig. 10.5) and strain alters the refractive index via the strain-optic effect. Now we know that, in a linearly birefringent fibre, only the two linear eigenmodes (i.e., those linearly polarized states which lie parallel to the birefringence axes) propagate without change of form. It is clear that a rotating linear polarization state cannot remain linearly polarized within such a fibre. It will emerge in an elliptical state which will depend not only on the bend birefringence but also on the way in which it has rotated along the fibre, i.e., on the current to be measured.

Now equation (10.7) indicates that the value of S is proportional to the difference between the two Wollaston outputs (I_1, I_2). However, this difference will also depend upon the axes of the polarization ellipse, i.e., on the linear bend birefringence, in addition to the circular birefringence due to the current to be measured. If, for example, the polarization ellipse happens to degenerate into a circle, S becomes zero even though the current is non-zero. Hence, this bend birefringence effect clearly will interfere with the measurement.

The effect can be quantified very conveniently with the aid of the Jones matrices discussed in section 3.8. The matrix for a polarization element which rotates the polarization through an angle ρ (i.e., which possesses circular birefringence 2ρ) in the presence of a linear birefringence δ, is given by [1]:

$$M = \begin{pmatrix} \alpha + i\beta & -\gamma \\ \gamma & \alpha - i\beta \end{pmatrix} \tag{10.8}$$

where, with $\Delta = \left(\rho^2 + \frac{1}{4}\delta^2\right)^{1/2}$ we have:

$$\alpha = \cos\Delta; \quad \beta = \frac{1}{2}\delta\left(\frac{\sin\Delta}{\Delta}\right); \quad \gamma = \rho\frac{\sin\Delta}{\Delta}$$

The matrix axes (Ox, Oy) are taken to be those of the linear birefringence fast and slow axes. This matrix is fairly complicated because the interaction between ρ and δ is quite subtle. The effect that δ has on the polarization state depends upon the state itself. For example it has no effect on a linear state parallel with one of the axes, but if that state is rotated through $45°$, it has a maximum effect; with the orientation changing continuously as a result of the rotation, ρ, it follows that the effect of δ also is changing continuously.

Suppose, then, that the linearly polarized input light to the fibre coil is launched with its polarization direction aligned with one of the birefringence axes, say Ox (fast). This can be represented by the column vector:

$$\begin{pmatrix} E \\ 0 \end{pmatrix}$$

where E is its electric field. The polarization state of the output light can now be determined as:

$$\begin{pmatrix} E_x \\ E_y \end{pmatrix} = M \begin{pmatrix} E \\ 0 \end{pmatrix}$$

using M from (10.8) (remembering that the Es are all complex numbers!). This allows S, from equation (10.7) to be constructed as:

$$S = 2\alpha\gamma = 2\rho\frac{\sin 2\Delta}{2\Delta} \tag{10.9}$$

which quantifies our measurement problem. For, suppose, first, that the bend-induced linear birefringence, δ, is very much larger than the current-induced circular birefringence 2ρ (and thus also very much greater than the rotation ρ). In this case

$$\Delta = \left(\rho^2 + \frac{1}{4}\delta^2\right)^{1/2} \approx \frac{1}{2}\delta$$

and thus from (10.9):

$$S \approx 2\rho\left(\frac{\sin\delta}{\delta}\right)$$

Hence, for $\delta > 0$, we see that $S < 2\rho$ and the measurement sensitivity is reduced. In fact, the situation is worse than a simple reduction of output, for the value of δ will be temperature dependent (via the temperature dependence of the strain-optic coefficient) and so S will be temperature dependent. Furthermore, δ can also be induced by vibration (since vibrational pressure also will strain the fibre) and hence S will suffer from vibrationally-induced A/C noise.

What can be done about these problems?

Firstly, the bend birefringence can be kept small by ensuring that the bend diameter is kept large ($> 0.5\,\text{m}$). Secondly, it is possible to use optical fibre which has an intrinsic circular birefringence (as has an optically active crystal such as quartz, for example). Circular birefringence can be induced in an optical fibre by several methods, one of which is by simply twisting the fibre about its own axis (section 3.7). In this latter case we now have:

$$2\rho \gg \delta$$

and hence:

$$\Delta \sim \rho$$

and

$$S \approx \sin 2\rho$$

(Physically, what has happened here is that the large value of intrinsic birefringence has caused the polarization state to rotate very rapidly along the fibre, thus averaging out the effects of the linear birefringence). However, ρ is now the sum of the intrinsic circular birefringence ($2\rho_0$) and the current-induced circular birefringence ($2\rho_\text{I}$), so that:

$$S \approx \sin(2\rho_0 + 2\rho_\text{I})$$

But from the discussion in Section 3.9(b) we know that there is a fundamental difference between these two components of circular birefringence: ρ_0 is reciprocal, while ρ_I is non-reciprocal. This means that if the light is back-reflected down the fibre, so that it performs a go-and-return passage through the coil around the conductor, the intrinsic reciprocal birefringence ($2\rho_0$) will be cancelled, while the current-induced non-reciprocal birefringence ($2\rho_\text{I}$) will be doubled. Hence on back-reflection in an arrangement such as that shown in Fig. 10.6 we shall have:

$$S \approx 4\rho_\text{I}$$

This has the added advantage of removing the temperature dependence of ρ_0 (which is due, again, to the temperature dependence of the strain-optic coefficient when ρ_0 is obtained from twist strain). Yet another advantage of this arrangement is that it has the convenience, for installation, of being single-ended (Fig. 10.6).

Hence, we can see that our detailed knowledge of polarization optics has allowed us to design a very satisfactory device, free from temperature and vibration effects, which is capable of making a very important measurement, cheaply and conveniently. Devices based on these principles have been used in the electricity supply industry, in various diagnostic and testing procedures where quick, easily installed devices have great advantages.

Figure 10.7 shows a particularly interesting application of optical-fibre current measurement, where it is not even necessary to twist the fibre since the intrinsic circular birefringence is not needed.

A fibre encloses a high-voltage transmission tower, and measures the current flowing into the ground when a short-circuit fault is struck between one of the high-voltage phase conductors and the earthed tower. This measurement is able to provide valuable information on the earth current which flows when such a fault occurs as a result, for example, of a direct lightning strike on the line. This current would be difficult to measure in any other way, and virtually impossible to measure using conventional current transformers. In this application the bend birefringence is not a problem because the coil diameter is very large ($\sim 10\,\mathrm{m}$); the vibration is not a problem because the measurement has been completed by the time that the mechanical stock of the fault propagates down the tower to the ground; the temperature dependence is not a problem because the temperature drift over such a short time is negligible. On the other hand, the advantages are that the bandwidth is large enough to ensure that the short period ($\sim 1\,\mu\mathrm{s}$) waveform is accurately reproduced; and the fibre is installed, and removed, in minutes. This is thus a good example of how the performance of an optoelectronic (or any other) system or device can be matched to the specific requirements with great advantage.

One final note on the bandwidth available with the optical-fibre current measurement device: the speed of the magneto-optic effect is not the limitation, but the bandwidth is limited by the time taken for the light to pass around the fibre loop. Clearly the measurement cannot take place in a time less than this, for the full rotation will not then have occurred. So perhaps just one loop is optimum? Not necessarily, because, as is evident from equation (10.6), the sensitivity of the measurement is proportional to the number of turns. Hence, we meet again the perennial problem of bandwidth versus sensitivity: their product usually is a constant for any given technique. Compromise (otherwise known as 'trade off') is central to the system or device designer's art!

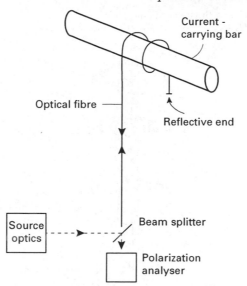

Fig. 10.6 Installation schematic for a single-ended optical-fibre current-measurement device.

10.4 THE INTEGRATED OPTICAL SPECTRUM ANALYSER

Topics needed:
   (i) Integrated optics (Chapter 8)
   (ii) Acousto-optic modulation (7)
   (iii) Fraunhofer diffraction (2)
   (iv) Piezo-electric effect (9)
   (v) Photodetection (7).

Spectral analysis of a given waveform is often required, for a number of reasons. By spectral analysis is meant the determination of the amplitudes and phases of the frequency components present in the waveform.

   The waveform to be analysed most often is in the form of an electrical signal. The spectral analysis may be necessary in order to determine the contribution being made by particular sources or processes with known frequency spectra, or to determine the way in which the different contributions vary with time. It is often required, in military applications, to identify aircraft, submarines or land vehicles via the characteristic spectral signatures which each emits in either acoustic or electromagnetic wave emissions. The spectrum may also be needed in order to determine the effect it will have on succeeding systems.

   In many of these applications the analysis must be done in real time, and in a time less than the time constant characterizing a significant change in the spectrum. Although electronic techniques can perform this task, they are often too slow to be acceptable.

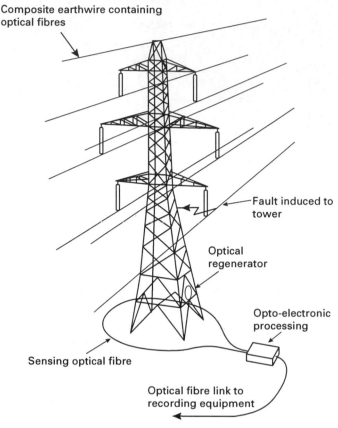

Composite earthwire containing optical fibres

Fault induced to tower

Optical regenerator

Opto-electronic processing

Sensing optical fibre

Optical fibre link to recording equipment

Fig. 10.7 'Tower-footing' optical-fibre current-measurement device.

In section 2.10 it was noted that the Fraunhofer diffraction pattern generated in the far field was just the Fourier transform of the aperture distribution function which gave rise to it (Appendix II). Now the Fourier transform (FT) is just what is needed, because it is a map of the amplitude and phase of all the frequencies present in a given function. It was seen in section 2.10 that a sinusoidal (for example) aperture distribution gives rise to just two lines in the far-field diffraction pattern, corresponding to the positive and negative 'frequencies' which define the amplitude and phase of the sine wave.

These ideas can be used to advantage in an optical spectrum analyser. Essentially what is needed is a diffraction aperture-distribution function which corresponds to the waveform to be spectrum-analysed.

Consider the arrangement shown in Fig. 10.8. The waveform to be analysed, which is assumed to be in the form of an electrical voltage of an appropriate magnitude (i.e previously voltage-amplified if necessary), is applied to an 'interdigital' transducer set into a slab of LiNbO$_3$ (lithium niobate) in what is now an integrated-optic (I/O) chip. This transducer, via its interlacing set of electrodes, uses the applied voltage to cause sympathetic expansions and contractions of the LiNbO$_3$ material, via the piezoelectric effect. (Recall that when an electric field acts on a crystalline material it will interfere with the interatomic electronic bonding which controls the intratomic spacing, so this spacing will change in sympathy with the field in certain preferred directions, depending on the particular crystal structure. This comprises the piezoelectric effect.) This sympathetic rarefaction and compression will launch a longitudinal acoustic wave into the crystal in a direction perpendicular to the interdigital electrodes.

Now as we know very well, from the study of the acousto-optic effect in section 7.3.3, this acoustic wave will, in turn, lead to a sympathetic variation of the refractive index of the medium, since the refractive index depends upon the density of medium. This acoustic wave, corresponding, as it does, to the waveform to be spectrum-analysed, thus comprises the diffractive aperture which is needed. For if (Fig. 10.8) plane, coherent light from a laser passes normally through the acoustic wave (just as for acousto-optic modulation in section 7.3.3), then the diffraction pattern generated in the far field comprises the spectral analysis required, in real time. The far field pattern is then allowed to fall on to a photodetector array (Fig. 10.9) which is arranged so that each detector in the array corresponds to the position to which a given frequency component will be diffracted. Hence the input from the array provides the required spectral analysis.

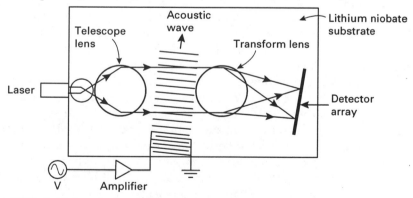

Fig. 10.8 An integrated-optical spectrum analyser.

We must remember the conditions, detailed in section 7.3.3, which determine whether the acousto-optic diffraction takes place in the Bragg or the Raman-Nath regime. At the lower frequencies the Raman-Nath operation will dominate, and there is no problem. However, as the frequency rises, the diffraction aperture changes its form before the light has fully crossed it, and we are into the Bragg regime. The Bragg regime is, of course, much more selective in respect to the angle of input, so that, at the

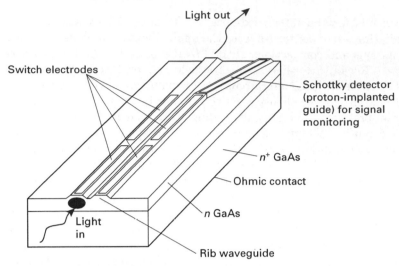

Fig. 10.9 An integrated-optical (I/O) signal monitoring switch.

higher frequencies careful further design is required, necessitating arrays of interdigital transducers set at different angles. However, we shall not delve into these complications, since they involve no essentially new principle.

The advantages, then, of this method for waveform spectral analysis are that the analysis is performed in parallel for each frequency component, and thus quickly in real time; and that it is performed on a small, compact, lightweight, rugged, I/O module which is not vulnerable to optical misalignment by shock, and which is readily manufacturable in large quantities. Clearly, all the other components needed are readily 'written' in to the I/O chip by the established methods of integrated optics. This optical processing function is only one of many which can be performed using optical/material interactions. As a group they form the subject known as 'optical information processing', which makes use of a large variety of both linear and non-linear optical effects in materials. This subject clearly is the precursor to ever more powerful processing operations using optical manipulations, and eventually leading to very powerful optical computers, with all their advantages of processing parallelism and high speed. Watch that space very carefully for future developments!

10.5 THE AUDIO COMPACT DISC (CD)

Topics needed:
    (i) Properties of laser light (Chapters 6, 7)
    (ii) Semiconductor lasers (7)
    (iii) Fresnel reflection (2)
    (iv) Optical interference (2)

(v) Diffraction (2)
(vi) Material dispersion (4)
(vii) Photodetection (7).

Our next view of optoelectronics in action is of a system which should be familiar to most readers as a result of the enormous impact it has made in recent years in improving the quality and convenience of music reproduction in the home and elsewhere. Such is the improvement which has been brought about that the audio compact disc (CD) is now well on the way towards replacing completely other methods of music reproduction (vinyl disc, magnetic tape) for the enthusiast.

The audio CD system is, therefore, a very suitable topic for this chapter, but in addition it illustrates quite a broad range of important principles.

To understand the operation of the CD system it, first, is necessary to understand the basic ideas involved in digital electronics. These ideas are properly the stuff of pure electronics (as opposed to optoelectronics) but they are useful to us, nevertheless, and they are quite straightforward.

The essential idea involved is that of transmission of information between two (or more) points by means of a series of 'yes or no' answers to questions. This is reminiscent of the well-known parlour game 'twenty questions', wherein a team is allowed to ask exactly twenty questions which can be answered only by yes or no, in order to identify a pre-chosen object. The main difference between the parlour game and digital electronics lies in the fact that, in the game, twenty questions might be asked in about five minutes, whereas in electronics up to $\sim 10^{11}$ questions can be asked per second!

A normal analogue signal may consist of a voltage which varies with time (Fig. 10.10) corresponding, in our present case, to the variation of the amplitude of a sound wave with time, translated into an electrical signal by means of a microphone. This voltage signal will need to be amplified, filtered and impedance-matched, to the point where it can activate a loudspeaker and thus reproduce the sound wave. All of these processes will add noise which will cause a reduction in the signal-to-noise ratio and thus a degradation in the quality of the sound.

Suppose, however, that the analogue voltage signal from the microphone is immediately digitized. By this is meant that the analogue waveform is sampled at certain time intervals (it must, in fact, be sampled at a sampling frequency which is equal to twice the maximum frequency present in the signal if the waveform is to be reproduced accurately: this follows from the so-called 'sampling theorem' (see Appendix IX)). The voltage level at each of the sampling points is then 'digitized', i.e., its value is expressed in digital code, as a series of pulses. In this series the presence or absence of a pulse correspondingly represents the presence or absence of a particular power of 2. For example, in a 3-bit (bit = *bi*nary dig*it*) system the pulse

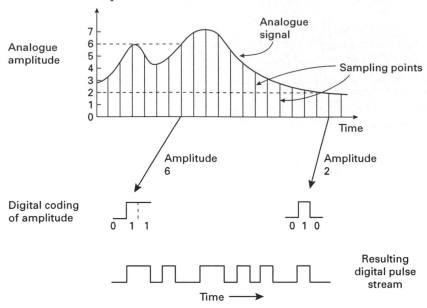

Fig. 10.10 Digital encoding of an analogue signal.

sequence 011 would be equivalent to $0 \times 2^0 + 1 \times 2^1 + 1 \times 2^2 = 6$. So the level for that sample is 6 on a range which runs from 0 (000) to 7 (111), thus comprising a total of $8(2^3)$ levels. Hence a 10-bit system, for example, will allow 2^{10}, or 1024 levels, and thus will allow the analogue amplitude to be defined to better than 0.1%.

The important advantage of this method of doing things is that all a detector system now has to do is to recognize whether a pulse is present or absent in a particular time slot: the answer is 'yes' it is, or 'no' it isn't. The magnitude of the pulse is unimportant provided only that it is above a threshold level, large enough for it to be distinguished unambiguously from noise.

Digital systems are thus much more resistant to noise. Their disadvantage is that they need more time to define the levels and this means more bandwidth (i.e., more variations (pulses) in unit time). Optical systems, with their enormous bandwidth ($\sim 10^{11}$ bits s^{-1}), effectively remove all problems of bandwidth, and thus almost all optical trunk communications systems are digital, in order to benefit from the signal-to-noise advantage. Furthermore, some very clever bandwidth compression techniques are possible when the signal is in digital form.

Let us now return to the sound wave. Clearly the wave can be digitized, as explained above, and thus turned into a pulse sequence. How can optics help in this case? For reproduction of the sound at any chosen time the pulse sequence has to be stored on a medium which can be 'read' at any time, so that the music (if music it be) can be reproduced. The medium

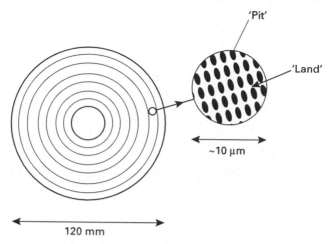

Fig. 10.11 Structure of the compact audio disc.

should allow a large number of bits (pulses) to be stored in a small space (it must be 'compact') and must be readable in a way which is convenient, and which does not degrade the information. For the latter, a non-contact method is preferable. (In a vinyl 'LP' the stylus makes contact and causes wear; in a magnetic tape the reading 'heads' do the same). Optics can satisfy the requirements in the form of a 'compact disc'. This is a disc which contains the pulse sequence as a series of pits, or 'lands' (absence of pits), in a path which spirals inwards from the outer circumference. The disc is a piece of plastic in which the pits have been 'punched' from a master disc (see Fig. 10.11). The disc is read, without mechanical contact, by a spot of light. The optical system used is shown in Fig. 10.12.

What are the design requirements of the optical system?

First, the spot of light must contain as much optical power as possible so as to provide a good, strong, reading signal; secondly, the spot must be as small as possible so that it can read small pits: the smaller is each pit, the greater the number that can be punched on to a single disc and, therefore, the greater is the amount of information which can be stored on the disc. Thus we need an intense, small spot and for this we turn naturally to a laser source. Furthermore, for its compactness, ruggedness and low operating voltage, we turn, specifically, to a semiconductor laser. This source emits intense, well collimated, quasi-monochromatic light which falls on to a plate beam-splitter. Fresnel reflection at the chosen angle ($\sim 45°$) ensures that about half of the light passes through a re-collimating lens followed by a focusing lens. The action of the collimating lens is to render the focusing process independent of the distance between the two lenses, and thus independent of vibrational and temperature disturbances. The focusing lens focuses the laser light down to a small-diameter spot. The size is limited by the wavelength spread of the source and the numerical

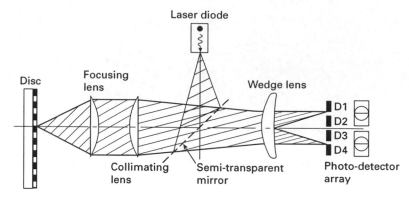

(a) Optical system for disc reader

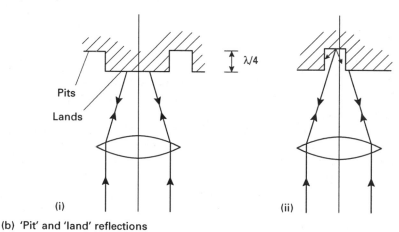

(b) 'Pit' and 'land' reflections

Fig. 10.12 Compact disc reading arrangements.

aperture of the lens. Let us examine this.

First, it is clear that the lens material will exhibit optical dispersion (section 4.3), i.e., its refractive index will vary with wavelength. Hence the focal length will vary with wavelength. For a well-defined spot we thus require a small wavelength spread. The laser provides this.

Secondly, the lens aperture will cause diffraction to occur, which will again limit the spot size. The best way to see this is to reverse the direction of the light rays and to consider the spot to be diffracting towards the lens. If the spot diameter is d and the wavelength (well-defined!) is λ, then we know (section 2.10) that the radiating spot will diffract into a half-angle $\sim \lambda/d$. If all of this light is to be collected by the lens, then it must subtend a half-angle at the spot position (i.e., the disc surface) equal to $\sim \lambda/d$. If D is the diameter of the lens and s its distance from the disc, then we have:

$$\frac{D}{2s} = \frac{\lambda}{d}; \quad d = \frac{2\lambda s}{D} = \frac{\lambda}{NA}$$

where NA is the numerical aperture. Let us calculate a typical value for d: we might have $\lambda = 850\,\text{nm}$ (GaAs semi-conductor laser) and $D = s$ (NA = 0.5). In this case we find that $d = 1.7\,\mu\text{m}$.

Clearly, what is now required is a 'pit' in the disc which has the same size as the laser spot size, so that the light can 'read' the disc unambiguously. If the pit has the same size ($1.7\,\mu\text{m}$), how many such pits can we punch on to a disc? For a typical disc of 120 mm diameter the answer will be $\sim [(120 \times 10^{-3})/(1.7 \times 10^{-6})]^2 \approx 5 \times 10^9$. To put this number into context, the contents of this book could be encoded into $\sim 5 \times 10^6$ bits [$\sim 10^6$ letters, spaces, punctuation marks etc., each letter requiring 5 bits ($2^5 = 32$)]. Hence a thousand books like this could easily be recorded on a single disc.

Clearly, if the focused spot falls on a polished 'land' the light will be well-reflected back, via the collimating lens system, to the detector lens system and the photodetector (Fig. 10.12(b)). The detection system is arranged so that any misalignment of the spot-focusing arrangement will generate an error signal which is fed back to stepper motors. These then correct the misalignment. If the spot falls into a pit it is necessary to ensure that as little light as possible is reflected back, so as to make it easy to distinguish between a '0' and a '1'. This is done by ensuring that:

(i) The depth of the pit is $\sim \frac{1}{4}\lambda$. This arranges that light reflected from the bottom of the pit interferes destructively with light reflected from a land, since they will be in antiphase.

(ii) The pit diameter is $\sim \lambda$ so that diffraction effects lead to large angle spreading of the light into the pit sides and outside the NA of the collecting lens. This requires a pit depth $\sim 0.2\,\mu\text{m}$.

Hence the pit-land (0-1) information on the disc now appears as an optical pulse stream at the photodetector which duly converts the optical stream into a stream of electric current pulses. These, via resistors, are converted into voltage pulses which are first digitally processed, and then converted back into an analogue signal via a digital-to-analogue converter (DAC). They are then amplified before passing to a loudspeaker for conversion into sound waves. These sound waves are now a very accurate, noise-free reproduction of the original sound.

The complete system is shown, diagrammatically, in Fig. 10.13. Optically, this is quite a simple system. However, it does illustrate well some basic advantages of laser optics (non-contact, high resolution, good signal-to-noise ratios) and some aspects of basic optical engineering which follow from the physical ideas (spot size, high-density information, dispersion, interference, diffraction, feedback control, etc.).

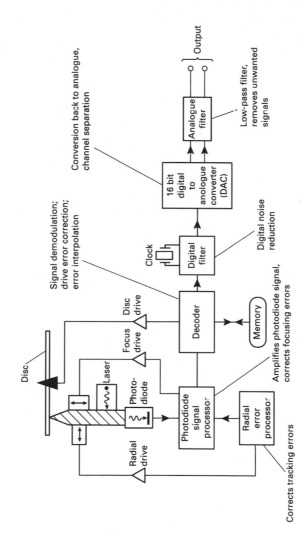

Fig. 10.13 Complete disc-reading system.

It also illustrates how optoelectronics can be used to improve the quality of life by enhancing the enjoyment of recorded music.

10.6 THE OPTICAL-FIBRE GYROSCOPE

Topics needed:
  (i) Wave interference (Chapter 2)
  (ii) Polarization properties of optical fibre (3)
  (iii) Mach-Zehnder interferometry (2)
  (iv) Optical Kerr effect (9)
  (v) Laser modulation (7)
  (vi) Integrated optics (8).

Gyroscopes are very important devices for navigation and automatic flight control. The conventional gyroscope based on the conservation of angular momentum in a spinning metal disc is highly developed, but contains parts which take time to be set in motion ('spin-up' time) and which wear. The device is also relatively expensive both to install and to maintain.

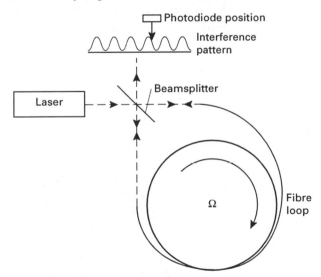

Fig. 10.14 Basic arrangement for the optical-fibre gyroscope.

The optical-fibre gyroscope overcomes all these problems (but, inevitably, has some of its own). Consider the arrangement shown in Fig. 10.14. Light from a laser is fed simultaneously into the two ends of a fibre loop, via the beamsplitter, so that two beams pass through the loop, in opposite directions. When the beams emerge at their respective ends they are brought together, again via the beamsplitter, and interfere on a receiving screen. This arrangement can be regarded as a special form of Mach-Zehnder interferometer, where the two arms of the interferometer lie within the same fibre, but the two signals traverse it in opposite directions. Clearly, under these conditions a sinusoidal interference pattern will be formed on the screen and, if the two beams are equal in intensity, the visibility will be 100%.

Suppose now that the complete system is rotating clockwise at an angular velocity Ω. In this case the clockwise propagating beam will view the end of the fibre receding from it as it travels, and it will thus have farther to go before it can emerge. Conversely, the anticlockwise rotating light will see its corresponding end approaching, and will have less far to go. The consequence is thus a relative phase shift between the two beams and a consequent shift in the interference pattern on the screen. (This is, in fact, a somewhat simplistic explanation of the physics involved; a rigorous explanation requires the help of the general theory of relativity, since rotating systems are accelerating systems, but the explanation given here is correct to the first order.) It follows that the change in the interference pattern can be used to measure the rotation, Ω, by placing a photodiode (for example) at a position on the screen where it can record a linear variation of received power with lateral shift (Fig. 10.14). The phase shift effect is known as the Sagnac effect, after the discoverer of the phenomenon (in 1913).

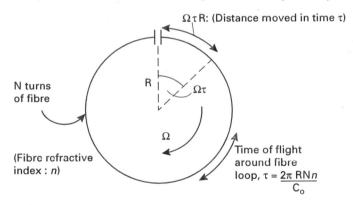

Fig. 10.15 Gyroscope geometry.

The phase shift caused by the angular rotation is readily calculated (Fig. 10.15). Suppose that there are N turns of fibre on the coil and that the coil radius is R. Then, in the absence of rotation, the time of flight around the coil will be given by

$$\tau = \frac{2\pi RNn}{c_0} \qquad (10.10)$$

where c_0 is the velocity of light in free space and n is the refractive index of the fibre material. If, now, the coil is rotated about an axis through its centre and perpendicular to its plane, with angular rotation Ω, then the fibre ends will have rotated through an angle $\Omega\tau$ while the light is propagating in the fibre, and thus through a distance $\Omega\tau R$. Hence the difference in distance travelled by the two counter-propagating beams will be twice this, i.e.:

$$dl = 2\Omega\tau R \tag{10.11}$$

Now the clockwise and anti-clockwise propagating light components no longer propagate with the same velocity when the coil is rotating, as viewed from the original stationary ('inertial') frame. We have to take account of the so-called 'Fresnel' drag, whereby light propagating in a medium which is moving with velocity v in the same direction as the light will propagate, relative to the stationary frame with a velocity given by (e.g., [2]):

$$c_v = \frac{c_0}{n} + v\left(1 - \frac{1}{n^2}\right)$$

where $\left(1 - n^{-2}\right)$ is called the Fresnel-Fizeau drag coefficient. (This is a direct consequence of special relativity.) Hence, in our case the two velocities around the loop are given by:

$$\begin{aligned} c_+ &= \frac{c_0}{n} + R\Omega\left(1 - \frac{1}{n^2}\right) \\ c_- &= \frac{c_0}{n} - R\Omega\left(1 - \frac{1}{n^2}\right) \end{aligned} \tag{10.12}$$

Consider now the difference between the times of arrival at the end of the fibre coil for the two counter-propagations. We have:

$$t = \frac{l}{v}; \quad dt = \frac{dl}{v} - \frac{l}{v^2}dv \tag{10.13}$$

and

$$dl = 2\Omega\tau R = \frac{4\pi R^2 Nn\Omega}{c_0} \tag{from 10.10}$$

$$v = \frac{c_0}{n}; \quad l = 2\pi RN$$

$$dv = 2R\Omega\left(1 - \frac{1}{n^2}\right) \tag{from 10.12}$$

Substituting the latter expressions into (10.13) we have:

$$dt = \frac{4\pi R^2 N\Omega}{c_0^2} \tag{10.14}$$

Note that this is *independent of the refractive index n* and therefore is independent of the fibre medium. (This is a common source of confusion in regard to the operation of the optical-fibre gyroscope). The phase difference between the two counter-propagations when the coil is rotating is now easily constructed from (10.14) as:

$$\Phi = \omega dt = \frac{2\pi c_0}{\lambda_0} dt = \frac{8\pi^2 R^2 N\Omega}{c_0 \lambda_0}$$

where λ_0 is the free space wavelength. This can also be written

$$\Phi = \frac{8\pi A\Omega}{c_0 \lambda_0} \tag{10.15a}$$

or

$$\Phi = \frac{2\pi L D\Omega}{c_0 \lambda_0} \tag{10.15b}$$

where A is the total effective area of the coil (i.e., the total area enclosed by N turns), L is the total length of the fibre and D is the diameter of the coil.

Let us now insert some numbers into (10.15b). Suppose that we use a wave-length of $1\,\mu\text{m}$ with a coil of length $1\,\text{km}$ and a diameter of $0.1\,\text{m}$. This gives:

$$\Phi = 2.1\Omega$$

For the earth's rotation of $15°\text{h}^{-1} (7.3 \times 10^{-5}\text{ radians s}^{-1})$ we must therefore be able to measure $\sim 1.5 \times 10^{-4}$ radian of phase shift. This can quite readily be done. In fact it is possible, using this device, to measure $\sim 10^{-6}$ radian of phase shift, corresponding to $\sim 5 \times 10^{-7}$ radians s^{-1} of rotation rate.

What, then, are the problems? First, since the fringe visibility will only be 100% if the two interfering beams have the same polarization, it is necessary to use polarization-maintaining monomode fibre: usually linearly birefringent (hi-bi) fibre is used. Secondly, there is a problem with the optical Kerr effect. The electric field of one beam will act, via the optical Kerr effect, to alter the phase of the other. The effect is small but then so is the phase difference which we are seeking to measure, at low rotation rates. This effect we can calculate, using the ideas discussed in Section 9.6. We know that, in fused silica, the non-linear electric polarization can be written:

$$P(E) = \chi_1 E + \chi_3 E^3 \tag{10.16}$$

to a good approximation.

The electric field, in this case, will be given by the sum of two counter-propagating waves:

$$E = E_+ \exp i(\omega t + kz) + E_- \exp i(\omega t - kz)$$

Substituting this value into (10.16) gives

$$P(E) = \chi_1 E + P_{N+} + P_{N-}$$

where:

$$P_{N+} = \chi_3(E_+^2 + 2E_-^2)E_+ \exp i(\omega t + kz)$$
$$P_{N-} = \chi_3(E_-^2 + 2E_+^2)E_- \exp i(\omega t - kz)$$

Now the refractive index is given in general by:

$$n^2 = \varepsilon = 1 + \chi = 1 + \frac{P(E)}{E}$$

Hence for the clockwise (+) propagating beam we have:

$$n_+ = (1 + \chi)^{1/2} = \left(1 + \chi_1 + \frac{P_{N+}}{E_+ \exp i(\omega t + kz)}\right)^{1/2}$$

or, using the binomial theorem,

$$n_+ = n_0 + \tfrac{1}{2}\chi_3(E_+^2 + 2E_-^2)$$

Similarly:

$$n_- = n_0 + \tfrac{1}{2}\chi_3(E_-^2 + 2E_+^2)$$

It follows that the non-linear (optical-Kerr-effect-induced) phase changes for each direction are:

$$\Delta\varphi_+ = \frac{2\pi}{\lambda_0} l \frac{\chi_3}{2} \left(E_+^2 + 2E_-^2\right)$$
$$\Delta\varphi_- = \frac{2\pi}{\lambda_0} l \frac{\chi_3}{2} \left(E_-^2 + 2E_+^2\right)$$

(10.17)

Clearly, these are not the same unless $E_+^2 = E_-^2$: the difference is, in fact,

$$\Delta\varphi_+ - \Delta\varphi_- = \left(E_-^2 - E_+^2\right) \frac{\pi}{\lambda_0} l \chi_3$$

A difference in optical power of just $1\,\mu\text{W}$ leads to a phase discrepancy of $10^{-6}\,\mu\text{rad}$, equivalent to a rotation of $0.01°\text{h}^{-1}$, whereas these devices are actually required to measure $\sim 0.01°\text{h}^{-1}$. This difference in optical power can easily result from the fibre attenuation, which produces inequalities of power, away from the centre of the coil.

The problem can be overcome by square-wave modulating the laser power. In this case each beam is influenced by the other for only half the time, so that, effectively, the cross-product term in (10.17) is reduced by a factor of 2 in each case, giving now:

$$\Delta\varphi_+ = \Delta\varphi_-$$

Other sources of noise are: Rayleigh backscatter in the fibre, which gives a coherent interfering signal, and drift in the value of the area of the fibre, due to temperature variation.

All these problems make it difficult for this device to compete successfully at the lowest rotation rates, although its simplicity and relatively low cost give it a distinct advantage for rotation rates $\sim 0.1°\mathrm{h}^{-1}$ and above. To achieve the highest sensitivity, the so-called minimum-configuration design is employed. This is shown in Fig. 10.16. It ensures that the common, reciprocal path comprises almost the entire system. Integrated optics is used to assist in this.

Polarization of the light and the beam-splitting are performed on the integrated-optical (I/O) chip. Also on the chip are a frequency shifter (acousto-optical) to allow the direction of rotation to be determined, and a phase modulator to ensure that the detection bias is maintained at its point of maximum sensitivity. The I/O chip also means that the device can be very compact, one having been built small enough to be enclosed in a sardine tin!

(I/O chips, such as this, which comprise both optical and electronic functions are sometimes referred to as 'optoelectronic integrated-optical circuits' or OEICs.)

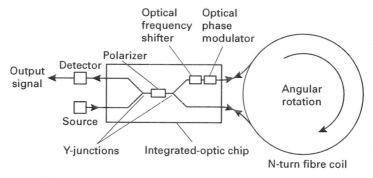

Fig. 10.16 'Minimum configuration' optical-fibre gyroscope.

Applications of the minimum configuration optical-fibre gyroscope range from ballistic missiles, through the location and control of oil-well drill tips, to motor vehicle navigation systems. This last application is a particularly promising growth area.

The optical-fibre gyroscope is in direct competition with the ring laser gyroscope (RLG), shown diagrammatically in Fig. 10.17, for many applications. This device uses the same Sagnac principle but does not use optical

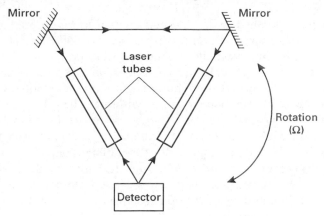

Fig. 10.17 Schematic ring laser gyroscope (RLG).

fibre; rather, it uses a triangular laser cavity, cut in a material such as quartz, which is filled with a laser gain medium (e.g., He-Ne). The RLG provides a difference frequency between the two counterpropagating laser modes, a difference resulting from the Sagnac effect - it is proportional to the rotation rate.

This RLG device is more highly developed at the present time than the optical-fibre gyroscope, although it still has its problems. The fibre gyroscope has the potential for equivalent performance and also for being a much cheaper alternative to the RLG. (It has been chosen for detailed description in this chapter largely because it embodies a broader range of optoelectronic ideas.)

10.7 HOLOGRAPHY

Topics needed:
    (i) Interference (Chapter 2)
    (ii) Coherence (5)
    (iii) Pulsed lasers: transverse and longitudinal modes (6,7)
    (iv) Diffraction (2).

Holography is a technique whereby three-dimensional images of objects can be reproduced by recording information on a two-dimensional photographic plate. This is valuable in a number of application areas: entertainment, advertising, industrial measurement and vibration analysis are a few, but it will be easier to appreciate the uses and potential uses of holography once we have understood how it works.

The three-dimensionality of an object is encoded in the light waves which reach our eyes from it, via the amplitudes and phases of the light waves which are reflected or refracted from each part of the object. When the object is photographed with a normal camera an image of the object is focused, by the camera lens, on to a plane photographic film at the back of the camera. This film can only respond to variations in the intensity of the light falling on it, since the chemical reactions, on which the recording relies, depend only on the numbers of photons which are intercepted by the film. Hence the phase information contained within the light waves is lost. This phase information relates to the distance which each individual light wave has travelled from the object, via its time delay; in other words it relates to the object's distribution in space, its three-dimensionality. Consequently a 'normal' photograph of an object appears only two-dimensional. The question which now arises is: how can we retain the phase information on our recorded film, and hence record the three-dimensionality of the object, with a view to regenerating it at will?

The answer is quite straightforward. We know that an interference pattern formed from two light waves of the same frequency is the result of their being added together with differing phase relationships. For example, in the two-slit interference pattern the sinusoidal variation of intensity produced on the screen is the result of a linearly increasing phase difference between the two waves as one moves in a direction normal to the slit lengths. The intensity pattern has a record, on the screen, of the phase information provided by the waves from the slits. Consider now the arrangement shown in Fig. 10.18. Here we have two plane waves incident on a photographic emulsion; one is normal to the plane of the emulsion, the other is incident at a small angle β to the normal. Clearly as in the two-slit case, the relative phase of the two waves will vary along a line parallel to the incident planes (Ox) within the emulsion. This can readily be quantified as follows. The two waves, at the plane of the emulsion, can be written as

$$E_1 = e_1 \cos \omega t$$

$$E_2 = e_2 \cos(\omega t + kx \sin \beta)$$

where, evidently, we have taken the phase of the first wave as our reference, and as always,

$$k = \frac{2\pi}{\lambda}$$

On superimposing the two waves we have a total field, E_T, which varies with x according to:

$$E_T(x) = e_1 \cos \omega t + e_2 \cos(\omega t + kx \sin \beta)$$

and hence the intensity of the resultant light will be proportional to the square of the amplitude of E_T, i.e.,

$$|E_T|^2 = e_1^2 + e_2^2 + e_1 e_2 \cos(kx \sin \beta)$$

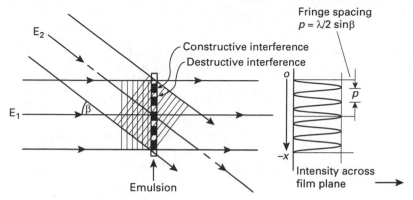

Fig. 10.18 Angled interfering waves in an emulsion.

Hence both the amplitude information and the phase difference between the waves are preserved in this intensity variation.

Suppose now that the screen is a photographic emulsion which records $|E_T|^2$. Suppose also that, having recorded it, the emulsion is put back into its place and is illuminated again by E_2. This illuminating wave will now have its amplitude modulated by the variations in the density of the emulsion which, of course, follow $|E_T|^2$. Hence the wave emerging from the emulsion will have the form:

$$E_R = E_2 |E_T|^2$$
$$= e_2 \cos(\omega t + kx \sin \beta)[e_1^2 + e_2^2 + 2e_1 e_2 \cos(kx \sin \beta)]$$

which, on expansion, gives:

$$E_R = \left(e_1^2 + e_2^2\right) e_2 \cos\left(\omega t + kx \sin \beta\right)$$
$$+ e_1 e_2^2 \cos(\omega t + 2kx \sin \beta) \qquad (10.17)$$
$$+ e_1 e_2^2 \cos \omega t$$

Look carefully at the three terms on the RHS of equation (10.17). They are what holography is all about.

The first term is just the 'straight-through' illuminating wave. The second term is a wave at approximately twice the angle which the reference beam makes with the emulsion's normal ($2kx \sin \beta \sim 2kx\beta$ if β is small). The third term, the all important one, reconstructs *the other wave (E_1) in correct amplitude and phase*, and multiplies it by e_2^2.

Consider now the arrangement shown in Fig. 10.19. A laser source is split into two (via a beam-splitting plate: Fresnel reflection!). One of the two resulting beams is allowed to fall on an object to be recorded, and the other falls directly (but at an angle) on to an emulsion. Reflected light from the object also falls on to the same emulsion. The result now

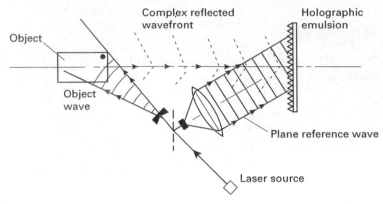

Fig. 10.19 Holographic recording.

is that each wave from the object interferes with the 'reference' wave to form an intensity interference pattern which preserves its amplitude and phase. The result is a complex sum-of-all-interference patterns recording which has preserved all the amplitude and phase information about the original object. This recording is a hologram. It is 'read' just as was the two wave pattern: by illuminating it with the reference wave. Then all the original object waves are reproduced, travelling in the same direction as before (Fig. 10.20). A viewer on that side of the hologram will then 'see' a virtual image of the object in the same relative position to the hologram as was the original object. Clearly the hologram can be taken anywhere, and the virtual image will reappear on illumination by the reference wave at the correct angle.

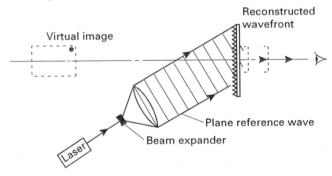

Fig. 10.20 Reconstruction of the virtual image.

In order to make good holograms (i.e., holograms which give a clear repro-duction of the object) several conditions have to be satisfied.

First, the interference pattern formed between object and reference waves must have good contrast. For this, as we know, the interfering waves must be coherent, that is: they must have a constant phase relationship and must have the same polarization state. This can be achieved by using a laser with a coherence length greater than the maximum dimension of the object and by ensuring that the difference in path length between object and reference waves at the emulsion does not exceed this coherence length. A coherence length of about 1 m is usually enough, but for this we must use a laser with only one longitudinal mode. For optimum operation the two optical paths should be exactly equal. An argon laser is often used. Secondly, if the optimum conditions are to be maintained over the full field of view, the laser also should be operating in a single transverse mode, otherwise interference conditions will vary with position on the emulsion. Thirdly, if mechanical vibrations are to be prevented from 'smearing' the interference pattern, either the whole set up must be very effectively vibration-isolated, or a pulsed laser should be used, with a pulse short enough to 'freeze' the pattern in the presence of the relatively low frequency mechanical waves: a pulse width of ~ 50 ns is usually sufficiently short.

Finally, the emulsion should be finely grained, in order to record the finely detailed interference patterns, representing as much detailed information about the object as possible. Grain sizes should be of the order of nanometres, rather than the micrometres used for conventional film. Clearly there will be fewer molecules per unit of information in the smaller grain size so that larger photon fluxes are required to provide given emulsion contrast. This means larger optical energies, either per pulse or per integrated (over time) CW laser light. Hence, most of the burden for good holograms falls on the laser source. It should be a high energy (~ 1 J) pulsed (~ 50 ns) single longitudinal mode (coherence length ~ 1 m), single transverse mode (coherence width ~ 1 m) laser, operating in the visible (where commercial emulsions have their highest sensitivity). For pulsed operations a ruby laser is often used, but with careful control. In addition to this the object must be firmly held, and of good reflectivity.

Industrial uses of holography revolve largely around hologrammetry (which involves the accurate measurement of specific features of the holographic image). This is most convenient when it is not possible or convenient to make direct measurements, owing to the hostility of the environment: the core of a nuclear reactor, and the base of an oil well are good examples. Sturdy holographic cameras record the object in these cases and careful measurements are then performed, at leisure, on the holographic image. One of the important additional advantages of such images is that they are not formed with the aid of lenses, so do not suffer from the variety of possible lens aberrations. Also, the images can be viewed from any angle within the aperture of the original recording.

Another very valuable industrial application is that of holographic interferometry. In this case a holographic image is compared directly, in situ, with the real object. If there are any differences between the two, then interference patterns are formed which reveal and quantify these differences. Hence the build-up of mechanical strains can be monitored. Also vibrational modes can be readily visualized by this method, since the hologram does not vibrate, while the real object does, creating a pattern of displacements which yields interference fringes. Dangerous resonances, among other deleterious effects, can be revealed in this way.

So far, our description of holographic image production has dealt only with *virtual* images. These are limited in their usefulness in that they can only be examined from a distance equal to at least that between the original object and the emulsion plate: this limits parallax and detail. It is much more useful to examine a *real* image, i.e., one that actually is formed in space and can be examined close up. It is also much more impressive for display and entertainment purposes, for the image appears to 'hang in the air', looking very real whilst totally insubstantial.

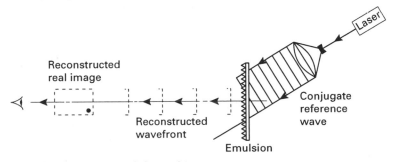

Fig. 10.21 Reconstruction of the real image.

A real image can be obtained from the hologram formed as previously described, simply by reversing the direction of the reference wave (i.e., using its 'phase conjugate'). What this does, essentially, is to reverse all the light waves previously generated, so that now the waves which appeared to come from the original object (to form the virtual image) now actually retrace their paths to form a real image in the original position of the object (Fig. 10.21). Unfortunately it is reversed back-to-front and left-to-right so that it is usually referred to as a 'pseudoscopic' image, but otherwise it appears just as the original. To understand what happens analytically in this case we need to extend slightly the analytical technique beyond that which has been used so far. Since we are concerned now also with *directions* of propagation it is necessary to express the waves in a way which makes the direction clear. Thus, returning to the case of two-wave interference, we see from Fig. 10.18 that the two waves can be written as

$$E_1 = e_1 \cos(\omega t - kz)$$

for propagation in the positive Oz direction, and

$$E_2 = e_2 \cos(\omega t - kz \cos \beta + kx \sin \beta)$$

for propagation at an angle β to the Oz direction in the negative $(-Ox)$ direction. Hence at z_0, the emulsion position, we have:

$$E_T(x, z_0) = E_1 + E_2$$

and

$$|E_T(x, z_0)|^2 = e_1^2 + e_2^2 + 2e_1 e_2 \cos(-kz + kz \cos \beta - kz \sin \beta)$$

This, then, represents the interference pattern which forms the hologram.

Let us now illuminate the hologram with the reference wave travelling in the *reverse* direction to the one which produced the hologram, i.e., with the wave:

$$E_2' = e_2 \cos(\omega t + kz \cos \beta - kx \sin \beta)$$

This wave will be modulated, as before, by the hologram, to produce the resultant wave complex:

$$
\begin{aligned}
E_R &= |E_T(x, z_0)|^2 E_2' \\
&= \left[e_1^2 + c_2^2 + 2e_1 e_2 \cos(-kz + kz \cos \beta - kx \sin \beta) \right] \\
&\quad \times \left[e_2 \cos(\omega t + kz \cos \beta - kx \sin \beta \right] \\
&= \left(e_1^2 + e_2^2 \right) e_2 \cos(\omega t + kz \cos \beta - kx \sin \beta) \\
&\quad + e_1 e_2^2 \cos(\omega t + 2kz \cos \beta - 2kx \sin \beta) \\
&\quad + e_1 e_2^2 \cos(\omega t + kz)
\end{aligned}
$$

The final term is the important one. This represents the other (object) wave travelling in the $-Oz$ direction, i.e., back to the object. Clearly, all such waves will converge on the object to produce the real image previously discussed.

In this description of holography we have been dealing with just one wavelength, so that the object would appear in just one colour. Clearly, however, other colours can be superimposed to form a true, coloured representation of the original object. And if all of this could be done quickly enough and in real time, we would have three-dimensional, moving, real images. This will require materials which can record, be read and erased in $\sim 50\,ms$ (to avoid eye flicker), however, and these do not yet exist. They will come! Three-dimensional television and cinematography (without special glasses) are on their way!

10.8 OPTICAL-TIME-DOMAIN REFLECTOMETRY (OTDR) AND ITS USE IN DISTRIBUTED OPTICAL-FIBRE RAMAN TEMPERATURE SENSING (DTS)

Points covered:

(i) Raman effect (Chapter 9)

(ii) Rayleigh scattering (4)

(iii) Backscatter in fibres (4, 9)

(iv) Avalanche photodiodes (7)

(v) Photon counting (7)

(vi) Non-linear effects in fibres (9).

In this section we shall consider another important type of optical-fibre sensor: the distributed optical-fibre sensor.

An optical fibre may be regarded (among many other things) as a one-dimensional measurement medium. This has a number of advantages. For example, it can be used to perform the 'line-integral' function of measuring the electric current in a conductor, by wrapping a coil around the conductor and measuring the line-integral of the magnetic field around the loop, via the Faraday magneto-optic effect. Since the line integral of the magnetic field around such a loop is equal to the current, this enables us to construct a convenient current measurement device, as was noted in section 10.3. Voltage can be measured (in principle) in a similar way, by line-integrating the electric field between two points of voltage difference.

Suppose, however, that we now consider *differentiating* the information on the light emerging from the fibre, instead of integrating it. If the differentiation is with respect to time, this could tell us about how a particular light property has evolved with time, and thus with distance (since its velocity is known) along the fibre. If this property were subject to external influences such as temperature, pressure, electric fields, magnetic fields, etc., then the differentiation could tell us what is the spatial distribution of that field along the length of the fibre, to within some prescribed spatial resolution interval. Since the fibre is very thin and flexible it would thus be possible to measure the distribution of, say, temperature, over a large structure such as an electrical generator, or a chemical boiler, and to note when dangerous 'hot spots' were developing (for example). Distributed optical-fibre sensing (DOFS) is potentially a very valuable measurement technology for large structures, and other examples of its use will be discussed at the end of this section. The technology is in its infancy, but one system for the distributed measurement of temperature is presently available commercially, and in order to illustrate the power of the technique (and cement our principles!) this system will now be described in some detail.

Most (but by no means all) distributed optical-fibre sensors make use of the principle of optical-time-domain reflectometry (OTDR). This is a valuable technique in itself for a number of diagnostic procedures in optical fibre communications systems. First, then, we must understand the principles of OTDR.

10.8.1 Optical-time-domain reflectometry (OTDR)

OTDR is essentially a one-dimensional radar (or 'lidar'!) along the fibre (Fig. 10.22). A short pulse of laser light is launched into the fibre at one end. As the pulse propagates it is continuously 'Rayleigh' backscattered by the small ($< \lambda$ in size) defects and inhomogeneities in the silica structure (section 4.2). These cause loss ($\sim \lambda^{-4}$), but they also backscatter some light to the launch end. If this light is detected, and the power level is differentiated with respect to time, then we have a mapping of the spatial distribution of the attenuation of the light along the length of the fibre. Clearly, for optical communications engineers this is extremely useful, because it will tell them just where there is a break in the fibre, a bad joint between fibres or perhaps a 'bad' piece of fibre (i.e.,a piece whose loss is higher than it should be according to the manufacturer!). Remedial action can then be taken. These ideas can readily be formalized as follows.

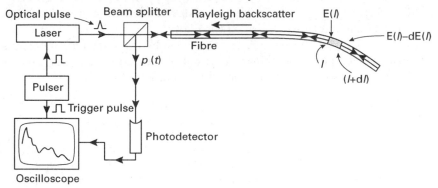

Fig. 10.22 Optical time-domain reflectometry (OTDR).

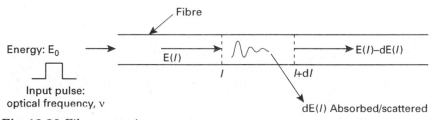

Fig. 10.23 Fibre scattering geometry.

Suppose we launch into a fibre a rectangular optical pulse of width τ, energy E_0 and optical frequency ν (Fig. 10.23). Consider the section of fibre which lies between l and $l + dl$ from the launch end. A fraction of the light energy that enters this section will fail to leave it as a result of loss, and this loss will be due to both scattering and absorption processes (section 4.2). This fraction will be given by Beer's law as:

$$\frac{dE(l)}{E(l)} = -\alpha(l)dl \qquad (10.18)$$

where $\alpha(l)$ is the loss coefficient, i.e., the loss per unit length at l, and will be the sum of the scattering and absorption coefficients, i.e.:

$$\alpha(l) = \alpha_s(l) + \alpha_s(l)$$

Integrating (10.18) we have:

$$E(l) = E_0 \exp\left(-\int_0^l \alpha(l) dl\right)$$

Now the portion of this light energy which is scattered in the element dl will be $\alpha_s(l)E(l)dl$. Not all of this will be guided back to the source. Suppose a fraction S is so guided (S will depend on the numerical aperture of the fibre, assuming the scatter is isotropic, and is, in fact, given by $3(NA)^2/8n_1^2$, where n_1 is the refractive index of the core material and NA is its numerical aperture; S is called the capture fraction). The light which is scattered back will suffer further attenuation by the same amount as was suffered by the light in reaching l, and hence the energy arriving back at the launch end due to scattering between l and $l + dl$ will be:

$$dE_{s,0}(l) = \alpha_s(l)E_0 \exp\left(-2\int_0^l \alpha(l) dl\right) dl \tag{10.19}$$

Suppose the light is launched at time zero. Then the light which arrives back at the launch end after time t will have travelled a distance $2l$ where:

$$2l = c_g t$$

where c_g is the group velocity of the pulse in the fibre.

Hence we can map distance into time and get:

$$\frac{dE_{s,0}}{dt} = \tfrac{1}{2}c_g\alpha_s(t)SE_0 \exp\left(-c_g \int_0^t \alpha(t) dt\right)$$

(Note that we have divided the LHS by $dl = \tfrac{1}{2}c_g dt$).

But $dE_{s,0}/dt = p(t)$, the optical *power* received at time t, so that:

$$p(t) = \tfrac{1}{2}c_g\alpha_s SE_0 \exp\left(-c_g \int_0^t \alpha(t) dt\right) \tag{10.20}$$

If α is constant along the fibre this becomes, simply

$$p(t) = \tfrac{1}{2}c_g\alpha_s SE_0 \exp\left(-c_g\alpha t\right)$$

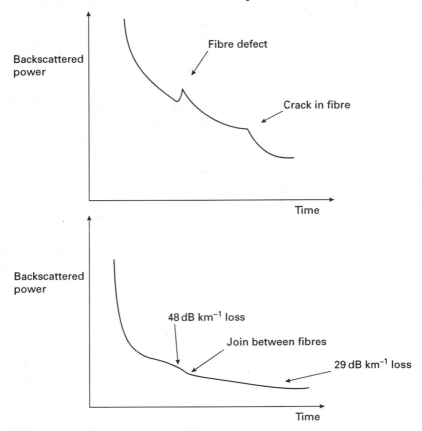

Fig. 10.24 Interpretation of 'typical' ODTR traces.

and can be seen to be an exponential decay with time.

If α is not constant, then we need to know how and where it changes so that we can find out why. This can be done by noting when, in time, the decay deviates from an exponential (see Fig. 10.24), and then using $l = \frac{1}{2}c_g t$ to find where. Provided that the fractional change in α is small over the length dl then it follows from differentiating (10.20) that:

$$\frac{1}{p(l)}\frac{dp(l)}{dl} = -2\alpha(l)$$

so that $\alpha(l)$ can be calculated as a function of l from a knowledge of $p(l)$, which is, of course, determinable from $p(t)$, using $2l = c_g t$.

10.8.2 Distributed optical-fibre Raman temperature sensing

OTDR, as described in the preceding section, effectively comprises a technique for the measurement of the distribution of attenuation along the length of a fibre. If, therefore, an external field had the effect of modify-ing the attenuation in a deterministic way, the strength of this field could be mapped correspondingly along the length of the fibre, using OTDR, thus comprising a distributed measurement of that field. Indeed, one such method was researched in 1986 [3], relying on the temperature dependence of absorption in rare-earth-doped fibres; but it had a number of disadvan-tages, not the least of which was that the sensitivity of the measurement in one section depended on how much absorption had occurred in another section (i.e., on its temperature).

A method which has been successful in measuring the distribution of temperature is based upon the non-linear Raman effect which we met in section 9.8. In discussing the origin of the Raman effect in silica, we noted that, provided the effect was spontaneous (i.e., not stimulated), then the anti-Stokes radiation depended on the population of the excited state from which the corresponding virtual level was drawn, and thus that the intensity of the anti-Stokes radiation varied as:

$$I_a \sim \exp\left(-\frac{E_1 - E_0}{kT}\right) \sim \exp\left(-\frac{h\nu_{10}}{kT}\right)$$

where $h\nu_{10}$ is the energy difference between the ground state and a rota-tional/vibrational excited state of the molecule. Remember also that this is true only if the medium is in thermal equilibrium at absolute temperature T. It is this dependence of I_a on temperature which is used to measure the temperature distribution along the fibre.

Suppose then that a narrow pump pulse is launched into one end of an optical fibre (monomode or multimode) with sufficient intensity to stim-ulate Raman scattering. The amount of this scattering will vary as λ^{-4} in accordance with scattering principles in an amorphous medium, and some of the isotropically scattered (again a result of the amorphous nature of the medium) radiation will be backscattered to the launch end of the fibre. Thus we shall be able to observe, as backscattered radiation, the normal Rayleigh scattering at the optical frequency of the pump, plus the broad spectrum of Stokes 's' and anti-Stokes 'a' Raman radiation at wave-lengths either side of the Rayleigh line. Furthermore, the variation in this backscattered spectrum with time will indicate the variation of the Raman scattering properties of the fibre with position, since, as in OTDR, any given time corresponds to a known position of the pulse along the fibre.

Suppose now that specially-defined narrow ($\sim 1\,\text{nm}$) bands of 'a' and 's' radiation are chosen, equally spaced from the pump frequency. Clearly the most convenient position for these will be close to the maxima in the 'a' and 's' spectra, which occur at $\pm 15\,\text{nm}$ from the pump line (see Fig. 9.11). The frequency difference, ν, between the pump and the chosen bands corresponds, of course, to a particular rotational or vibrational energy level.

Thus we have:

$$\nu_a = \nu_p + \nu$$

$$\nu_s = \nu_p - \nu$$

where ν_p is the pump optical frequency. Since the Stokes radiation is not temperature dependent we can use this as a convenient reference level to cancel the effects of source variation and fibre attenuation. Hence we measure the backscattered levels of both I_a at $\nu_p + \nu$, and I_s at $\nu_p - \nu$ (using a monochromator, i.e., a diffraction grating!) and construct the ratio:

$$R_T = \frac{I_a}{I_s} = \left(\frac{\nu_a}{\nu_s}\right)^4 \exp\left(-\frac{h\nu}{kT}\right) \tag{10.21}$$

R_T is clearly a direct measure of the temperature, T, since all other parameters are constants. Hence a measurement of R_T as a function of time will map the temperature as a function of position along the fibre. We have a convenient distributed temperature measurement sensor. Furthermore it will measure *absolute* temperature, and will be independent of the fibre material. To process the signal we can simply take the logarithm of equation (10.21):

$$\ln R_T(t) = 4\ln\left(\frac{\nu_a}{\nu_s}\right) - \frac{h\nu}{kT(t)}$$

and with a reference section at temperature ϑ, the first term on the RHS can be eliminated, for we have

$$\ln R_\vartheta = 4\ln\left(\frac{\nu_a}{\nu_s}\right) - \frac{h\nu}{k\vartheta}$$

and thus, subtracting the two above equations:

$$\frac{1}{T(t)} = \frac{1}{\vartheta} - \frac{k}{h\nu}\ln\left[\frac{R_T(t)}{R_\vartheta}\right]$$

A schematic of the practical arrangement for implementing these ideas is shown in Fig. 10.25 with the corresponding measured temperature distribution in Fig. 10.26. Note that an avalanche photodiode (APD) is used as the photodetector. This can be used in a photon-counting mode, since each incoming photon will yield a burst of electrons when the APD is suitably biased. It will be instructive to consider in a little more detail how a photon-counting system might be designed for this measurement system, using approximate quantities.

First, the pump power must not be too high, for we must not allow stimulated Raman gain to occur: if it did, thermal equilibrium would no longer prevail, and the Raman backscatter would not then measure the

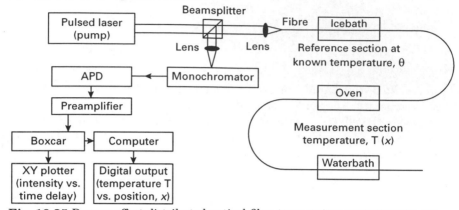

Fig. 10.25 Raman-effect distributed optical-fibre temperature measurement system.

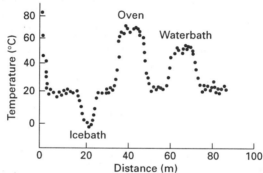

Fig. 10.26 Measured temperature distribution (for Fig. 10.25).

temperature of the medium. Suppose the peak pump power is $\sim 1\,\mathrm{W}$. The duration of the pulse will fix the spatial resolution of the measurement: suppose that we require this to be 1 m. This corresponds to a pulse duration of 10 ns (remember that $2l = c_g t$ and $c_g \sim 2 \times 10^8\,\mathrm{m\,s^{-1}}$ in fibre) and hence a pulse energy of 10 nJ. If the optical wavelength of the pump is $1\,\mu\mathrm{m}$, then the number of photons in 10 nJ is:

$$n_\mathrm{p} = \frac{E}{h\nu} = \frac{E\lambda}{hc} = \frac{10^{-8} \times 10^{-6}}{6.6 \times 10^{-34} \times 2 \times 10^8} = 7.6 \times 10^{10}$$

Of these only ~ 1 in 10^6 will be Raman scattered, i.e., $\sim 10^5$. Of these, only $\sim 10\%$ will be anti-Stokes scattered, since Stokes scattering is much the stronger process, leaving $\sim 10^4$ for the anti-Stokes scatter. Of these only a fraction $\sim 5 \times 10^{-3}$ will be guided back to the front end, giving ~ 50. Taking now the losses in the fibre, the beamsplitter and the monochromator, and the quantum efficiency of the detector, we can expect only ~ 5 photons to be detectable per resolution length of 1 m. In section 1.7 it was noted that the signal-to-noise ratio for a count of N photons was

$N^{1/2}$. If we require a measurement accuracy of 1% (i.e., a signal-to-noise ratio of 100), then N must be at least 10^4. Hence we must integrate over $\sim 2,000$, i.e., $10^4/5$ pump pulses to achieve this accuracy. For a total fibre length of 1 km this means waiting for a time $\sim 2000 \times 10^{-5} = 20$ ms for the measurement to be made. This is usually acceptable for temperature measurements on large structures, since thermal time constants on these are usually large. However, photon counting techniques are expensive to implement and commercially available 'DTS' (Distributed Temperature Systems) presently do not use them. They use an APD in analogue mode with a transimpedance amplifier. The best performance with such a system at the time of writing was $\pm 1°C$ accuracy over 4 km of fibre with 1 m spatial resolution and 10 s standard measurement time. The system continues to evolve. What is needed is a smaller resolution interval. However reducing this interval means reducing the pulse width, and this means smaller energy, since the peak pulse power cannot be increased (stimulated Raman!), and that means a reduced signal level. It also means a faster photodetector (since the pulse is shorter), which means more noise (greater bandwidth) and thus increased measurement time, or shorter total length. Such, as usual, are the many problems of system design - the art of compromise and trade-offs.

10.8.3 Distributed optical-fibre measurement in general

Optical-fibre methods of measurement sensing offer many important advantages for industrial use: the fibre is a flexible, insulating, dielectric medium which can readily be installed in industrial plant without significant disturbance of the measurement environment; the range of measurands which is accessible to measurement by optical-fibre techniques is very large, since the propagation of light within an optical fibre is sensitive to a wide variety of physical influences external to it.

Optical-fibre distributed measurement sensing is a technique which utilizes the one-dimensional nature of the optical fibre as a distinct measurement feature. It is possible, in principle, to determine the value of a wanted measurand continuously as a function of position along a length of a suitably-configured optical fibre, with arbitrarily large spatial resolution. The normal temporal variation of the distribution is determined simultaneously.

Such sensing systems are normally referred to as fully distributed systems, to distinguish them from the quasi-distributed systems, which possess the capability of sensing the measurand only at a number of discrete, predetermined points.

The fully distributed facility opens up an enormous number of possibilities for industrial application. For example, it would allow the spatial and temporal strain distributions in large critical structures, such as multistorey buildings, bridges, dams, aircraft, pressure vessels, electrical generators, etc., to be monitored continuously. It would allow the temperature distributions in boilers, power transformers, power cables, aerofoils, office blocks, etc. to be determined, and thus heat flows to be computed. Electric and magnetic field distributions could be mapped in space so that electro-magnetic design problems would be eased and sources of electromagnetic interference would be quickly identifiable.

There are two important definable reasons for requiring the informa-tion afforded by distributed optical-fibre measurement sensors. The first is that of providing continuous monitoring so as to obtain advance warning of any potentially damaging condition in a structure, and thus to allow alleviative action to be taken in good time. The second is that this spatial and temporal information allows a much deeper understanding of the be-haviour of large (or even quite small) structures, with many implications for improvements in their basic design.

Conventional industrial measurement sensor technology doesn't provide this facility. When measurand distributions of any kind are vital in a given situation the solution usually is to festoon the structure with a multitude of thermocouples, or strain gauges, or whatever. This then presents problems of multiplexing, logging and calibration, and, in any case, relies on the choice of position for each of the many sensors being the correct one - a choice which cannot properly be made without a prior knowledge of the very distribution one is seeking to measure! This 'solution' is thus expensive, tedious and usually broadly inadequate.

The optical fibre, on the other hand, can be readily installed in industrial plant (retrospectively if necessary), produces minimal disturbance of the measurement environment, is cheap, passive and electrically insulating, acts as its own telemetering channel, can easily be rearranged in accordance with acquired knowledge and allows a choice of any or all measurement points along its length within the limits of the spatial resolution interval. If such a technique can be made to work satisfactorily for a number of measurands, a new dimension appears in the field of industrial measurement. These possibilities are being actively explored.

10.9 MEASUREMENT OF VERY SHORT OPTICAL PULSES: THE AUTOCORRELATOR

Topics needed:
   (i) Non-linear susceptibility (Chapter 9)
   (ii) Second harmonic generation (9)
   (iii) Correlation (5)
   (iv) Photodetection response time (7).

Very short, sometimes called 'ultrashort', laser pulses, are in increasing demand in all kinds of application areas. Such pulses lie in the femtosecond

(1 fs $= 10^{-15}$ s) range and are difficult to time-resolve with accuracy, since few detection systems respond sufficiently quickly. Ultrashort laser pulses are required for research applications such as time-resolved spectroscopy, where the dynamics of fast chemical reactions are studied, and in high-speed switching for analysis of charge transfer effects in semiconductors, and also for practical instrumentation systems such as laser radars (e.g., OTDR, lidar), optical computing and storage, and high-speed sampling oscilloscopes.

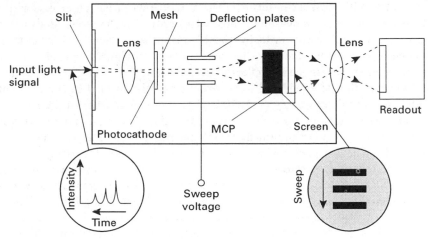

Fig. 10.27 Schematic for a streak camera.

In order to understand properly the data which these narrow pulses can provide it is necessary to have a good knowledge of the amplitude and phase profile of the pulse, i.e., it is necessary to know something of its (complex) shape, in time. We cannot, for example, interpret the results of an experiment in which an ultrashort laser pulse stimulates a chemical reaction, in terms of the dynamics of the reaction, unless the temporal pulse shape is known. When it is known, the results can be 'unwound' by pulling out the pulse shape, a process which is known as 'deconvolution'.

For pulses with widths down to about 10 ps, high-speed photodiodes can be used in combination with fast oscilloscopes to achieve 'direct' detection. If the pulse is repetitive, sampling oscilloscopes are used; in this case a small part of the pulse is observed for any one pulse, and the complete pulse shape is built up by scanning the sampling time across the pulse shape progressively for a number of successive pulses. For pulse widths in the range 500 fs to 10 ps a device known as a 'streak camera' can be used. In this, the optical pulse is allowed to impinge on to a photocathode and thus to generate photo-electrons, as in the photomultiplier. The electrons then are accelerated towards an anode (again as in the photomultiplier) but at the same time they are rapidly deflected sideways by a strong transverse electric field, and then allowed to fall upon a phosphorescent screen. The result is a 'streak' across the screen whose intensity variation displays the time evolution of the original pulse. The streak camera is, however, an elaborate and expensive device (Fig. 10.27).

A device which is much cheaper than a streak camera, and also allows time resolutions down to a few tens of femtoseconds, is the autocorrelator.

The autocorrelator relies on two sets of ideas previously discussed: autocorrelation (unsurprisingly) and second harmonic generation (SHG). Figure 10.28(a) illustrates the action of the autocorrelator used for pulse width measurement. The (quasi-monochromatic) optical pulse to be analysed is split into two identical parts by the beamsplitter. Each portion is then passed to a reflector which, in this case is a 'roof' prism. (The roof prism is a simple prism, arranged for reflection as shown in the diagram; its function is to ensure that the reflected light returns exactly parallel to the incident light, and simple geometry (Fig. 10.28(b)) shows that this will be so even for small rotations of the prism about its axis. The advantage of this is that accurate angular alignment is not necessary; this is especially useful when the reflector position needs to be variable, as in this case.) One of the roof prisms is variable in position as shown in the diagram, so that a variable time delay can be inserted between the two identical halves of the pulse. This is a very convenient method for controlling time delays of order the width of these ultrashort pulses: a change of position of $5\,\mu$m (for example), which is readily achievable, corresponds to a delay of ~ 33 fs.

After the prescribed delay the two halves of the pulse are recombined (so that the complete system up to this point forms what is essentially a Michelson interferometer). The resultant collinear light is then fed into a non-linear crystal, via a lens and an optical filter. The lens ensures that the light is focused properly into the crystal, and the filter acts to exclude all unwanted light (for example, if the pulse originated from a solid state laser, there might be a second harmonic component in the primary beam; clearly this should be excluded).

As the light passes into the SHG crystal at the appropriate angle with respect to the crystal axes, and with the appropriate polarization direction (see section 9.4), it will generate the second harmonic of the fundamental optical frequency. The level of second harmonic generated will depend upon the relative delay between the two halves of the pulse: crudely (a proper analysis will follow), this can be understood by considering that if there is no delay we have just the original pulse before beamsplitting, and the second harmonic level will be proportional to the square of its intensity, i.e., proportional to I^2 (equation (9.5)); if the delay is greater than the pulse width, then the maximum intensity is $\frac{1}{2}I$, so the second harmonic level will be proportional to $\frac{1}{4}I^2$. Hence the delay necessary to take the SHG level from I^2 to $\frac{1}{4}I^2$ will be equal to the pulse width. The SHG component thus generated by the crystal is passed to a photomultiplier

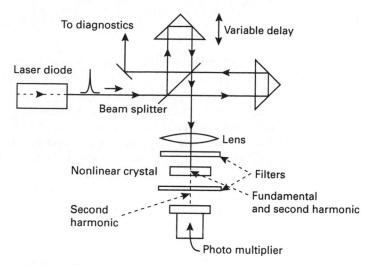

(a) Basic arrangement for an autocorrelator

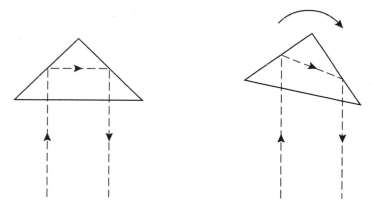

(b) Action of roof prism in maintaining direction of reflected light

Fig. 10.28 The optical SHG autocorrelator.

(PMT) via a filter which blocks the fundamental. If, as is usually the case, the SHG component is generated with the orthogonal linear polarization compared with the fundamental, then this last filter can be a simple linear polarizer (e.g., a Nicol or a Wollaston prism). Hence it is clear that this arrangement comprises a fairly straightforward method for measurement of the pulse width. For a full analysis of the action of this autocorrelator it is necessary to turn to the theory of autocorrelation.

Suppose that the pulse to be analysed is written as an intensity variation in time, $I(t)$. This intensity will correspond to an electric field variation

$E(t)$ where:

$$|E(t)|^2 = KI(t)$$

and K is a universal constant. Note that $E(t)$ is a complex quantity since it contains information on both the optical amplitude and the phase. The intensity function contains no phase information, and the autocorrelator presently being described cannot, therefore, recover the phase. (This is not a serious disadvantage since in most practical applications the systems under investigation respond only to intensity, i.e., to a photon flux.)

In the autocorrelator (Fig. 10.28) the pulse is split into two equal components each of field amplitude $e(t)$ say. It follows that:

$$2|e(t)|^2 = |E(t)|^2 = KI(t)$$

so that

$$e(t) = \frac{E(t)}{\sqrt{2}} \tag{10.22}$$

If one of the components is delayed by τ (via the roof prism) and the two are then recombined, the resulting electric field will be:

$$e_T(t) = e(t) + e(t + \tau)$$

From equation (9.1) we know that the second harmonic component is generated from the square law term in the expression for the nonlinear electric polarization, so that we can write:

$$P_{\text{SHG}} = \chi_2 e_T^2(t) = \chi_2 \left[e(t) + e(t + \tau) \right]^2$$

$$= \chi_2 \left[e^2(t) + e^2(t + \tau) + 2e(t)e(t + \tau) \right]$$

From the discussions in section 9.4 we know that each of these three terms represents a second harmonic field amplitude, so they will lead to an SHG intensity given by

$$I_{\text{SHG}} = K' \left[|e^2(t)|^2 + |e^2(t + \tau)|^2 + |2e(t)e(t + \tau)| \right]^2$$

Hence, from (10.22):

$$I_{\text{SHG}} = KK' \left[\tfrac{1}{4}I^2(t) + \tfrac{1}{4}I^2(t + \tau) + I(t)I(t + \tau) \right]$$

(Note that the last term equivalence follows from the fact that the modulus of a product of two complex numbers is equal to the product of their moduli.)

Now since a simple temporal delay does not cause any variation in the intensity, it follows that

$$I^2(t) = I^2(t + \tau)$$

and hence

$$I_{\text{SHG}} = \tfrac{1}{2}KK' \left[I^2(t) + 2I(t)I(t + \tau) \right]$$

The photodetector, in this case a photomultiplier, cannot respond in a time of the order of the pulse width. If it could, there would be no need for the autocorrelator, since direct detection could be used! It follows that the photomultiplier output is the integral, effectively over all time, of the signal SHG, i.e., its response is

$$R = \int I_{\text{SHG}}dt = \tfrac{1}{2}KK' \left[\int I^2(t)dt + 2\int I(t)I(t + \tau)dt \right]$$

The response thus consists of two terms, one of which varies with time delay τ, and one which does not. If this latter (easily recognizable experimentally) 'DC' term is used to normalize the output we have a normalized response:

$$R' = 1 + \frac{2\int I(t)I(t + \tau)dt}{\int I^2(t)dt} \tag{10.23}$$

The second term in this expression can be recognized as the autocorrelation function of $I(t)$.

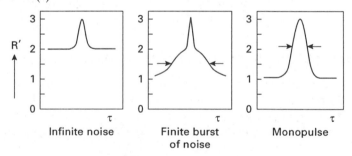

(a) Examples of normalized autocorrelation functions

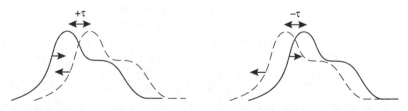

(b) Each of these states, symmetrical about $\tau = 0$, must give the same autocorrelation signal; hence the latter is always symmetrical

Fig. 10.29 Autocorrelator signals.

It is clear that the autocorrelation term varies from a value of 2 when $\tau = 0$ to a value zero when τ is greater than the pulse width, in which case $I(t)$ and $I(t + t)$ will be completely independent. This result is thus in general accordance with the brief discussion earlier in this section.

Hence R' varies from a maximum of 3 to a minimum of 1. The delay, τ, which takes it from 3 to 1 will be equal to the width of the pulse, which can thus be experimentally determined. Some examples of measured results for R' are shown in Fig. (10.29), in the presence of noise. However, the variation of R' with τ does not allow $I(t)$ to be uniquely determined in shape, for any given autocorrelation function will correspond not to a unique temporal profile but to a class of profiles. Furthermore, the autocorrelation function will always be symmetrical, no matter how asymmetrical is the actual pulse. This is clear from Fig. 10.29. The sum of the two pulses always follows the same form either side of $\tau = 0$.

Notwithstanding these limitations, however, the autocorrelator is an extremely valuable device for determining the widths of femtosecond pulses, most of which are, in practice, symmetrical to first order in any case. The autocorrelator is relatively simple and cheap to construct, once the crystal is available: lithium iodate and potassium niobate are commonly used crystals. And it does not require fast-response photodetection. Furthermore, there are other more sophisticated (interferometric) versions which can recover both the full shape and the phase information, if so desired. These are too complex for our present requirements but the interested reader can discover the details of these in ref. [4].

10.10 TOPICS IN OPTICAL-FIBRE COMMUNICATIONS

The basic principles of optical-fibre telecommunications systems were described in sections 8.5 and 8.6 and there have been other allusions elsewhere. These principles will be expanded here as a prelude to a treatment of four quite advanced topics which lie at the forefront of present-day research in the subject.

This is done for several reasons. First, in keeping with our present purpose, they represent excellent examples of optoelectronics in action. Secondly, they give a flavour of the front-line activity in the subject and, hopefully, this will whet the reader's appetite to continue his or her studies of the subject. And, thirdly, since optical telecommunications is such an important present-day application area for optoelectronics (probably the most important) it is a fitting way to end this book.

Again in keeping with the style of this text, the topics will not be dealt with in great analytical detail: physical principles and mechanisms will be paramount.

The elements of a conventional optical-fibre communications system are shown in Fig. 10.30. An optical source is modulated by the information

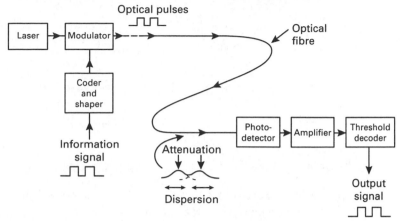

Fig. 10.30 Basis subsystems for optical-fibre telecommunications.

signal, and the light is fed into a fibre for guided transmission. While propagating in the fibre the light suffers from attenuation and dispersion: the former reduces the magnitude of the signal, and the latter distorts and degrades it. The fall in magnitude is unacceptable beyond a certain point since noise will always be present in the detector, and unless the signal is larger than the noise level the signal will not be adequately recognizable. The distortion is unacceptable beyond a certain point since distortion represents a loss of information, and beyond this point the signal again becomes unrecognizable. For long distance communication, amplifying repeaters are used, i.e., the signal is detected, amplified, reshaped and regenerated for onward transmission.

Most optical-fibre communications systems are digital: the information is coded into a series of pulses which represent 'yes' or 'no' to the question: 'are you present in this particular time slot (see section 10.5)?' In this (prevailing) case the condition on magnitude is that the pulses should be just above the noise level. As long as they are above the defined level (threshold) the answers to the above question will be correct to a remarkable degree (only ~ 1 in 10^9 pulses (or pulse absences) will be incorrectly read by the photodetector), but below that threshold the 'bit error rate' (BER) rises rapidly. It is the robustness of the system above the threshold which characterizes a digital system.

The condition on dispersion also is quite straightforward in the digital case. Dispersion broadens the pulses so that there comes a point (see Fig. 10.31) where the noise prevents a clear distinction being made between successive pulses. The question 'present or absent?' can no longer be answered confidently and the BER rises precipitously.

From sections 8.5 and 8.6 we know that both attenuation and dispersion increase linearly with distance along the fibre. The attenuation does so because it is a consequence of absorption and scattering by atoms and

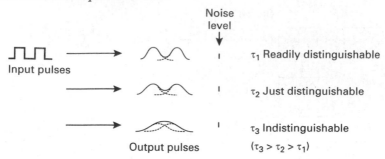

Fig. 10.31 Effect of increasing fibre dispersion on pulse distinguishability.

molecules, so it is proportional to the total number of these which it has encountered, i.e., proportional to total distance in a fibre of fixed cross-section and uniform composition. The dispersion does so because it is a consequence of the differing velocities at which different wavelength components travel, and the displacement between the components thus increases linearly with distance.

Advanced research in optical communications is largely concerned with overcoming the effects of attenuation and dispersion, in order to transmit larger and larger quantities of information at higher and higher speeds over greater and greater distances. The large quantity at high speed implies large bandwidth, and the great distance implies good signal strength at the output. And we saw in section 8.6.2 that the two requirements are largely in opposition: as one increases, the other decreases. For example, a low dispersion can be achieved with a narrowband source, but it is difficult to squeeze optical power into a narrow spectral source-width. Other examples can readily be generated. Consequently, a given optical-fibre communications system can usually be characterized by the product of the signal bandwidth and the distance over which that signal can be transmitted without degradation: this is the bandwidth-distance product and it is usually (for trunk systems) expressed in units of $Gb\,s^{-1}$ km (gigabits per second × kilometres). The best systems at the present time have product $\sim 1000\,Gb\,s^{-1}$ km, so that we can use them for $10\,Gb\,s^{-1}$ bandwidths over 100 km, or $1\,Gb\,s^{-1}$ bandwidths over 1000 km, for example. At the limit of the range, if we need to transmit a further distance, the signal must be amplified and regenerated in some way by an 'amplifying repeater'.

Present-day research is aimed at increasing this bandwidth-distance product by a variety of stratagems, including some which involve highly efficient in-line amplification. We shall begin with one of these.

10.10.1 The optical-fibre amplifier

Topics needed:
   (i) Rare earth doping (Chapter 7)
   (ii) Optical pumping (6)

(iii) Inverted population (6)

(iv) Stimulated emission (6)

(v) Semiconductor lasers (7).

In Chapter 6 the basics of laser action were treated. It was learned there that the laser essentially was an optical amplifier, with positive feedback to cause it to oscillate. It follows that the same principles can be used to construct an optical amplifier; essentially all that is necessary is the removal of the positive feedback.

The first requirement for optical amplification is to generate a population inversion. In order to do this, as section 6.2.4 has detailed, it is necessary to 'pump' (i.e., to populate) appropriate energy levels. The silica of which an optical fibre is made (together with a few other low-concentration dopants) does not possess a suitable energy level structure: it is an amorphous medium with a range of bond energies, leading to broad, overlapping energy levels which do not lend themselves to the definite transitions which useable population inversion requires.

This problem can be overcome by doping the fibre material with 'rare-earth' ions. The rare-earth elements comprise a series in the chemical periodic table whose members are chemically very similar (and they are, as a consequence, very difficult to separate). The 15 elements in question range from lanthanum (atomic number 57) to lutetium (71). The reason for their chemical similarity is that their outer-electron structures are identical, and it is, of course, these structures which determine the chemical behaviour, by interacting with the outer structures of other atoms. As one progresses up the series the atomic number (number of protons in the nucleus) increases and more electrons must be added to compensate. The particular feature which distinguishes the rare-earth elements (and some other series in the periodic table) is that these electrons are added to the inner atomic shells since these, unusually, are of lower energy than the outer shells. In other words, there are available energy levels below those of the outer electrons.

How does all of this help us to produce optical amplifiers?

If a rare-earth element is used as a dopant in silica, then it sits in the amorphous silica lattice with its outermost electrons interacting with all the various bond strengths in the lattice. In fact, for this particular lattice the rare earth element finds it energetically convenient to 'lose' some of its outermost electrons, and become a trivalent ion, e.g., $Er^{3+}, Nd^{3+}, Pr^{3+}$. But the really important point is that these outer shell reconfigurations leave the unfilled inner shells substantially (though not entirely) unaffected, so that these inner energy levels remain quite sharp and can be used for population inversion. Moreover, even though the outer structures are very similar for all the rare-earth elements, the inner levels can be quite different, providing a range of available energies among all these elements, and thus a range of optical frequencies which can be amplified.

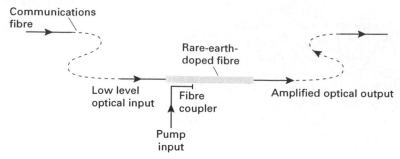

Fig. 10.32 Schematic for an in-line optical-fibre communications amplifier.

Let us look now at how all of this information can be used to construct an optical amplifier. The basic arrangement for an 'in-line' optical fibre amplifier is shown in Fig. 10.32. This allows the signal propagating in a trunk telecommunications fibre (for example) to be amplified without leaving the fibre, and thus without suffering any of the coupling losses associated with electronic repeater-amplifiers where the signal is detected, amplified electronically, and then used to modulate a new a laser source for onward transmission.

Referring again to Fig. 10.32 we note that, at the amplifier point, 'pump' power is coupled into the fibre medium, which is a section of link-compatible fibre doped with a suitable (for the optical wavelength to be amplified) rare-earth ion.

The action of the pump is to generate an inverted population whilst propagating coaxially with the signal to be amplified.

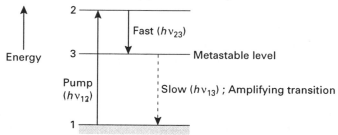

Fig. 10.33 Basic energy level dynamics for the erbium-doped fibre amplifier (EDFA).

Consider, for example, the energy-level diagram for erbium (Er) doping (a few parts per million (ppm)) in a silica lattice (Fig. 10.33). The pump power excites Er^+ ions from level 1 to level 2, where they decay quickly to level 3. This level (3) is a metastable level (its decay to level 1 is 'forbidden' by the quantum rules), so that the ions linger there for a relatively long time (tens of microseconds), thus allowing the population of ions in level 3 to exceed that of level 1: the population is thus inverted, and photons of energy $h\nu_{13}$ will stimulate in-phase photons to be generated by the downward $3 \rightarrow 1$ transition. Hence, the incoming photons at $h\nu_{13}$ will be amplified: all of this we know from Chapter 6. There are several special features relating to the optical-fibre amplifier which are new, however, and we shall now deal with these.

First, what do we use for a pump, and how is its light to be coupled into the fibre?

It is clear that the pump wavelength must be smaller than that of the wavelength to be amplified since the energy $h\nu_{12}$ must be greater than $h\nu_{13}$ (level 2 is higher than level 3).

Taking as an example the Er^+ fibre amplifier operating at 1550 nm (this being close to the wavelength of minimum attenuation in silica), the energy-level diagram shows that a convenient pump wavelength for populating the metastable level is 800 nm. This is especially convenient since there is a very readily available GaAs semiconductor laser at this wavelength. However, there is a problem: the phenomenon known as excited-state absorption (ESA). This is illustrated in Fig. 10.34. The pump radiation at 800 nm also is able to raise the ions, in the metastable, amplifying state (3), to a still higher level (4). This, clearly, reduces the population of level 3 and reduces the efficiency of the amplifier (efficiency = optical gain per watt of pump power). To overcome this problem there are two solutions. First, use a different wavelength: 980 nm and 1480 nm are possible. There are no good pump sources available at 980 nm. At 1480 nm there are sources but the wavelength is getting uncomfortably close to 1550 nm, and separation of the two wavelengths becomes difficult. Nevertheless, 1480 nm sources are often used when large gain is required. Secondly, the Er^+ can be doped into a different glass structure: fluoride glasses are being researched for this purpose. Because the lattice interactions are different, so also are the energy levels, and 800 nm radiation can now be used without the encumbrance of ESA. These glasses are more difficult to fabricate than silica glasses, however.

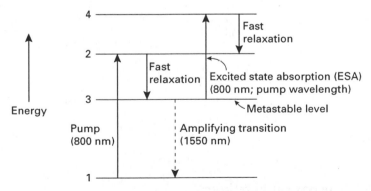

Fig. 10.34 Mechanism for excited-state absorption in the erbium-doped fibre amplifier.

Of course, the most widely used wavelength for installed communications fibres presently is 1300 nm. Attenuation is low at this wavelength in silica (although not quite as low as at 1550 nm) and group velocity dispersion is at a minimum there. A basic problem with producing silica fibre amplifiers at 1300 nm, however, is that the photons at this wavelength couple very effectively to phonons (sound vibrations) at this frequency, and the energy transition is therefore non-radiative. This problem can be overcome by doping fluoride fibre with praseodymium. However, the performance of praseodymium-doped fluoride fibre amplifiers (PDFFAs) remains inferior to that of erbium-doped silica fibre amplifiers (EDFAs).

What of the practicalities of this performance?

With a typical EDFA we might obtain a gain spectrum such as that shown in Fig. 10.35. This shows a not-quite-flat gain of 20 to 40 dB (10^2 to 10^4 in power amplification) over a wavelength range of ~ 40 nm. This corresponds to an optical frequency range of ~ 5000 GHz. The reason for this broad gain bandwidth lies in the chaotic nature of the fused-silica lattice. Although the optically active levels lie below the outer levels, they are affected to some extent by the bonding between the erbium atoms and the silica molecules, and this bonding is highly variable in strength, leading to a spread of the linear Er^{3+} levels as a consequence. This, of course, leads to a range of photon transitions, and hence to gain over a range of wavelengths. With narrow-linewidth semiconductor sources having spectral widths ~ 0.1 nm this means that several different carrier wavelengths can be used in a single fibre, with the same amplifier. One might have up to 20 such carriers for a typical EDFA-based system. Such systems are called wavelength-division-multiplexed (WDM) systems and clearly allow the bandwidth-distance product for any given fibre to be increased by at least an order of magnitude.

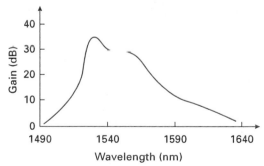

Fig. 10.35 Gain spectrum for an EDFA.

10.10.2 The optical-fibre laser

Topics needed:

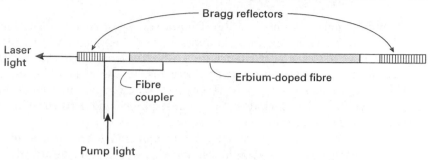

Fig. 10.36 A fibre-laser using Bragg reflectors.

(i) Laser action (Chapter 6)
(ii) Photosensitivity (9)
(iii) Optical fibre waveguides (8).

It is a short step from the optical-fibre amplifier to the optical-fibre laser. Positive feedback must be applied to the amplifier in order to cause it to oscillate. This can be done with external mirrors, mirrors coated on to the fibre ends or by 'writing' reflective Bragg gratings into the fibre, using the photorefractive effect discussed in section 9.10 (Fig. 10.36). This latter is the preferred reflective method, at least for one of the reflectors, because it is very wavelength selective and, since the wavelength which it selects can be varied by varying the Bragg-grating spacing (via an applied stress or a temperature variation), this means that the laser is wavelength tunable over a limited range.

Optical-fibre lasers clearly are very convenient for optical-fibre communication since the laser can be 'fused' directly on to the end of the transmission fibre to provide a very efficient launch arrangement.

10.10.3 Optical waveguide couplers and switches

Topics needed:
(i) Optical waveguides (Chapter 8)
(ii) Coupled oscillators (6)
(iii) Electro-optic effect (7, 9)
(iv) Integrated optics (8)
(v) Mach-Zehnder interferometer (2)
(vi) Evanescent waves (2).

There are many occasions, in optical communications and elsewhere, where it is convenient, sometimes necessary, to couple light from one waveguide to another. This might be coupling between planar or cylindrical guides. For example, an important problem in WDM systems is that of coupling light into a single transmission fibre from a variety of sources, each independently modulated, and then 'de-multiplexing' (i.e., separating them all again) when they emerge at the far end. Another example is that for the devices discussed in the previous two sections; it was in fact the second of the two questions asked about the optical pump source for the optical amplifier or laser, i.e., how does one couple pump light into the fibre so that it propagates coaxially with the signal?

The fibre coupler is a special example of an arrangement which allows coupling between any two waveguides, and this will now be described.

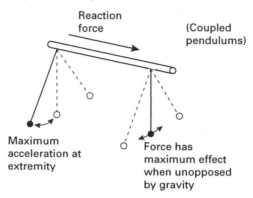

Fig. 10.37 Quadrature relationship between coupled oscillators.

When two oscillators are allowed to interact i.e., to stimulate each other's oscillations, power will transfer periodically between them. The most straightforward illustration of this is that of two, identical pendulums (Fig. 10.37), coupled via a common support. The coupled system is a complete entity and can be described in composite terms. There are two eigenmodes of oscillation, i.e., two states of oscillation which do not change with time (assuming that there are no dissipative forces, such as air resistance). The first of these is that where both pendulums oscillate in phase with the same amplitude: call this mode 1, with oscillation frequency f_1. The second is where they oscillate in antiphase with same amplitude: this takes place at frequency f_2. Suppose now that with both pendulums initially at rest, one is set swinging. It interacts with the other via a force which, when the first pendulum is accelerating (maximum acceleration is at the extremities of swing), is transmitted along the support. This starts the second pendulum swinging. Now when a periodic force drives an oscillator at resonant frequency (identical pendulums!) it does so with a phase difference of $\frac{1}{2}\pi$, that is to say the amplitude of the driving force is in quadrature with the amplitude of the oscillator's displacement. This is because the force is most effective in increasing the velocity of the point mass on which it is acting when it is unopposed by the restoring force of the oscillator, i.e., at the oscillator's mid-point of swing (neutral position). This means that force and oscillator are thus in quadrature. In the case of the two pendulums, the driving force from one pendulum (a maximum at the extremities of the swing) has maximum effect on the other at the latter's mid-point of swing, and hence the second pendulum oscillates $\frac{1}{2}\pi$ out of phase with the first (Fig. 10.37). As it increases its amplitude in response to the motion of the first, the second pendulum starts to react back on the first. Again it does so $\frac{1}{2}\pi$ out of phase and therefore in anti-phase ($\frac{1}{2}\pi + \frac{1}{2}\pi = \pi$!) with the first. The result is that the first pendulum has its swing amplitude reduced, while that of the second increases, until the first is stationary and the second has maximum amplitude. The process then continues in reverse, so that energy of swing is continuously transferring between the two pendulums. The frequency at which the transfer takes place is $f_1 - f_2$. Clearly $f_1 - f_2$ depends upon the strength of the coupling between the pendulums. The stronger the coupling, the higher will be $f_1 - f_2$. (For zero coupling $f_1 = f_2 = f$, the frequency of each independent oscillation, and $f_1 - f_2 = 0$.)

Let us now transfer these ideas to coupled optical waveguides.

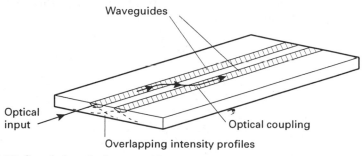

Fig. 10.38 Coupled optical waveguides.

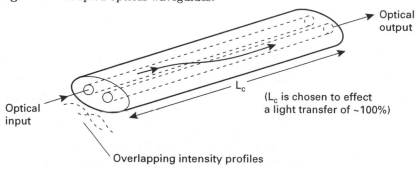

Fig. 10.39 A fused-cladding fibre coupler.

Suppose that there are two identical, parallel waveguides close enough together for their evanescent fields to overlap (Fig. 10.38). In this case, if an optical wave is propagating in just one of them, some of its light will leak into the second: the two waveguides are coupled. Now the coupling occurs as a result of the optical electric field of the first guide acting on the atomic oscillators of the second to set them into motion (classical physics approach!). This will happen with a relative phase of $\pi/2$, just as for the pendulums. As the waves progress down the pair of guides, the wave in the second guide increases in strength and acts back on the first in antiphase, reducing its amplitude, again just as for the pendulums. Hence as the propagation progresses, the light power transfers back and forth between the two guides, with a spatial frequency (i.e., over a waveguide distance) which depends upon the coupling strength (e.g., how close the waveguides are). It follows that, if the coupling strengths and the distance are chosen appropriately, all the light power will have transferred from one waveguide to the other after that distance, so if the guides are separated at that point we shall have constructed an effective waveguide coupler, coupling light from one guide to the other.

Suppose then that we have two fibres with claddings fused together over a length which just allows this complete coupling to occur (Fig. 10.39). If one of these fibres is part of a communications link, and a section of it is doped with erbium, the other can be used to couple in the pump power from, say, a 1480 nm semiconductor laser.

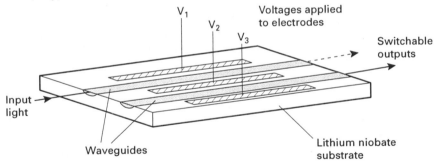

Fig. 10.40 A switchable planar optical coupler.

We can become even cleverer with these ideas. Consider the integrated-optical arrangement shown in Fig. 10.40. In this case we have two coupled planar waveguides written into $LiNbO_3$, an electro-optic material. Also we have electrodes allowing an external electric field to be applied across the waveguides. The effect of an applied voltage is, in the electro-optic material, to alter the refractive indices of the waveguide materials. This alters the propagation constants for the light signals in the waveguides and thus alters the frequencies of the system's eigenmodes (it alters f_1 and f_2 in the pendulum analogy). Hence the voltage controls $f_1 - f_2$, the spatial frequency at which coupling takes place between the guides. Therefore at zero voltage (say) all the light will emerge from guide 1. At voltage V, over that same distance, all the light will emerge from guide 2. We have produced a photonic switch! Light can be switched between waveguides at

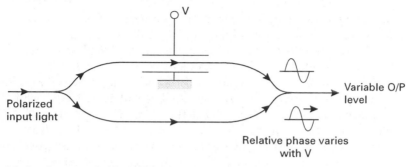

(a) An electro-optic Mach–Zehnder modulator

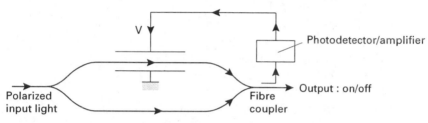

(b) A Mach–Zehnder two-state memory switch

Fig. 10.41 Electro-optic Mach-Zehnder optical switches.

GHz rates (limited only by the speed with which the material responds to the voltage, i.e., to the speed of the electro-optic effect). Clearly this is a very useful device for telecommunications, and many other, purposes.

We have dealt only with identical waveguides. Asymmetrical waveguides also can be coupled and these have various interesting properties. A full analysis of these coupling effects is quite complex, involving the solution of second-order differential coupled-mode equations. A complete understanding of this analysis clearly is essential for anyone wishing to design devices based on these effects and this can be obtained from other texts (e.g. [5]).

Another type of electro-optic integrated-optical (I/O) waveguide switch is shown in Fig. 10.41(a). This is a waveguide example of the Mach-Zehnder interferometer discussed in section 2.9. Linearly polarized light enters into the single-mode waveguide at the front end and is split into two equal components at the first Y-junction. The two components propagate separately in the two arms and are brought together again at the second Y-junction. Clearly, if the two optical paths are exactly equal and the same polarizations are maintained, then the components will be in phase at the second junction and hence they will reinforce to give the original input level again at the output. If, however, a voltage is applied to one arm of the Mach-Zehnder in the electro-optic material so as to alter the phase of the light in that arm by π, then the two components will be in antiphase on recombination and the output from the I/O module will be zero. Thus the voltage activates the switch between maximum and zero outputs, with all levels between these accessible by varying the applied voltage.

It is even possible to arrange that the electric field required for the switching is provided by a light beam, so that light switches light, via the *optical* electro-optic effect. Moreover, if the switching light is derived from the guide output (Fig. 10.41(b)), the switch will remain 'on' whilst the input remains above a certain level; below that level it turns 'off'. Thus we have a two-state switch which can separate different light levels. This is an example of an optical 'logic gate' which can be used in optical signal processing and, in more advanced forms, in optical computing.

10.10.4 Coherent systems

Topics needed:
  (i) Photon statistics (Chapter 1)
  (ii) Interference (2)
  (iii) Coherence (5)
  (iv) Photodetection (7)
  (v) Laser spectra and stability (6, 7)
  (vi) Laser modes (6)
  (vii) Polarization in optical fibres (3, 8)
  (viii) Polarization control (3).

The optical communications systems which have been discussed so far have all been of the type called, in conventional parlance, amplitude-modulation (AM) systems. By this it is meant that the power level of the source is varied in sympathy with the signal modulation. Since an optical source is more readily categorized by (among other things) its output power, in the case of optical communications we should perhaps speak of 'power modulation' or, since the power propagates in a fibre of fixed cross-section 'intensity modulation' (IM), rather than amplitude modulation. The receiver in this case has a relatively simple task: to provide an electrical current proportional to the power it receives. It is a 'direct detector (DD)' and such systems are often referred to as IM/DD systems.

Such systems have the important advantage of simplicity, thus requiring relatively unsophisticated (and therefore cheap and reliable!) components. However, they also have a number of disadvantages.

The first is that they are relatively insensitive. This means that it is difficult to obtain good receiver signal-to-noise ratios (SNRs) over long distances, thus necessitating frequent amplifying repeaters. The reason for the insensitivity is that the optical signal level is small after a long distance transmission, and the quantum noise is such as to allow a maximum SNR (section 1.7):

$$\text{SNR} = \left(\frac{P_s}{Bh\nu} \right)^{1/2} \tag{10.24}$$

P_s being the received signal power, B the bandwidth, h the quantum constant and ν the optical frequency. In other words the SNR is smaller, the smaller is the received power. Any subsequent amplification in the detection system can only degrade the SNR further, since it will add noise.

The second disadvantage is that of the large optical bandwidth. A typical multimode semiconductor laser has a spectral width $\sim 5\,\text{nm}$, which corresponds to $\sim 1,000\,\text{GHz}$ of bandwidth. Since spacing between channels in a wavelength-division-multiplexed (WDM) system needs to be $\sim 10\times$source width (to avoid cross-talk between channels) this means that each channel effectively occupies $\sim 10,000\,\text{GHz}$. Such a large frequency spread does not allow more than one or two channels in either of the $1,300\,\text{nm}$ and $1,550\,\text{nm}$ transmission 'windows' in silica fibre (Fig. 4.3).

The $\sim 1,000\,\text{GHz}$ spectral width of the multimode semiconductor laser means that it is little more than an optical noise source (in communications terms) which, in digital AM systems, is simply switched on and off.

The development of radio and microwave techniques has shown that, with spectrally pure sources, much better system performance can be achieved by modulating the frequency, phase or polarization state, rather than simply the amplitude. In order to do this, the modulation parameter (e.g., frequency in FM) must be stable to better than $\sim 1\%$ of the modulation bandwidth if the signal is not to be distorted. Hence the requirement is for high power sources with narrow linewidth, and good frequency/phase stability. Such sources are clearly going to possess a high degree of coherence (section 5.1) and the systems based on them are thus known generally as 'coherent' systems.

Let us examine some of these ideas in the context of optical-fibre communications.

Suppose that we were to have available a 'pure' high power optical source (i.e., a laser) whose output at the far, receiving, end of the fibre could be characterized in terms of its optical electric field as, effectively, a pure sinusoid:

$$E_s = e_s \cos(\omega_s t + \varphi_s) \tag{10.24}$$

Suppose, first, that this source is used in an IM/DD system with the information signal directly modulating the mean optical power level.

When this signal is directly detected, the detector provides an output current proportional to the input optical power (Section 7.4). The optical power is proportional to the intensity of the wave, i.e., to the square of its amplitude, averaged over the response time of the photodetector:

$$P_s = C\langle e_s^2 \cos^2(\omega_s t + \varphi_s)\rangle = \tfrac{1}{2}Ce_s^2\langle(1 + \cos 2(\omega_s t + \varphi_s))\rangle$$

where C is a constant.

Now since the detector cannot respond to frequencies as high as $2\omega_s$, the current which flows is proportional only to the first term, i.e.,

$$i_d \sim P_s = \tfrac{1}{2}Ce_s^2 \tag{10.25}$$

The current will therefore follow a modulation of the power level of the source, $P_s(t)$, up to the maximum speed of response of the photodetector.

Suppose now that we have available yet another pure optical source but at a different optical frequency. Let us describe its output by

$$E_L = e_L \cos(\omega_L t + \varphi_L) \tag{10.26}$$

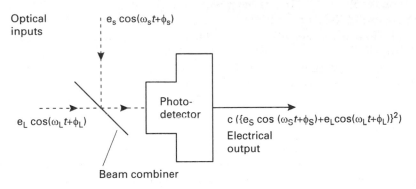

Fig. 10.42 Basics of coherent optical detection.

Let us allow the two sources described by equations (10.24) and (10.26) to fall simultaneously on to the photodetector (Fig. 10.42). Now the input power and the resulting output current will be proportional to the time-averaged value of the square of the total electric field of the two waves, so that, assuming that they have the same polarization (but see later) the optical power will now be represented by:

$$P_D = C\langle (E_s + E_L)^2 \rangle$$

$$= C\langle [e_s \cos(\omega_s t + \varphi_s) + e_L \cos(\omega_L t + \varphi_L)]^2 \rangle$$

$$= C\langle \tfrac{1}{2} e_s^2 [1 + \cos 2(\omega_s t + \varphi_s)] + \tfrac{1}{2} e_L^2 [1 + \cos 2(\omega_L t + \varphi_L)]$$

$$+ e_s e_L \cos[(\omega_s - \omega_L)t + \varphi_s - \varphi_L] + e_s e_L \cos[(\omega_s + \omega_L)t + \varphi_s + \varphi_L] \rangle$$

Again, the detector cannot respond to frequencies as high as $2\omega_s, 2\omega_L$ or $(\omega_s + \omega_L)$. It can, however, respond to $(\omega_s - \omega_L)$ if this is low enough, say $< 1\,\text{GHz}$. In this case we shall have a detector current given by:

$$i_d' \sim \tfrac{1}{2} e_s^2 + \tfrac{1}{2} e_L^2 + e_s e_L \cos[(\omega_s - \omega_L)t + \varphi_s - \varphi_L]$$

Since

$$P_s \sim \tfrac{1}{2} e_s^2, \quad P_L \sim \tfrac{1}{2} e_L^2$$

and writing

$$\omega_s - \omega_L = \omega_{IF}$$

$$\varphi_\text{s} - \varphi_\text{L} = \varphi_\text{IF}$$

we can simplify the expression for i'_d to:

$$i'_\text{d} \sim P_\text{s}(t) + P_\text{L} + 2[P_\text{s}(t)P_\text{L}]^{1/2}\cos(\omega_\text{IF}t + \varphi_\text{IF}) \qquad (10.27)$$

ω_IF is called the 'intermediate frequency' and will be familiar to readers who have experience of 'superhet' radio techniques. We call this 'coherent' detection for reasons which will be clearer shortly.

Look carefully at equation (10.27). It tells us that we have two 'DC' terms (P_s and P_L) followed by a term which represents a frequency which is now in the electronic (as opposed to the optical) range. Furthermore, its amplitude is dependent upon the product of the power levels (P_s, P_L) of the two optical sources.

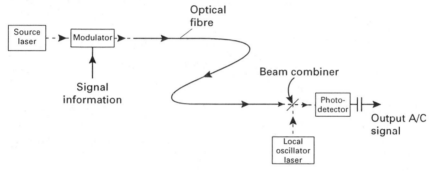

Fig. 10.43 Schematic coherent optical communication system.

Consider now the optical-fibre communications system shown in Fig. 10.43. Here the 'pure' signal source is intensity modulated in the usual way. At the detector it is 'mixed' on the photodetector surface with the signal from the other pure source: the 'local oscillator (LO)'. With DC filtering in the detector circuit the only term which remains in the current from the detector is:

$$i''_\text{d} \sim 2[P_\text{s}(t)P_\text{L}]^{1/2}\cos(\omega_\text{IF}t + \varphi_\text{IF}) \qquad (10.28)$$

This has two advantages. First, it is of narrow bandwidth, fixed by the spectral widths of the two sources ($\omega_\text{IF} = \omega_\text{s} - \omega_\text{L}$), of which more later. Secondly, the signal amplitude $P_\text{s}(t)$ is boosted by the $P_\text{L}^{1/2}$ term which effectively provides optical amplification, since the local oscillator contains no signal information, does not pass through the fibre, and we are free to make P_L as large as we like. This optical amplification has the effect of increasing the received power, over the direct detection arrangement, by a factor [see equations (10.25) and (10.28)] of

$$2\frac{(P_\text{s}P_\text{L})^{1/2}}{P_\text{s}} = 2\left(\frac{P_\text{L}}{P_\text{s}}\right)^{1/2}$$

Since the SNR increases as the square root of the received power (equation (4.16)), this means that the SNR rises by a factor of

$$\sqrt{2}\left(\frac{P_L}{P_s}\right)^{1/4}$$

Let us take an example. Suppose that both sources have an output power of 1 mW. The signal source suffers attenuation in the fibre of, say, 50 dB, so the optical source power emerging is $10^{-5} \times 10^{-3} = 10^{-8}$ W. This is mixed with 1 mW from the local oscillator source to give amplification by a factor of

$$2\left(\frac{P_L}{P_s}\right)^{1/2} = 2\left(\frac{10^{-3}}{10^{-8}}\right)^{1/2} = 632.5$$

The SNR increases by:

$$\sqrt{2}\left(\frac{10^{-3}}{10^{-8}}\right)^{1/4} = 25.14 = 14\,\mathrm{dB}$$

A 14 dB improvement in SNR is well worth having! Up to about 20 dB improvement can be obtained with this technique. This is a valuable advantage since it easily could be equivalent to another 100 km of communications distance.

What is the price to be paid for this improvement?

Fairly clearly one price is that of the required purity of the lasers used for signal and local oscillator sources. Looking again at equation (10.27), it is evident that ω_{IF} must have a value which does not stray by more than about 1% of the bandwidth of the modulation signal $P_s(t)$. If it did, it would corrupt and distort $P_s(t)$ and thus effectively introduce noise into the system. Since $\omega_{IF} = \omega_s - \omega_L$, this means, in turn, that neither ω_s nor ω_L can stray by more than about 0.5% of the signal bandwidth. Look yet again at equation (10.27): φ_{IF} also needs to be stable for the same reason as ω_{IF}, and $\varphi_{IF} = \varphi_s - \varphi_L$, so the phases need to be locked together to the same kind of accuracy.

Let us now insert some numbers into all of this. Suppose that the signal modulation bandwidth is ~ 1 GHz (a 1 Gb s^{-1} digital system for example). Each laser now has to be stable to 0.5%, i.e., to at least 5 MHz, or 1 part in 10^8 of the optical frequency. Single longitudinal mode semiconductor lasers can be fabricated using distributed feedback (DFB) reflectors for mirrors (i.e., mirrors which are Bragg gratings and reflect one very narrow band of wavelengths). Using external distributed Bragg grating reflectors (DBR) laser linewidths as narrow as 10 kHz have been obtained. However, the output frequency and phase drift with temperature in these devices, so they must be loop-stabilized. Figure 10.44(a) shows schematically how this can be done. The source laser is bled by a small amount and compared with an ultra-stable frequency from, for example, an atomic secondary standard, to provide an error signal proportional to the drift. This signal is fed to a temperature or current controller to correct the drift. A similar arrangement is used at the detector end (Fig. 10.44(b)), only this time it is the much lower frequency IF which must be maintained constant. A stable electronic oscillator is now required for the reference, but phase error, in addition to frequency error, is required, so as to maintain both ω_{IF} and φ_{IF} constant.

With this kind of stability in the channel frequency it follows that the 1 GHz signal needs only about 10 GHz of channel separation as opposed to 10,000 GHz in the IM/DD case. Hence it becomes possible to run ~ 1000 channels in each of the two silica windows, thus increasing the bandwidth distance product by a factor ~ 1000.

Clearly, the requirements on the stability of the lasers are stringent. The lasers need to be not only of stable, narrow linewidth, but also tunable, to allow the locking action for frequency and phase. Such lasers are expensive and temperamental, and more work needs to be done before their performances are wholly satisfactory for coherent communications links.

There is just one more problem, unfortunately. (This has not been mentioned before in order not to over-complicate the ideas for readers encountering them for the first time, but it is important for coherent optical-fibre communications designers). The problem is this: equation (10.27) is valid only if the two signals have exactly the same polarization. It is not difficult to arrange that each laser has a stable, linearly polarized output but, as we know (section 8.7), the fibre can play a variety of polarization tricks on light propagating within it. Consequently, the signal light falling on the photodetector is likely to have a varying polarization and hence the IF signal will be subject to fading.

There are two possible types of solution to this problem. The first is to lock the polarization of the signal by means of a feedback loop and polarization controller using, for example, the electro-optic effect and/or the magneto-optic effect; using, in fact, the same principles as for polarization modulation (sections 7.3.1, 7.3.2). The reference polarization in this case can be that of the local oscillator. The second is a 'diversity' arrangement (Fig. 10.45) where the two orthogonal, linear components are separated (by a Wollaston prism for example) and IF-detected separately. The two IF signals can then be added electronically. Such systems can be made to work satisfactorily but, again, they add complexity, cost and unreliability.

The full, reliable implementation of coherent systems with their significant advantages of high channel selectivity (and therefore ease of multiplexing) and of high sensitivity (+20 dB) will come. But at the time of writing (1996) the components necessary for them are not quite ready.

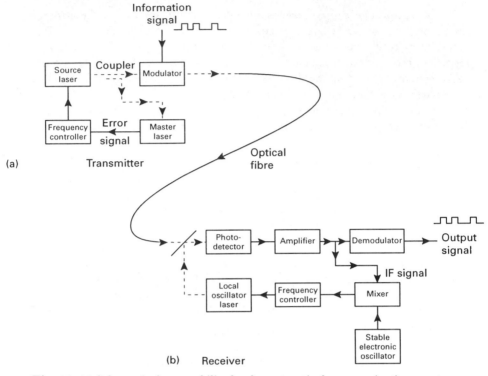

Fig. 10.44 Schematic for a stabilised coherent optical communications system.

10.11 CONCLUSIONS

In this chapter we have seen many examples of optoelectronics in action. There are, of course, very many others, but it is hoped that the ones chosen have served to illustrate the power, flexibility and utility of optoelectronic device and system design.

It is hoped also that, by viewing the ideas in action, the ideas will themselves have become clearer to the point where new applications and new researches may suggest themselves to the reader.

PROBLEMS

10.1 A glass microscopic slide having a refractive index 1.55 is to be coated with a magnesium fluoride film (refractive index 1.38) in order to increase the transmission, at normal incidence, of yellow light at wavelength 550 nm. What is the minimum film thickness that should be deposited? What is the next lower wavelength which will be maximally transmitted with that thickness?

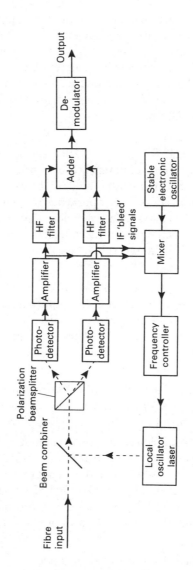

Fig. 10.45 Schematic for a coherent optical detector with polarization diversity.

10.2 Describe in detail an arrangement for measuring the electric current in a high voltage conductor, making use of the Faraday magneto-optic effect in a single-mode optical fibre. Discuss the advantages and disadvantages of the arrangement.

A length of single-mode fibre is wound ten times around a former of radius 1 m which encloses a copper conductor. Linearly polarized light of wavelength 633 nm is launched into one end of the fibre. What magnitude of rotation of the polarization direction has occurred when the light emerges from the fibre at the other end if a current of 100 A is passing through the conductor? What is the maximum current which can be measured without ambiguity? (the Verdet constant for the fibre material $= 5.3 \times 10^{-6}$ radians A^{-1} at 633 nm.)

10.3 An integrated-optic RF spectrum analyser is to be constructed from a $LiNbO_3$ substrate with a tilted interdigital transducer. Calculate the angle of tilt necessary for optimum operation in the Bragg regime, for a wavelength of 633 nm, at a transducer operating frequency of 500 MHz. If the transducer bandwidth is 200 MHz how wide may the acoustic beam be before Bragg selectivity limits the diffraction efficiency? (The refractive index of $LiNbO_3$ is 2.2 at 633 nm and the acoustic velocity is $6.57 \times 10^3 \, m\,s^{-1}$.)

10.4 A compact disc system uses a He-Ne laser, wavelength 633 nm, and a lens with a numerical aperture of 0.45. Approximately how many bits of information can be stored on a compact disc of diameter 30 cm? If each sample of the analogue waveform for storage on this disc is represented by 4 bits, how accurately can the signal be reproduced by the disc on playback?

10.5 Draw a diagram of a minimum-configuration fibre interferometer gyroscope. Label the system components and briefly describe, with reasons, the desirable features and main function of each component in the system.

For a fibre gyroscope consisting of a circular coil of N turns of fibre, derive a formula expressing the optical phase shift $\Delta\varphi$ produced by a rotation rate Ω in terms of the coil area A and source wavelength λ_0. Use your formula to calculate the unambiguous range of rotation rate measurable by a gyroscope employing an 830 nm source and 700 m of fibre wound on to a coil of 10 cm diameter. How could this range be extended?

10.6 Describe the processes involved in holography. What are the experimental arrangements necessary for producing virtual and real (pseudoscopic) images. Derive the relevant mathematical expressions and interpret them in physical terms.

Describe three practical applications of holography.

10.7 Describe the principles of optical time-domain reflectometry (OTDR). Show that the power received at time t from a pulse of energy E_0 launched at time $t = 0$ is given by:

$$p(t) = \tfrac{1}{2}c_g\alpha_s SE_0 \exp(-c_g\alpha t)$$

where c_g is the group velocity of the pulse, α_s is the scatter coefficient, α is the total loss coefficient, and S is the backscatter capture fraction.

How can this equation be used to measure $\alpha(l)$, the variation of α with distance along the fibre? Discuss how these ideas can be used in distributed optical-fibre sensing (DOFS).

10.8 A short optical pulse has the form:

$$f(t) = a_0 \exp\left(-\frac{t^2}{2\tau^2}\right)$$

What is its Fourier spectrum?

How would you measure a_0 and τ if (i) $\tau \sim 1\,\mu s\ (10^{-6}\,s)$ and (ii) $\tau \sim 1\,fs$ $(10^{-15}\,s)$?

10.9 A communications link operating at a wavelength of $1.55\,\mu m$ consists of 50 km of monomode fibre whose loss characteristic is $0.2\,dB\,km^{-1}$. This 50 km is comprised of ten lengths of fibre joined by fusion splices, the loss at each splice being 0.1 dB. The front-end laser diode launches 2.5 mW into the fibre. The emerging signal is to be boosted to a power of 1 mW by an Er^+-doped fibre amplifier which has a gain of $0.25\,dB\,m^{-1}$. What length of this latter fibre is required?

10.10 A fabrications facility is turning out directional couplers in an I/O planar geometry, in Ti-diffused $LiNbO_3$. The inter-waveguide gaps vary with fabrication conditions, but the following data were obtained from various guides: gap (in μm) 4, 5 and 6 and the corresponding complete coupling length (in mm) 7, 7.5 and 9, respectively.

Calculate the length required for a 50% coupling when the gap is $7\,\mu m$.

10.11 Discuss the relative advantages and disadvantages of coherent optical detection compared to direct detection. Draw a block diagram of an optical fibre communications system employing coherent detection and briefly describe the nature and function of the major system components.

Derive a formula showing the relation between the output current, I, and the received optical power P_c at an optical frequency ω_c for a coherent detector having a local oscillator of power P_0 and optical frequency ω_0. If the DC level of the photocurrent in a coherent receiver is $1\,\mu A$ with the local oscillator turned off and $100\,\mu A$ when the oscillator is working, what is the optical power gain for the coherent signal?

10.12 In a coherent communications system, an incoming optical carrier signal can be represented by an electric field $E_c(t) = A_c \cos(\omega_c t + \varphi_c)$, where A_c and ω_c are the field amplitude and radian frequency respectively, and φ_c is a constant. Similarly, the local oscillator field $E_0(t) = A_0 \cos(\omega_0 t + \varphi_0)$.

Derive an expression for the photocurrent generated by a detector of responsivity R receiving the sum of these fields, in terms of the mean powers of the input fields. The carrier power P_c can be taken to be in units such that $P_c = A_c^2$, and for the local oscillator power, $P_0 = A_0^2$.

Explain the implications of the resulting formula.

10.13 An optical wave with maximum electric field amplitude E_0 is passing at velocity c through a medium whose dielectric constant, at the frequency of the wave, is ε. Show that the optical power passing through unit area normal to the wave direction is given by:

$$I = \tfrac{1}{2} c e E_0^2$$

Two optical waves, each with maximum electric field amplitude E_0, are passing collinearly through this medium and are, together, incident normally on a photodetector of area $A\,m^2$, and sensitivity $s\,AW^{-1}$. The two waves have the same frequency and are linearly polarized in the same direction, but differ in phase by $\tfrac{1}{3}\pi$. Show that the resulting current provided by the photodetector is given by:

$$i = \tfrac{3}{2} c \varepsilon s A E_0^2 \text{ Amperes}$$

If the polarization direction of one of the waves is rotated through $\tfrac{1}{4}\pi$, what is the new value for this current?

10.14 The frequency stability required for an amplitude modulated (ASK) optical coherent communications system is $\pm 10\,MHz$. If the local oscillator laser emits at $1.55\,\mu m$ wavelength and its frequency drifts at a rate of $14\,GHz\,K^{-1}$ calculate:
(i) the fractional stability necessary for the laser;
(ii) the maximum temperature change permissible in the absence of external control;
(iii) the maximum transmission bandwidth allowed by the frequency stability.

REFERENCES

[1] Jones, R. C. (1948) A new calculus for the treatment of optical systems, *J. Opt. Soc. Am.* 38(8): 671-85.
[2] Richtmeyer, F. K., Kennard, E. H., Lauritsen, T. (1955) *Introduction to Modern Physics*, McGraw-Hill, 5th edn, chap. 2.

[3] Farries, M. C. *et al.*   (1986) Distributed temperature sensor using Nd^+-doped optical fibre. *Elect. Lett.* 22; 418-419.

[4] Ippen, E. P. and Shank, C. V. (1977) in *Techniques for Ultra-Short Light Pulse Measurements*, ed. S. L. Shapiro, Springer, New York, pp. 83-122.

[5] Syms, R. and Cozens J. (1992) *Optical Guided Waves and Devices*, McGraw-Hill, chap. 10.

FURTHER READING

Andonovic, I. and Uttamchandani, D. (1989) *Principles of Modern Optical Systems*, Artech House (Optical information processing, storage media, integrated optics, polarimeters.)

Bjarklev, A. (1993) *Optical Fiber Amplifiers*, Artech House.

Blaker, J. W. and Rosenblum, W. B. (1993) *Optics: An Introduction for Students of Engineering*, Macmillan (New York) (Instruments, interferometry and holography.)

Dakin, J.P. (1990) *The Distributed Fibre-optic Sensing Handbook*, Springer-Verlag.

Grattan, K. T. V. and Meggitt, B. T. (1995) *Optical-Fibre Sensor Technology*, Chapman and Hall (Interferometric sensors, optical-fibre current measurement, distributed optical-fibre sensors.)

Keiser, G. (1991) *Optical Fiber Communications*, McGraw Hill.

Lefevre, H. (1993) *The Fiber-Optic Gyroscope*, Artech House.

Ryan, S. (1995) *Coherent Lightwave Communications Systems*, Artech House.

Ref [5] above, for waveguide couplers and switches.

11
Epilogue: And what of the future...?

We have been dealing, throughout the preceding chapters, with the properties of photons, of electrons and of their various interactions with and within matter.

It has been noted that, although a full understanding of the quantum nature of the fundamental processes presently is lacking, enough is understood about behaviour to construct a self-consistent framework which can be used not only to design optoelectronic systems and devices, but also to predict new phenomena: we have constructed an 'heuristic' theory which serves our present technological purposes very well.

Optoelectronics technology is progressing rapidly at the present time (1996) largely as a result of the information revolution. Larger and larger quantities of information are being gathered, conveyed and processed to meet the requirements of industry, business, commerce, entertainment, education and government. The increased accessibility and flow of information is changing the very fabric of society, almost entirely for the better.

The requirements of the information revolution have led to the harnessing of optoelectronics for each of the above-mentioned processes: optical sensing for gathering; optical communications for conveyance; optical signal processing for data reduction. For the future, it is not too difficult to forecast the developments in the short term (~ 5 years) from the present trends. Information gathering will take advantage of more advanced optoelectronic sensing, with a strong emphasis on distributed optical-fibre sensing, which provides information in space and time to enable the behaviour of structures to be understood, monitored and controlled. 'Smart' materials technology, in conjunction with optoelectronics technology, will lead to self-adjusting, self-compensating artefacts ranging from continuously variable aircraft shape, to earthquake-resistant buildings and bridges.

In order to use and share, in real time, all the information which is gathered in this way, in addition to all the information which must be conveyed for purposes of business, commerce, industry, education, media dissemination, entertainment and so on, the present-day optical-fibre communications systems will become more and more sophisticated, and much faster. Already $40\,\mathrm{Gb\,s^{-1}}$ systems are being researched and $140\,\mathrm{Gb\,s^{-1}}$ systems are under consideration. Coherent systems and soliton systems undoubtedly will quite soon be contributing to these further developments.

In order to switch the multi-channels of these communications systems, and to process the information at the far end into the required form for use or action, fast optical signal switching and processing devices will be required. Presently, these are all relatively slow, the limitations on speed being largely the result of the limited availability of materials. These limitations comprise something of a bottleneck in development at the present time; we shall return to this problem shortly.

The short-term predictions, then, can be fairly confident: there will be major developments in the collection, accessibility and flow of information which will lead to a great many societal conveniences in entertainment (interactive), education (software-based), medicine (instant access to the best advice/treatment), law (ditto), and so on. In short, these developments will replace the flow of people (e.g., to and from school, workplaces, hospitals/surgeries, law courts) by the flow of information. Our lives will be facilitated beyond our present states, but society would still be recognizable from our present position. Suppose now, however, that we attempt to look further into the future. Suppose we consider what might be possible if a range of new optoelectronic materials were to become available and entirely novel transmission, switching networking and storage stratagems were developed to make use of them.

Glimpses of what might be possible here can be snatched by considering the present research into such materials. New types of optical fibre will allow low dispersion, low attenuation communications to span the globe, with in-built fibre amplifiers performing the 'repeater' functions. But the real breakthroughs are likely to come from the 'tailoring' of new material properties at the quantum level. Multiple-quantum-well (MQW) materials presently are being researched: in these the structure of the material is laid down in layers of quantum dimensions (nm), in order to 'write' a given band structure (or other properties) into the material, as desired. Thus we may imagine the availability of materials which can store, switch, process and control information at great speed via very fast (ps?), sensitive nonlinear phenomena such as the photorefractive effect, the electro-optic effect, the magneto-effect and, certainly, some effects yet to be discovered.

Ultimately, we might reasonably expect that a 'bit' of information will correspond to a change of state in a single electron, allowing, at least, as many bits to be stored in a piece of material as the number of atoms the piece contains. This would allow, for example, the complete works of Shakespeare to be stored on the head of a pin!

Clearly, there is a long way to go before information can be stored and accessed with such densities, but it is very likely that, when this does become possible, it will be via optoelectronic processes. For photons have the very big advantage of being able to interact with electrons within materials without interacting with each other (in the linear optical regime!) and hence many photon-electron processes can occur simultaneously, by using many non-interacting streams of photons within a photorefractive material, for example.

These processes open up the possibility of photonic 'parallel processing' wherein many processing functions occur simultaneously, leading to an enormous increase in computing speed. Indeed, it is clear that the human brain works in this way; it is for this reason that parallel processing systems are sometimes referred to as 'neural networks'.

The individual bits of information in the human brain are handled by biological switches called 'neurons', and the enormous capacity of the human brain, indeed human intelligence itself, appears to be due not so much to the speed of these switches but to the great complexity with which they are interconnected. Of course, there is a very great deal yet to be understood about the workings of the human brain, but if our rudimentary understanding is broadly correct, the possibility is open for the construction of a very fast optoelectronic neural computing network, perhaps of higher intelligence (whatever that may truly mean) than human beings; perhaps, also, making use of organic optoelectronic materials.

Whatever the moral, religious, ethical, philosophical or medical implications of such possibilities might be, it is clear that even to go some way down this path is to progress society well beyond its present complexion, with optoelectronic robots to perform most of the repetitive, functional tasks, and optoelectronic information storage/access flow removing the necessity for the transport of either people or objects. Society could be transformed by optical computers alone.

All this could happen over the next 50 years. If such advances are to be properly managed for the benefit of humankind, optoelectronics knowledge should be widely dispersed throughout the population at large. I hope that this book has helped in this. The subject of optoelectronics bears heavy responsibilities as, now, do you!

12

Appendices

APPENDIX I: MAXWELL'S EQUATIONS

Maxwell's equations may be expressed in the vectorial form:

1. div $\mathbf{D} = \rho$ (Gauss' theorem)

2. div $\mathbf{B} = 0$ (no free magnetic poles)

3. curl $\mathbf{E} = -\dfrac{\partial \mathbf{B}}{\partial t}$ (Faraday's law of induction + Lenz's law)

4. curl $\mathbf{H} = \dfrac{\partial \mathbf{D}}{\partial t} + \mathbf{j}$ (Ampere's circuital theorem + Maxwell's

displacement current)

where ρ is the density of electric charge, $\mathbf{j}$ is the current density, $\mathbf{B} = \mu\mathbf{H}$, and $\mathbf{D} = \varepsilon\mathbf{E}$.

In free space: $\mu = \mu_0; \varepsilon = \varepsilon_0, \rho = 0$ and $\mathbf{j} = 0$, so that the above equations become:

1. div $\mathbf{E} = 0$

2. div $\mathbf{H} = 0$

3. curl $\mathbf{E} = -\mu_0 \dfrac{\partial \mathbf{H}}{\partial t}$

4. curl $\mathbf{H} = \varepsilon_0 \dfrac{\partial \mathbf{E}}{\partial t}$

Taking the curl of equation (3) we have the mathematical identity:

$$\text{curl curl } \mathbf{E} = \text{grad div} \mathbf{E} - \nabla^2 \mathbf{E}$$

so that

$$\text{curl curl } \mathbf{E} = \mu_0 \text{curl} \frac{\partial \mathbf{H}}{\partial t} = -\mu_0 \frac{\partial}{\partial t} \text{curl } \mathbf{H}$$

and thus, since div $\mathbf{E} = 0$:

$$\nabla^2 \mathbf{E} = \varepsilon_0 \mu_0 \frac{\partial^2 \mathbf{E}}{\partial t^2} \qquad (\text{I.1})$$

This is a wave equation for $\mathbf{E}$ with wave velocity:

$$c_0 = \frac{1}{(\varepsilon_0 \mu_0)^{1/2}}$$

There will clearly be a similar solution for $\mathbf{H}$, from symmetry.

A sinusoidal solution for $\mathbf{E}$ is:

$$E_x = E_0 \exp[i(\omega t - kz)]$$

In this case, we have from (3), with the resolution $\mathbf{E} = E_x \mathbf{i} + E_y \mathbf{j} + E_z \mathbf{k}$:

$$\text{curl } \mathbf{E} = \begin{vmatrix} \mathbf{i} & \mathbf{j} & \mathbf{k} \\ \frac{\partial}{\partial x} & \frac{\partial}{\partial y} & \frac{\partial}{\partial z} \\ E_x & E_y & E_z \end{vmatrix} = \mathbf{j} \frac{\partial E_x}{\partial z} = -\mu_0 \frac{\partial \mathbf{H}}{\partial t} \qquad (\text{I.2})$$

(Since $E_y = E_z = 0$; $\partial E_x / \partial y = 0$).

Thus $\mathbf{H}$ can have only a y component ($\mathbf{j}$ vector) and we have:

$$H_y = H_0 \exp[i(\omega t - kz)]$$

as the corresponding value for H_y.

Moreover, using (I.2):

$$\frac{E_0}{H_0} = \left(\frac{\mu_0}{\varepsilon_0} \right)^{1/2} = Z_0$$

Z_0 is called the electromagnetic impedance, of free space in this case.

Quite generally:

$$\frac{|\mathbf{E}|}{|\mathbf{H}|} = \left(\frac{\mu}{\varepsilon} \right)^{1/2} = Z \qquad (\text{I.3})$$

The energy stored per unit volume in an electromagnetic wave is given (from elementary electomagnetics) by:

$$U = \tfrac{1}{2}(\mathbf{D}.\mathbf{E} + \mathbf{B}.\mathbf{H}) = \tfrac{1}{2}(\varepsilon \mathbf{E}^2 + \mu \mathbf{H}^2)$$

From (I.3) we have:

$$U = \tfrac{1}{2}\varepsilon \mathbf{E}^2 = \tfrac{1}{2}\mu \mathbf{H}^2$$

so that the energy stored in each of the two fields is the same. The energy crossing unit area per second in the Oz direction for components E_x and H_y will be:

$$cU = \frac{1}{(\varepsilon\mu)^{1/2}} \tfrac{1}{2} \left(\varepsilon E_x^2 + \mu H_y^2 \right)$$

$$= \left(\frac{\varepsilon}{\mu} \right)^{1/2} E_x^2 = \left(\frac{\mu}{\varepsilon} \right)^{1/2} H_y^2 = E_x H_y = \mathbf{E} \times \mathbf{H}$$

This quantity, the vector product of $\mathbf{E}$ and $\mathbf{H}$, is the Poynting vector, i.e.,:

$$\mathbf{\Pi} = \mathbf{E} \times \mathbf{H}$$

and represents the flux of energy through unit area in the direction of wave propagation. Its mean value over one cycle of the optical wave therefore represents the mean power per unit area in the optical propagation and is thus equal to the intensity (or irradiance) of the wave.

APPENDIX II: THE FOURIER INVERSION THEOREM

The Fourier inversion theorem states that: if $A(\alpha)$ is the Fourier transform (FT) of $f(x)$, then $f(x)$ is the inverse FT of $A(\alpha)$.

The proof is straightforward. If

$$A(\alpha) = \int_{-\infty}^{\infty} f(x) \exp(-i\alpha x)\mathrm{d}x$$

then the inverse FT of $A(\alpha)$ is

$$\int_{-\infty}^{\infty} A(\alpha) \exp(i\alpha x')\mathrm{d}\alpha$$

$$= \int_{-\infty}^{\infty} \int_{-\infty}^{\infty} f(x) \exp(-i\alpha x)\mathrm{d}x \exp(i\alpha x')\,\mathrm{d}\alpha$$

$$= \int_{-\infty}^{\infty} \int_{-\infty}^{\infty} f(x) \exp[-i\alpha(x - x')]\mathrm{d}x\mathrm{d}\alpha$$

Integrating w.r.t. α, we obtain:

$$\int_{-\infty}^{\infty} f(x) \left\{ \frac{\exp[-i\alpha(x - x')]}{-i(x - x')} \right\}_{\alpha=-\infty}^{\alpha=\infty} \mathrm{d}x$$

Now the function within the square brackets can be written as:

$$\lim_{\alpha \to \infty} \left[\frac{2 \sin \alpha(x - x')}{x - x'} \right]$$

and this is clearly the δ-function (see Appendix VI) $2\pi\delta(x - x')$.

Hence:

$$\int_{-\infty}^{\infty} A(\alpha)\exp(i\alpha x')d\alpha = 2\pi \int_{-\infty}^{\infty} \delta(x - x')f(x)dx$$

By definition, the δ-function is non-zero only at $x = x'$; hence:

$$\int_{-\infty}^{\infty} A(x)\exp(i\alpha x')dx = 2\pi f(x')$$

and the proposition is proved apart from the factor of 2π. It is for this reason that the two sides are often divided by $\sqrt{2\pi}$, so that

$$A'(\alpha) = \frac{A(\alpha)}{\sqrt{2\pi}}$$

$$f'(x) = \sqrt{2\pi}f(x)$$

and then:

$$A'(\alpha) = \frac{1}{\sqrt{2\pi}} \int_{-\infty}^{\infty} f(x)\exp(-i\alpha x)dx$$

$$f'(x) = \frac{1}{\sqrt{2\pi}} \int_{-\infty}^{\infty} A(\alpha)\exp(i\alpha x)d\alpha$$

so each is now a true inverse transform of the other. This relationship is often expressed by use of the notation

$$A'(\alpha) \rightleftharpoons f'(x)$$

APPENDIX III: SYMMETRY OF THE PERMITTIVITY TENSOR

Let us begin with a well-known result (proved in any text on elementary electrostatics) that the energy of a capacitor of capacitance C when charged to a voltage V is

$$U = \tfrac{1}{2}CV^2 \tag{III.1}$$

Suppose that this capacitor is of the parallel plate variety with area A and plate separation s. Then we have:

$$C = \frac{\varepsilon\varepsilon_0 A}{s}$$

and the electric field E between the plates is given by:

$$E = \frac{V}{s}$$

The capacitor's energy is, of course, stored in this field and hence

$$U = \frac{1}{2}\frac{\varepsilon\varepsilon_0 A}{s}(Es)^2 = \frac{1}{2}As\varepsilon\varepsilon_0 E^2$$

or

$$u = \frac{1}{2}\varepsilon\varepsilon_0 E^2$$

where u is now the energy per unit volume of field. Hence for a small change of field $\mathrm{d}\mathbf{E}$ the change in energy density will be:

$$\mathrm{d}u = \varepsilon\varepsilon_0\mathbf{E}.\mathrm{d}\mathbf{E} = \mathbf{E}.\mathrm{d}\mathbf{D} \tag{III.2}$$

If ε is a tensor, then $\mathbf{E}$ and $\mathbf{D}$ are not necessarily in the same direction. In fact we write, in general:

$$D_i = \varepsilon_0\varepsilon_{ij}E_j$$

where D_i and E_j are now vectors and ε_{ij} is the permittivity tensor. From (III.2) the work done in establishing extra displacement $\mathrm{d}D_i$ is given by:

$$\mathrm{d}u = \frac{1}{\varepsilon_0\varepsilon_{ij}}D_j\mathrm{d}D_i$$

or

$$\frac{1}{\varepsilon_{ij}} = \varepsilon_0\frac{\partial^2 u}{\partial D_i\partial D_j}$$

But u is a function of the state, since it is a property which is independent of the system's history: hence u is a perfect differential. It follows that:

$$\frac{1}{\varepsilon_{ij}} = \varepsilon_0\frac{\partial^2 u}{\partial D_i\partial D_j} = \varepsilon_0\frac{\partial^2 u}{\partial D_j\partial D_i} = \frac{1}{\varepsilon_{ji}}$$

Hence $\varepsilon_{ij} = \varepsilon_{ji}$ and the tensor is thus symmetrical.

APPENDIX IV: THE POLARIZATION ELLIPSE

When referred to rectangular cartesian axes Ox, Oy, the two electric field components of any polarized optical wave may be written:

$$\begin{aligned}E_x &= e_x\cos(\omega t - kz + \delta_x)\\ E_y &= e_y\cos(\omega t - kz + \delta_y)\end{aligned} \tag{IV.1}$$

It is straightforward to eliminate $(\omega t - kz) = \tau$, say, from these equations, as follows:

$$\frac{E_x}{e_x} = \cos\tau\cos\delta_x - \sin\tau\sin\delta_x$$

$$\frac{E_y}{e_y} = \cos\tau\cos\delta_y - \sin\tau\sin\delta_y$$

so that:

$$\frac{E_x}{e_x}\sin\delta_y - \frac{E_y}{e_y}\sin\delta_x = \cos\tau\sin\delta$$

$$\frac{E_x}{e_x}\cos\delta_y - \frac{E_y}{e_y}\cos\delta_x = \sin\tau\sin\delta$$

where $\delta = \delta_y - \delta_x$. Squaring and adding these gives:

$$\left(\frac{E_x}{e_x}\right)^2 + \left(\frac{E_y}{e_y}\right)^2 - 2\frac{E_xE_y}{e_xe_y}\cos\delta = \sin^2\delta \qquad \text{(IV.2)}$$

which is the polarization ellipse referred to E_x, E_y (Fig. IV.1).

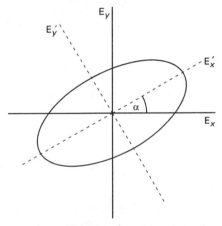

Fig. IV.1 Co-ordinate systems for the polarization ellipse.

To find the ellipticity and orientation of this ellipse we may cast it into the standard form

$$\frac{x^2}{a^2} + \frac{y^2}{b^2} = 1$$

by a rotation of the axes through an angle α. The new field components E'_x, E'_y are related to E_x, E_y by:

$$E_x = E'_x\cos\alpha - E'_y\sin\alpha$$

$$E_y = E'_x\sin\alpha + E'_y\cos\alpha$$

Substituting these into (IV.2) we have

$$E_x'^2 \left(\frac{\cos^2 \alpha}{e_x^2} + \frac{\sin^2 \alpha}{e_y^2} - \frac{\sin 2\alpha \cos \delta}{e_x e_y} \right)$$

$$+ E_y'^2 \left(\frac{\sin^2 \alpha}{e_x^2} + \frac{\cos^2 \alpha}{e_y^2} + \frac{\sin^2 \alpha \cos \delta}{e_x e_y} \right) \qquad \text{(IV.3)}$$

$$- E_x' E_y' \left(\frac{\sin 2\alpha}{e_x^2} - \frac{\sin 2\alpha}{e_y^2} + \frac{2 \cos^2 \alpha \cos \delta}{e_x e_y} - \frac{2 \sin^2 \alpha \cos \delta}{e_x e_y} \right) = \sin^2 \delta$$

Now to cast this into the required standard form, the coefficient of the cross-product term $E_x' E_y'$ is equated to zero, giving the value of α as:

$$\tan 2\alpha = \frac{2 e_x e_y \cos \delta}{(e_x^2 - e_y^2)}$$

If we now define an angle β such that:

$$\frac{e_y}{e_x} = \tan \beta$$

then:

$$\tan 2\alpha = \tan 2\beta \cos \delta \qquad \text{(IV.4)}$$

Substituting this into (IV.3) and defining a new angle χ such that

$$\tan \chi = \pm \frac{b}{a}$$

we find that

$$\sin 2\chi = - \sin 2\beta \sin \delta \qquad \text{(IV.5)}$$

Hence the orientation (β) and the ellipticity b/a of the ellipse are now determinable from the earlier parameters e_x, e_y, and δ.

Taking now the original axes E_x, E_y, arbitrarily chosen for measurement of the Stokes parameters:

$$S_0 = I(0°, 0) + I(90°, 0)$$
$$S_1 = I(0°, 0) + I(90°, 0)$$
$$S_2 = I(45°, 0) - I(135°, 0)$$
$$S_3 = I(45°, \pi/2) - I(135°, \pi/2)$$

where, as described in section 3.8, $I(\vartheta, \varepsilon)$ denotes the intensity of the incident light passed by a linear polarizer set at angle ϑ to E_x, after the E_y component has been retarded by angle ε as a result of the insertion (or not) of a quarter wave plate with its axes parallel with E_x, E_y (see Fig. 3.14(a)). Using the original expressions for E_x, E_y from (IV.1) it is clear that:

$$S_0 = |E_x|^2 + |E_y|^2 = e_x^2 + e_y^2$$
$$S_1 = |E_x|^2 - |E_y|^2 = e_x^2 - e_y^2$$
$$S_2 = \tfrac{1}{2}|E_x + E_y|^2 - \tfrac{1}{2}|E_x - E_y|^2 = 2e_x e_y \cos\delta$$
$$S_3 = \tfrac{1}{2}|E_x + iE_y|^2 - \tfrac{1}{2}|E_x - iE_y|^2 = 2e_x e_y \sin\delta$$

From (IV.4) and (IV.5) it now follows that:

$$\frac{S_3}{S_0} = \sin 2\chi$$

and

$$\frac{S_2}{S_1} = \tan 2\alpha$$

with, also:

$$S_0^2 = S_1^2 + S_2^2 + S_3^2$$

Hence the measurement of the Stokes parameters provides a quick and convenient method for complete specification of the polarization ellipse.

The degree of polarization, in the case of partially polarized light, is given by:

$$\eta = \frac{S_1^2 + S_2^2 + S_3^2}{S_0^2}$$

APPENDIX V: RADIATION FROM AN OSCILLATING DIPOLE

Consider Maxwell's equations for charge density ρ and current density $\mathbf{j}$ in free space:

$$1.\ \operatorname{div}\mathbf{D} = \frac{\rho}{\varepsilon_0}$$

$$2.\ \operatorname{div}\mathbf{B} = 0$$

$$3.\ \operatorname{curl}\mathbf{E} = -\frac{\partial\mathbf{B}}{\partial t}$$

$$4.\ \operatorname{curl}\mathbf{H} = \varepsilon_0\frac{\partial\mathbf{E}}{\partial t} + \mathbf{j}$$

Now define a vector $\mathbf{A}$ such that $\mathbf{B} = \operatorname{curl} \mathbf{A}$ where $\mathbf{A}$ is called the 'vector potential'.

From (3):

$$\operatorname{curl} \left(\mathbf{E} + \frac{\partial \mathbf{A}}{\partial t} \right) = 0$$

Hence:

$$\mathbf{E} + \frac{\partial \mathbf{A}}{\partial t} = -\operatorname{grad} \varphi$$

(since $\operatorname{curl}(\operatorname{grad} \varphi) = 0$) and also:

$$\frac{\partial \mathbf{E}}{\partial t} = -\frac{\partial^2 \mathbf{A}}{\partial t^2} - \operatorname{grad} \left(\frac{\partial \varphi}{\partial t} \right)$$

Inserting this in (4) gives:

$$\operatorname{curl} \mathbf{B} = \operatorname{curl} \operatorname{curl} \mathbf{A} = \operatorname{grad} \operatorname{div} \mathbf{A} - \nabla^2 \mathbf{A}$$

$$= \varepsilon_0 \mu_0 \left[-\frac{\partial^2 \mathbf{A}}{\partial t^2} - \operatorname{grad} \left(\frac{\partial \varphi}{\partial t} \right) \right] + \mu_0 \mathbf{j}$$

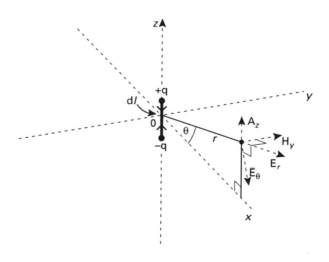

Fig. V.1 Geometry for the fields generated by an oscillating dipole.

Now impose the condition:

$$\operatorname{div} \mathbf{A} + \varepsilon_0 \mu_0 \frac{\partial \varphi}{\partial t} = 0$$

and the result is:

$$\nabla^2 \mathbf{A} = \frac{1}{c_0^2} \frac{\partial^2 \mathbf{A}}{\partial t^2} - \mu_0 \mathbf{j}$$

This differential equation for $\mathbf{A}$ can be solved in many practical cases, giving $\mathbf{B}, \varphi$, and $\mathbf{E}$ for those cases. When $\mathbf{j}$ is time-varying the solution for $\mathbf{j}$ is of the form:

$$\mathbf{j} = \mathbf{j} \left(t - \frac{r}{c} \right)$$

and

$$\mathbf{A} = \frac{\mu_0}{4\pi} \iiint \frac{[\mathbf{j}] \, dV}{r} \qquad (V.1)$$

where $[\mathbf{j}]$ here is what is known as the 'retarded' value of $\mathbf{j}$. This is the value of $\mathbf{j}$ at time $t - r/c$ rather than t. Hence the integral must be evaluated with this in mind.

Consider now a Hertzian dipole where a charge q oscillates through a distance dl at frequency ω (Fig. V.1). Then the dipole moment is given by:

$$\mathbf{P}(t) = q(t)\mathbf{dl}$$

and, since the electric current magnitude is given by

$$i_c = \frac{dq}{dt}$$

we have

$$i_c \mathbf{dl} = \dot{\mathbf{P}}(t)$$

So in this case:

$$\mathbf{A} = \frac{\mu_0}{4\pi} [i_c] \frac{\mathbf{dl}}{r} = \frac{\mu_0}{4\pi} \frac{[\dot{\mathbf{P}}]}{r}$$

from (V.1).

Suppose now that the dipole oscillation takes place in direction Oz, one of a set of Cartesian axes Ox, Oy, Oz (Fig. V.1). Then it is true in this case that:

$$A_z = \frac{\mu_0}{4\pi} \frac{[\dot{\mathbf{P}}]}{r}$$

Consider, first, just the xz plane and a point, in that plane, distant r from the origin, the corresponding radius vector making an angle ϑ with Ox (Fig. V.1).

We have

$$\mathbf{B} = \operatorname{curl} \mathbf{A}$$

so that

$$\mathbf{H} = \frac{1}{4\pi\mu_0} \operatorname{curl} \frac{[\dot{\mathbf{P}}]}{r}$$

Hence with (mathematical identity)

$$\operatorname{curl} A_z \mathbf{k} = \begin{vmatrix} \mathbf{i} & \mathbf{j} & \mathbf{k} \\ \dfrac{\partial}{\partial x} & \dfrac{\partial}{\partial y} & \dfrac{\partial}{\partial z} \\ 0 & 0 & A_z \end{vmatrix}$$

we have:

$$H_y = -\frac{1}{4\pi}\frac{\partial}{\partial r}\left(\frac{[\dot{P}_z]}{r}\right)\frac{\partial r}{\partial x} = -\frac{1}{4\pi}\frac{\partial}{\partial r}\left(\frac{[\dot{P}_z]}{r}\right)\cos\vartheta$$

or

$$H_y = \frac{1}{4\pi}\frac{\partial}{\partial r}\left(\frac{[\dot{P}_z]}{r^2} + \frac{[\ddot{P}_z]}{c_0 r}\right)\cos\vartheta$$

Hence, we have derived the transverse component of **H** for *any* vertical plane which contains the oscillating dipole. **E** can now be derived, as follows:

$$\mathbf{E} = -\operatorname{grad}\varphi - \frac{\partial \mathbf{A}}{\partial t}$$

and

$$\operatorname{div}\mathbf{A} + \varepsilon_0\mu_0\frac{\partial\varphi}{\partial t} = 0$$

so:

$$\frac{\mu_0}{4\pi}\frac{\partial}{\partial z}\frac{[\dot{\mathbf{P}}]}{r} + \varepsilon_0\mu_0\frac{\partial\varphi}{\partial t} = 0$$

Hence:

$$\frac{\mu_0}{4\pi}\frac{\partial}{\partial z}\frac{[\dot{\mathbf{P}}]}{r} + \varepsilon_0\varphi = 0$$

This is an equation for φ and hence also for **E**. Solving this gives:

$$E_r = \frac{1}{2\pi\varepsilon_0}\sin\vartheta\left(\frac{[P_z]}{r^3} + \frac{[\dot{P}_z]}{c_0 r^2}\right)$$

and

$$E_\vartheta = \frac{1}{4\pi\varepsilon_0}\cos\vartheta\left(\frac{[P_z]}{r^3} + \frac{[\dot{P}_z]}{c_0 r^2} + \frac{[\ddot{P}_z]}{c_0^2 r}\right)$$

Terms in $1/r^3$ comprise the electrostatic field of the dipole; terms in $1/r^2$ are the induction field terms; the only term of significance for large r (say $r \gg \lambda$) is:

$$E_\vartheta = \frac{1}{4\pi\varepsilon_0}\cos\vartheta\frac{[\ddot{P}_z]}{c_0^2 r} \qquad\qquad (V.2)$$

This is the E-field normal to the radius vector and in the xz plane: it is the transverse E component of the radiation field. The surviving term in H_y for large r, correspondingly will be:

$$H_y = \frac{[\ddot{P_z}]}{4\pi c_0 r} \cos \vartheta$$

Clearly, H_y and E_ϑ comprise the mutually orthogonal components of the propagating electromagnetic wave generated by the oscillating dipole. The Poynting vector (flux of energy across unit area in the direction of wave propagation) will be:

$$\Pi_r = |\mathbf{E}_\vartheta \times \mathbf{H}_y| = \frac{1}{16\pi^2 \varepsilon_0} \frac{1}{c_0^3} \frac{1}{r^2} [\ddot{P_z}]^2 \cos^2 \vartheta$$

This is seen to vary as $\cos^2 \vartheta$, which is Lambert's law; it also is seen to obey the inverse square law with distance.

The total power radiated by the dipole will be given by:

$$P_t = 2 \int_0^{\pi/2} \Pi_r 2\pi r^2 \cos \vartheta d\vartheta \qquad (V.3)$$

(Note that the r^2 term now cancels out.)

Suppose now that we consider the specific case of an oscillating dipole. A charge q oscillates sinusoidally with angular frequency ω along a distance $\pm z_0$, so that at any time t, its distance from the origin is given by

$$z = z_0 \cos \omega t$$

The dipole moment at any time will be

$$P_z = q z_0 \cos \omega t$$

Hence

$$[\dot{P_z}] = -\omega q z_0 \sin \omega \left(t - \frac{r}{c_0} \right)$$

and

$$[\ddot{P_z}] = -\omega^2 q z_0 \cos \omega \left(t - \frac{r}{c_0} \right) = -\omega^2 p_0 \cos \omega (t - \frac{r}{c_0})$$

where p_0 is the maximum value of the dipole moment.

Hence, in this case, the Poynting vector's amplitude becomes:

$$\Pi_r = \frac{1}{32\pi^2 \varepsilon_0} \frac{1}{c_0^3} \frac{1}{r^2} \omega^4 p_0^2 \cos^2 \vartheta$$

when averaged over one period of oscillation.

Correspondingly, the total radiated power for this case, on performing the integration (V.3), becomes:

$$P_t = \frac{\mu_0 \omega^4 p_0^2}{12\pi c_0} = \frac{4\pi^3 c_0 p_0^2}{3\epsilon_0 \lambda^4}$$

This now embodies the λ^{-4} dependence which is characteristic of Rayleigh scattering by small ($< \lambda$) dipoles.

APPENDIX VI: THE δ-function

Consider a rectangular optical pulse of duration τ and height h. The Fourier transform (frequency spectrum) of this pulse is given by:

$$a(\omega) = \int_{-\infty}^{\infty} f(t)\exp(i\omega t)\mathrm{d}t = \int_{-\pi/2}^{\pi/2} h\exp(i\omega t)\mathrm{d}t = \alpha\left(\frac{\sin\frac{1}{2}\omega\tau}{\frac{1}{2}\omega\tau}\right) \quad \text{(VI.1)}$$

where $\alpha = h\tau$ is the 'area' of the pulse. This is the standard 'sinc' function and is shown in Fig. VI.1(a).

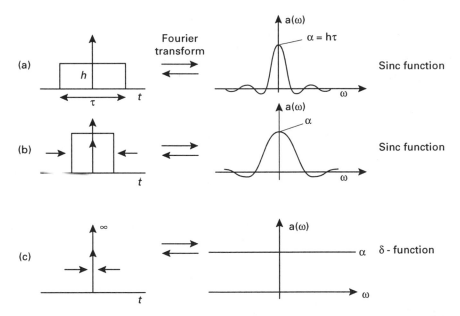

Fig. VI.1 The δ-function as the limit of a sinc function of increasing width.

Suppose now that the pulse is made narrower while maintaining its 'area', α, constant. As $\tau \to 0, h \to \infty$ if α is to remain constant (Fig.

VI.1(c)) and, in the limit, the pulse becomes a δ-function formally defined by

$$\alpha\delta(t) = 0, t \neq 0$$

$$\alpha \int_{-\infty}^{\infty} \delta(t) = \alpha$$

(i.e., $\int_{-\infty}^{\infty} \delta(t) = 1$ for the generalized δ-function). Fig. VI.1 illustrates the limit process.

From (VI.1) it is clear that, as $\tau \to 0$,

$$a(\omega) \to \alpha$$

so all frequencies from $-\infty$ to ∞ are present, with equal amplitudes α and all are in phase at $t = 0$. Hence they all add in phase at $t = 0$ to give infinite amplitude there, and average to zero, by mutual destructive interference, everywhere else.

The δ-function is a very useful mathematical device for expression of limiting cases such as this. It was introduced by Paul Dirac in the 1920s, and is often called the 'Dirac δ-function'.

In section 5.1 we deal with a set of randomly positioned δ-function optical pulses. It is clear that all frequencies will be present in equal amounts and that, owing to the random positions of the pulses, the phases will be uncorrelated.

APPENDIX VII: THE FERMI-DIRAC FUNCTION

The derivation of the Fermi-Dirac (FD) function for particles which possess anti-symmetrical wave functions (fermions) requires the ideas of statistical mechanics. These ideas have been rigorously and beautifully developed in detail in various texts, so they will only be summarized here.

Electrons are fermions and they thus obey FD statistics. It is for this reason that the FD function is so important for us.

Suppose that the electrons can occupy any energy level in a range $\varepsilon_1, \varepsilon_2 ... \varepsilon_i ... \varepsilon_s$, and that the i-th level has degeneracy g_i, i.e., there are g_i possible states (wave functions) with that energy (ε_i). Since we are dealing with anti-symmetrical particles, only one electron can occupy any one state. This is a salient property of fermions.

Suppose that n_i of the g_i states at energy ε_i are occupied by electrons; then $(g_i - n_i)$ states will be unoccupied. The g_i states are all different, so there are $g_i!$ ways of arranging them in sequence. However, it is of no physical significance how the electrons are arranged among the n_i occupied states since all electrons are equivalent; neither does it matter how the $(g_i - n_i)$ empty states are arranged. Hence the total number of physically different ways of arranging the n_i electrons among g_i states:

$$W_i = \frac{g_i!}{n_i!(g_i - n_i)!}$$

Considering now all the i possible energy levels, the total number of ways of arranging all the electrons amongst all possible states is:

$$W = \prod_{i=1}^{s} \frac{g_i!}{n_i!(g_i - n_i)!}$$

where $\prod$ denotes the continued product over all possible values of i.

However, the implicit assumption has, so far, been that there is a particular number of electrons, n_i, for each ϵ_i, hence for complete generality the total number of arrangements (complexions) must include all other possible sets of n_i, so that:

$$W_T = \sum_{\text{all possible sets of } n_i} \left(\prod_i \frac{g_i!}{n_i!(g_i - n_i)!} \right)$$

Now the greater the number of ways of arranging the electrons for a given configuration, the greater will be the chance of finding the electrons in that configuration and thus the greater will be the possibility of their taking up that configuration in practice. It is easy to show that for one particular set of n_i the value of W is very much greater (by a factor of $\sim \exp 10^8$) than any other. Hence we need only consider this maximum value set and ignore all others:

$$W_T = \prod_{i=1}^{s} \frac{g_i!}{n_i!(g_i - n_i)!} \tag{VII.1}$$

The condition that must be imposed on this function in order to determine the nature of the electron distribution in terms of other physical parameters is that it must be a maximum against variations in the n_i and hence the condition to be imposed is:

$$\frac{\partial W_T}{\partial n_i} = 0$$

In order to handle this condition mathematically it is necessarily firstly to transform the continued product into a sum. This can be done by taking natural logarithms (denoted as 'log' rather than the more usual 'ln' in order to avoid confusion with the n_i).

$$\log W_T = \sum_i [\log(g_i!) - \log(n_i!) - \log(g_i - n_i)!]$$

The next step is to use Stirling's approximation, which states that:

$$\log n! = n \log n - n$$

when n is very large. Hence, since all our n_i, g_i are very large:

$$\log W_T =$$

$$\sum_i [g_i \log g_i - g_i - n_i \log n_i + n_i - (g_i - n_i) \log (g_i - n_i) + (g_i - n_i)]$$

Now if W_T is at a maximum then so is $\log W_T$ since a monotonic relationship exists between them. Hence

$$\frac{\partial (\log W_T)}{\partial n_i} = \sum_i [-\log n_i + \log (g_i - n_i)] = 0 \qquad \text{(VII.2)}$$

is now the condition to be imposed for our purposes.

The next step is to recognize that there are two other conditions to be imposed on the system: the total number of electrons must remain constant; and the total energy of the system must remain constant. Hence:

$$\sum_{i=1}^{s} n_i = N$$

$$\sum_{i=1}^{s} n_i \varepsilon_i = E \qquad \text{(VII.3)}$$

The expressions given by (VII.2) and (VII.3) are used in conjunction with a mathematical device known as Lagrange's method of undetermined multipliers to solve the problem. We use constants α and β such that:

$$\alpha \sum_i dn_i = 0$$

$$\beta \sum_i \varepsilon_i dn_i = 0$$

and then construct the legitimate equation:

$$\sum_{i=1}^{s} [-\log n_i + \log (g_i - n_i) + \alpha + \beta \varepsilon_i] \, dn_i = 0$$

However the dn_i variations are not independent, owing to (VII.3). If now α and β are fixed by particular variations dn_k and dn_j, then:

$$-\log n_k + \log (g_k - n_k) + \alpha + \beta \varepsilon_k = 0$$

and

$$- \log n_j + \log (g_j - n_j) + \alpha + \beta \varepsilon_j = 0$$

This allows all other $(s - 2)$ values of i to become independent, with these defined values of α and β. Hence it follows that, for all values of i, including k and j:

$$- \log n_i + \log (g_i - n_i) + \alpha + \beta \varepsilon_i = 0$$

comprising s equations which fix now the s values of n_i. Hence we have:

$$\log \left(\frac{g_i - n_i}{n_i} \right) + \alpha + \beta \varepsilon_i = 0$$

or

$$\frac{g_i}{n_i} = 1 + \exp \left[-(\alpha + \beta \varepsilon_i) \right]$$

i.e.,

$$\frac{n_i}{g_i} = \frac{1}{1 + \exp \left[-(\alpha + \beta \varepsilon_i) \right]} \tag{VII.4}$$

Now if $g_i \gg n_i$, i.e., there are very many more states available than there are electrons to fill them, then the electrons are not restricted by their quantum nature, and their distribution approaches the classical case described by Boltzmann statistics. Hence, for this case:

$$\frac{n_i}{g_i} \ll 1$$

and thus

$$\exp \left[-(\alpha + \beta \varepsilon_i) \right] \gg 1$$

so that:

$$n_i \approx g_i \exp(\alpha + \beta \epsilon_i)$$

However, we know from Boltzmann statistics that, where it is valid:

$$n_i - N_0 \exp \left(-\frac{\epsilon_i}{kT} \right)$$

where k is Boltzmann's constant and T is the absolute temperature.
It follows that $\beta = -1/kT$
Defining now, for convenience, an energy ε_F such that

$$\alpha = \frac{\varepsilon_F}{kT}$$

(note that $\varepsilon_F \sim kT$), we may finally write for n_i/g_i, which is, of course, the fraction of available states which is occupied by electrons:

$$F(\varepsilon) = \frac{n_i}{g_i} = \frac{1}{1 + \exp \dfrac{\epsilon - \varepsilon_F}{kT}}$$

This is equation (6.18a) and comprises the Fermi-Dirac distribution function.

APPENDIX VIII: SECOND HARMONIC GENERATION

The generation of the second harmonic from a fundamental optical wave, when it is passing through a nonlinear optical material, must properly be treated with Maxwell's equations as the starting point.

For an insulating medium of permittivity ε and permeability μ we can write Maxwell's equations in the form:

1. $\operatorname{div} \mathbf{E} = 0$

2. $\operatorname{div} \mathbf{H} = 0$

3. $\operatorname{curl} \mathbf{E} = -\mu\mu_0 \dfrac{\partial \mathbf{H}}{\partial t} = -\dfrac{\partial \mathbf{B}}{\partial t}$

4. $\operatorname{curl} \mathbf{H} = \varepsilon\varepsilon_0 \dfrac{\partial \mathbf{E}}{\partial t} = \dfrac{\partial \mathbf{D}}{\partial t}$

Now assuming that the non-linear electric polarization vector lies in the same direction as the applied electric field, the first two terms of its expansion can be written in the form

$$\mathbf{P} = \varepsilon\chi_1\mathbf{E} + \varepsilon_0\chi_2|\mathbf{E}|\mathbf{E}$$

Hence we have

$$\mathbf{D} = \varepsilon_0\mathbf{E} + \mathbf{P} = \varepsilon_0\mathbf{E} + \varepsilon_0\chi_1\mathbf{E} + \varepsilon_0\chi_2|\mathbf{E}|\mathbf{E}$$

or

$$\mathbf{D} = \varepsilon\varepsilon_0\mathbf{E} + \varepsilon_0\chi_2|\mathbf{E}|\mathbf{E}$$

(since $\varepsilon = 1 + \chi$). Substituting this expression for $\mathbf{D}$ in Maxwell's equation (4), we have

$$\operatorname{curl} \mathbf{H} = \varepsilon\varepsilon_0 \frac{\partial \mathbf{E}}{\partial t} + \varepsilon_0\chi_2 \frac{\partial(|\mathbf{E}|\mathbf{E})}{\partial t} \tag{VIII.1}$$

Taking now the curl of Maxwell's equation (3) and using the mathematical identity

$$\operatorname{curl}\operatorname{curl} \mathbf{E} = \operatorname{grad}\operatorname{div} \mathbf{E} - \nabla^2\mathbf{E}$$

we find that

$$\nabla^2\mathbf{E} = \mu\mu_0 \frac{\partial(\operatorname{curl} \mathbf{H})}{\partial t}$$

(since div $\mathbf{E} = 0$). Using equation VIII.1 we now have that:

$$\nabla^2 \mathbf{E} = \mu\mu_0\varepsilon\varepsilon_0\frac{\partial^2 \mathbf{E}}{\partial t^2} + \mu\mu_0\varepsilon\varepsilon_0\chi_2\frac{\partial^2(|\mathbf{E}|\mathbf{E})}{\partial t^2} \qquad \text{(VIII.2)}$$

Suppose now that we consider a solution for this equation, which consists of the fundamental and a second harmonic, of the form:

$$\mathbf{E}(z,t) = \mathbf{E}_1(z)\exp\left[i(\omega_1 t - k_1 z)\right] + \mathbf{E}_2(z)\exp\left[i(2\omega_1 - k_2 z)\right]$$

This can now be substituted into equation (VIII.2) in order to determine the relationship between the first and second harmonic components. Essentially this relationship will tell us how the second harmonic is 'generated' from the fundamental. It is clear that this generation can only result from the square of the fundamental component since only that leads to the correct frequency. It is, therefore necessary only to deal with terms which oscillate with frequency $2\omega_1$; all other terms will operate independently.

We assume that all vectors again are parallel, and orthogonal to the propagation direction Oz.

We shall deal with each side of equation (VIII.2) in turn. Remembering that we are only concerned with terms in $2\omega_1 t$ the right-hand side becomes, on substituting $E(z,t)$:

$$(\nabla^2 E) = -\mu\mu_0\varepsilon\varepsilon_0 4\omega_1^2 E_2(z)\exp[i(2\omega_1 t - k_2 z)$$

$$-\mu\mu_0\varepsilon_0\chi_2 4\omega_1 1^2 E_1^2(z)\exp[i(2\omega_1 t - k_1 z)]$$

where μ is assumed constant for both components, and for the left-hand side:

$$\nabla^2 E(z,t) = \frac{\partial^2}{\partial z^2} E_2(z)\exp[i(2\omega_1 t - k_2 z)]$$

$$-\left[k_2^2 E_2(z) + ik_2\frac{\partial E_2(z)}{\partial z}\right]\exp[i(2\omega_1 t - k_2 z)]$$

where, for this last expression, it has been assumed that:

$$\left|2k_2\frac{\partial E_2(z)}{\partial z}\right| \gg \left|\frac{\partial^2 E_2(z)}{\partial z^2}\right|$$

i.e., that $\partial E_2(z)/\partial z$ is sensibly constant over one second harmonic wavelength. Equating the right-hand side and left-hand side of the equation (VIII.2) and cancelling the common factor $(\exp(2i\omega_1 t))$ we obtain:

$$4\mu\mu_0\varepsilon_2\varepsilon_0\omega_1^2 E_2(z)\exp(-ik_2 z)$$

$$+4\mu\mu_0\varepsilon_0\chi_2\omega_1^2 E_1^2(z)\exp(-2ik_1 z)$$

$$= k_2^2 E_2(z) \exp(-ik_2 z)$$

$$+ik_2 \frac{\partial E_2(z)}{\partial z} \exp(-ik_2 z)$$

Now we know that:

$$c_2^2 = \frac{1}{\mu\mu_0\varepsilon_2\varepsilon_0}; \qquad \frac{2\omega_1}{k_2} = c_2$$

(since the permittivity and permeability constants will refer to the propagation for which they are the coefficients, in this case the second harmonic).

It follows that the first terms on each side cancel out to give:

$$ik_2 \frac{\partial E_2(z)}{\partial z} \exp(-ik_2 z)$$

$$= 4\mu\mu_0\varepsilon_0\chi_2\omega_1^2 E_1^2(z) \exp(-2ik_1 z)$$

or

$$\frac{\partial E_2(z)}{\partial z} = \frac{-2i\chi_2\omega_1}{c_2\varepsilon_2} E_1^2(z) \exp[-i(2k_1 - k_2)z] \qquad \text{(VIII.3)}$$

This is the 'generator' equation, showing the relationship between the spatial growth of $E_2(z)$ as a result of $E_1^2(z)$. In order to determine the value of $E_2(z)$ after a length L of generation in the nonlinear crystal, equation (VIII.3) must be integrated:

$$E_2(L) = \frac{-2i\chi_2\omega_1}{c_2\varepsilon_2} E_1^2(L) \frac{\exp[-i(2k_1 - k_2)L] - 1}{-i(2k_1 - k_2)}$$

Hence the intensity of $E_2(L)$ will be proportional to:

$$E_2(L)E_2^*(L) = \frac{4\chi_2^2\omega_1^2 L^2 E_1^4(L)}{c_e^2\varepsilon_2^2} \left[\frac{\sin\left(k_1 - \frac{1}{2}k_2\right)L}{\left(k_1 - \frac{1}{2}k_2\right)L} \right]^2$$

This is seen to have the same form as equation (9.5), which was derived from (largely) physical intuition. It is instructive to compare the two equations in detail, and this will be left as an exercise for the reader.

APPENDIX IX: THE SAMPLING THEOREM

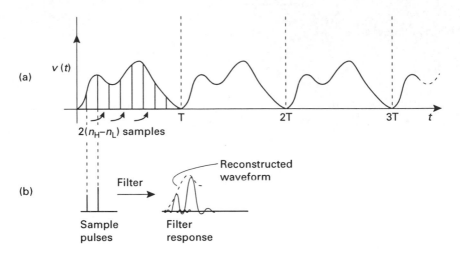

Fig. IX.1 The sampling theorem.

The sampling theorem may be stated in the following form: 'If an analogue waveform has bandwidth Δf, then it can be specified completely by sampling its value at a rate $2\Delta f$'. This means that if the waveform's value is taken at specific points in time at that rate, then enough information will be available to reproduce the waveform exactly.

The proof is quite simple. Suppose that the waveform $v(t)$ has a duration T. Suppose now that we consider the waveform to be repeated at intervals of T (Fig. IX.1(a)). Let the bandwidth of the waveform be Δf.

From Fourier theory, the waveform, periodic in T, can be represented as the sum of harmonics of the fundamental frequency, $2\pi/T$. The waveform can, therefore, be represented by:

$$v'(t) = \sum_n \left(a_n \cos \frac{2\pi nt}{T} + b_n \sin \frac{2\pi nt}{T} \right) \qquad (IX.1)$$

where a_n, b_n are constants which specify the amplitude and phase of the n-th harmonic of the fundamental frequency.

Suppose that the lowest frequency present in the waveform is f_L and the highest f_H. Then the lowest and highest values of n necessary are given by

$$\frac{n_L}{T} = f_L; \qquad n_L = f_L T$$

$$\frac{n_H}{T} = f_H; \qquad n_H = f_H T$$

where n_L and n_H are, of course, positive integers. Hence the total number of harmonics necessary to define $v'(t)$ in (IX.1) will be $(n_H - n_L)$.

Since there are two unknowns, a_n, b_n, for each value of n, it follows that $2(n_H - n_L)$ independent samples of $v'(t)$ will be sufficient to set up $2(n_H - n_L)$ linear equations for the determination of a_n and b_n for each n. But

$$2(n_H - n_L) = 2(f_H - f_L)T \qquad (IX.2)$$

If the $2(n_H - n_L)$ samples are taken in the time T they will be sufficient to define the waveform completely since it merely repeats after the interval T. To take $2(n_H - n_L)$ samples in time T implies, from (IX.2), a sampling rate:

$$f_s = \frac{2(n_H - n_L)}{T} = 2(f_H - f_L) = 2\Delta f$$

which proves the theorem.

Note that if the waveform's spectrum extends down to DC, then the sampling rate will be $2f_H$, i.e. twice the highest frequency present in the signal waveform.

Maximum accuracy will be achieved for sampling at regular intervals within T, for this will provide the best 'conditioning' for the matrix (a_n, b_n). Indeed it is possible for the accuracy of determination of some of the a_n, b_n to collapse to zero under some conditions. For example, if one of the frequency components, say $2\pi n/T$, were sampled at intervals equal to the period T/n then only one piece, rather than two pieces, of information is available for it and hence a_n and b_n cannot be separately determined. Provided, however, that the sampling interval is equal to $T/2n_H$, i.e. sampling is done uniformly at a frequency equal to the highest frequency present, then the above cannot occur, for any lower frequency, and the matrix is 'well-conditioned'.

The waveform can be reconstructed, in practice, from the sample pulses, by passing the resulting pulse train through a filter of the same bandwidth, Δf. In this case the sum of the filter's responses to the pulses reproduces the waveform (Fig. IX.1(b)). Clearly, a knowledge of the a_n, b_n also allows the waveform to be computed directly.

Evidently, f_s represents a minimum rate and the accuracy can be improved by sampling at a rate greater than f_s, a procedure known as 'oversampling'.

APPENDIX X: THE SEMICONDUCTOR EQUATION

In Chapter 6 the energy distribution of electrons in solids was considered. It was shown there that the distribution was the product of two factors: the number of states which were available at a given energy, and the actual probability that any state at a given energy is occupied by an electron.

The number of states available at a given energy is known as the density of states function (sometimes the degeneracy function) and is conditioned by the restrictions placed upon the electrons by their being bound within the atomic lattice of the solid (Fig. X.1(c)). These ideas culminated in equation (6.16) for the density of states at energy E:

$$g(E) = \frac{4\pi}{h^3}(2m_e^*)E^{1/2}dE$$

where m_e^* is the 'effective' electron mass.

Electrons are fermions, so they obey Fermi-Dirac statistics. The probability that any given state is occupied by an electron is given by the Fermi-Dirac function derived in Appendix VII. Hence the actual density distribution (i.e. number of electrons per unit volume with energy between E and $E+dE$) is given by the product of the two functions (Fig. X.1(d)):

$$n(E)dE = g(E)F(E)dE \qquad (X.1(a))$$

$$= \frac{4\pi}{h^3}(2m_e^*)^{3/2}E^{1/2}\frac{dE}{1+\exp\dfrac{E-E_F}{kT}} \qquad (X.1(b))$$

where E_F is, of course, the Fermi energy. Consider now a semiconductor material with energy E_c at the bottom of the conduction band and E_v at the top of the valence band. From equation (6.25) we know that the Fermi level for this lies somewhere near the middle of the band gap (i.e., $E_F \sim (E_c + E_v)/2$, but see later), so at absolute zero of temperature all the states in the valence band are full, and all those in the conduction band are empty (Fig. X.1(a)). As the temperature rises, the Fermi-Dirac function tells us that there is a non-zero probability that some of the states in the conduction band will now be occupied by electrons, according to $F(E)$ at $T > 0$.

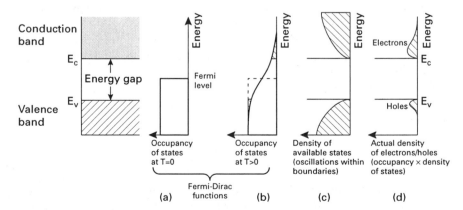

Fig. X.1 Occupancies and densities amongst energy levels in a semiconductor.

The probability that states in the valence band will not be occupied by electrons will correspondingly be $1 - F(E)$, so this will be the probability that these states will now be 'occupied' by holes (Fig. X.1(b)). For the conduction band the correct energy, E_c, represents the effective zero of energy for electrons trapped in the band, so from (X.1), their energy distribution becomes:

$$n_c(E)dE = \frac{4\pi}{h^3}(2m_e^*)^{3/2}(E - E_c)^{1/2}\frac{dE}{1 + \exp\dfrac{E - E_F}{kT}}$$

The total number density of electrons in the conduction band is thus given by:

$$n = \int_{E_c}^{E_t} n(E)dE$$

where E_t is the energy at the top of the band. This is a difficult integral but is manageable with the help of two reasonable, simplifying assumptions:

1. The lower levels of the band, only, are significant, so E_t is effectively at infinity.
2. The value of E_c is sufficiently in excess of E_F to render:

$$exp\left(\frac{E_c - F_F}{kT}\right) \gg 1$$

so that the '1' in the denominator of the Fermi-Dirac function can be ignored for all $E > E_c$.

Under these conditions the integral becomes:

$$n = \frac{2}{h^3}(2\pi m_e^* kT)^{3/2}\exp\left(-\frac{E_c - E_F}{kT}\right) \qquad \text{(X.2(a))}$$

The same process can now be performed for the holes in the valence band, the only differences being that $1 - F(E)$ must be used in place of $F(E)$ in equation (X.1(a)), and that m_h^* must replace the effective electron mass m_e^*. For this case we obtain, using the same assumptions 1 and 2 (now in the negative energy direction), the total number density of holes in the valence band as:

$$p = \frac{2}{h^3}(2\pi m_h^* kT)^{3/2}\exp\left(-\frac{E_F - E_v}{kT}\right) \qquad \text{(X.2(b))}$$

Writing X.26(a) and X.26(b) for convenience as

$$n = n_c\exp\left(-\frac{E_c - E_F}{kT}\right)$$

$$p = n_p\exp\left(-\frac{E_F - E_v}{kT}\right)$$

where n_c and n_p are constants (known as the 'effective density of states') at a given temperature, we see that the product of n and p is given by:

$$np = n_c n_p \exp\left(-\frac{E_c - E_v}{kT}\right) \qquad \text{(X.3)}$$

This is the required result, the 'semiconductor equation', for it shows that np is independent of the Fermi energy and depends only on the band gap $(E_c - E_v)$ and the temperature. Equation (X.3) thus is always true, even when the Fermi level is altered by the addition of impurity donors or acceptors, in an extrinsic semiconductor. Physically, what happens in this latter case is that donor atoms, for example, provide electrons to the conduction band and these must then come into general equilibrium in accordance with equation (X.1(a)). The result is that some electrons fall to the valence band to annihilate holes and thus to maintain the constancy of the product pn. In an intrinsic semiconductor the number of holes must equal the number of electrons, so:

$$p = n = n_i$$

and thus

$$pn = n_i^2 = n_c n_p \exp\left(-\frac{E_c - E_v}{kT}\right)$$

In an extrinsic semiconductor

$$pn = n_i^2$$

but

$$p \neq n$$

Instead, with donor and acceptor atom densities N_d and N_a we now have, under the condition of charge neutrality: $n + N_a = p + N_d$, assuming that all the impurity atoms are fully ionized. This relation together with $pn = n_i^2$ allows p and n to be determined, if N_a and N_d are known.

If, as is usually the case, in an n-type material, for example

$$N_d \gg N_a, n_i$$

then

$$n \approx N_d$$

and

$$p \approx \frac{n_i^2}{N_d}$$

Similarly for a p-type material

$$N_a \gg N_d, n_i$$

$$p \approx N_{\mathrm{a}}$$

and

$$n \approx \frac{n_i^2}{N_{\mathrm{a}}}$$

Note that, with N_{d} or $N_{\mathrm{a}} \gg n_{\mathrm{i}}$, the minority carrier density is, in each case, considerably depressed below that of the intrinsic material.

Finally, another important result can be derived from the equations (X.2) for n and p. They can be used to express the Fermi energy E_F in terms of n and p. Taking the logarithm of the ratio of p and n and re-arranging we obtain:

$$E_{\mathrm{F}} = \tfrac{1}{2}(E_{\mathrm{v}} + E_{\mathrm{c}}) - \tfrac{1}{2}kT \ln \frac{p}{n} - \tfrac{3}{4}kT \ln \frac{m_{\mathrm{e}}^*}{m_{\mathrm{h}}^*}$$

which is equation (6.26), and shows that, for an intrinsic material where $p = n$, then $E_{\mathrm{F}} = \tfrac{1}{2}(E_{\mathrm{v}} + E_{\mathrm{c}})$, i.e., half way up the band gap, except for the extent to which m_{e}^* differs from m_{h}^*.

13

Answers to numerical questions

Chapter 1

1.3   Gamma-ray intensity: $6.626 \times 10^{-6} \, \mathrm{W \, m^{-2}}$.

1.4   0.1 s

1.5 3.88×10^{-12} m.

1.6 $1.99 \times 10^{-24} \, \mathrm{kg \, m \, s^{-1}}$.

Chapter 2

2.1   Frequency range: 7.5×10^{14} Hz to 4.3×10^{14} Hz. Wavenumber range: $1.57 \times 10^7 \, \mathrm{m^{-1}}$ to $8.98 \times 10^6 \, \mathrm{m^{-1}}$.

2.2 $(a^2 \cos^2 \varphi + b^2 \sin^2 \varphi)^{1/2}$;

$\tan^{-1} \left(\dfrac{b}{a} \tan \varphi \right) ; (a^2 \cos^2 \varphi + b^2 \sin^2 \varphi)$;

one is the square of the other.

2.3 $E_x = \dfrac{10}{\sqrt{2}} \exp \left[i\omega t - \dfrac{k}{\sqrt{2}}(x+y) \right]$

$E_x = \dfrac{10}{\sqrt{2}} \exp \left[i\omega t - \dfrac{k}{\sqrt{2}}(x+y) \right]$

Intensity in free space: $0.133 \, \mathrm{W \, m^{-2}}$

Intensity in medium: $0.202 \, \mathrm{W \, m^{-2}}$

2.4   0.162 (parallel); 0.24 (orthogonal).

2.5   59.24°.

2.6   0.341.

2.7   2.65 mm.

2.8 $2\pi/7$; $7 \times$ each individual amplitude.

2.9   (i) 14.13, (ii) 0.22 radians, (iii) 1.25×10^{-11} m, (iv) 5.65×10^5.

Chapter 3

3.1   0.765; 0° (i.e., parallel to the major axis).

3.2 $e = 0.70$; $\gamma = -10.42°$ (w.r.t. Ox).

3.4 $\begin{pmatrix} -2i \\ -1 \end{pmatrix}$

Chapter 4

4.3 $313 \, \mu\mathrm{W}$ (3.13×10^{-4} W).

4.4 $18.7 \, \mathrm{pW} (1.87 \times 10^{-11}$ W).

Chapter 5

5.1 9.65×10^{-8} radian (5.53×10^{-6} deg).

5.3   5.89 mm.

5.4 2.75×10^9 m.

Chapter 6

6.2   1960 K.

6.3 $2.64 \times 10^5 \, \text{m}^{-3}$.

6.5   0 K: 0; 1.

300 K: 0.4; 0.87.

6.7   0.014 eV below the mid-point of the band.

Chapter 7

7.3   2.93 m; 51.2 MHz; 0.1 ns.

7.4   5.8 mW.

7.5   0.82.

7.7 $0.17°; 0.34°; 80 \, \text{MHz}$.

7.8   11.3 kΩ.

7.9   58.5 µA; Yes, 8.6 dB.

7.10   9.95 kHz; 12.7 pA; $2.8 \times 10^{-13} \, \text{W} \, \text{Hz}^{-1/2}$.

Chapter 8

8.1   20.13°.

8.2   2.89 µm; 0.44.

8.3   1.67 µm; 290 modes; 10 ns.

8.4   Modal dispersion is the limit, at 6 MHz.

8.5 $4.55 \, \text{dB} \, \text{km}^{-1}$.

Chapter 9

9.5   −82.03 dB; 81.77 dB.

9.6 $6.237 \times 10^3 \, \text{m} \, \text{s}^{-1}$.

Chapter 10

10.1   99.6 nm; 183.3 nm.

10.2 5.3×10^{-3} radian (0.304°); 29.6 kA.

10.3   0.627°; 7.5 mm.

10.4 4.6×10^{10}; 6.25%.

10.5 $0.89 \, \text{radians} \, \text{s}^{-1} (51° \, \text{s}^{-1})$.

10.9   27.68 m.

10.10   4.14 mm.

10.11   10 dB.

10.14   (i) 1.93 in 10^7. (ii) 7×10^{-4} K. (iii) 50 MHz.

Index

Page numbers representing figures are in **bold**; tables are in *italics*